Understanding Disaster Risk

Understanding Disaster Risk
A Multidimensional Approach

Edited by

Pedro Pinto Santos
Ksenia Chmutina
Jason Von Meding
Emmanuel Raju

ELSEVIER

Elsevier
Radarweg 29, PO Box 211, 1000 AE Amsterdam, Netherlands
The Boulevard, Langford Lane, Kidlington, Oxford OX5 1GB, United Kingdom
50 Hampshire Street, 5th Floor, Cambridge, MA 02139, United States

Notices
Knowledge and best practice in this field are constantly changing. As new research and experience broaden our
understanding, changes in research methods, professional practices, or medical treatment may become
necessary.

Practitioners and researchers must always rely on their own experience and knowledge in evaluating and using
any information, methods, compounds, or experiments described herein. In using such information or methods
they should be mindful of their own safety and the safety of others, including parties for whom they have a
professional responsibility.

To the fullest extent of the law, neither the Publisher nor the authors, contributors, or editors, assume any liability
for any injury and/or damage to persons or property as a matter of products liability, negligence or otherwise, or
from any use or operation of any methods, products, instructions, or ideas contained in the material herein.

Library of Congress Cataloging-in-Publication Data
A catalog record for this book is available from the Library of Congress

British Library Cataloguing-in-Publication Data
A catalogue record for this book is available from the British Library

ISBN: 978-0-12-819047-0

For information on all Elsevier publications
visit our website at https://www.elsevier.com/books-and-journals

Publisher: Candice Janco
Acquisitions Editor: Marissa LaFleur
Editorial Project Manager: Amy Moone
Production Project Manager: Kumar Anbazhagan
Cover Designer: Christian J. Bilbow

Typeset by SPi Global, India

Contents

SECTION 2 PREVENTION

Contributors

Oshienemen Albert
University of Huddersfield, Global Disaster Resilience Centre, Huddersfield, United Kingdom

Dilanthi Amaratunga
University of Huddersfield, Global Disaster Resilience Centre, Huddersfield, United Kingdom

José Leandro Barros
Centre for Social Studies, Universidade de Coimbra, Coimbra, Portugal

Melissa Bedinger
School of Energy, Geoscience, Infrastructure and Society, Heriot-Watt University, Edinburgh, United Kingdom

Lindsay C. Beevers
Institute for Infrastructure and Environment; School of Energy, Geoscience, Infrastructure and Society, Heriot-Watt University, Edinburgh, United Kingdom

Rainer Bell
Department of Geography, University of Innsbruck, Innsbruck, Austria

Fronika de Wit
Institute of Social Science, University of Lisbon, Lisboa, Portugal

Yaella Depietri
Faculty of Architecture and Town Planning, Technion—Israel Institute of Technology, Haifa, Israel

Rahul Dewan
WeLoveTheCity, Rotterdam, The Netherlands

Constanza Espinoza-Valenzuela
Department of Architecture, Technical University Federico Santa María, Valparaiso, Chile

Jessica Fullwood-Thomas
Resilience and Fragility Advisor, International Programme Team, Oxfam GB, Oxford, United Kingdom

Ricardo A.C. Garcia
Centre for Geographical Studies, Institute of Geography and Spatial Planning, University of Lisbon, Lisbon, Portugal

Andrea Garcia Tapia
Stevens Institute of Technology, Hoboken, NJ, United States

Andries Geerse
WeLoveTheCity, Rotterdam, The Netherlands

Mario A. Giampieri
Department of Urban Studies and Planning, Massachusetts Institute of Technology, Cambridge, MA, United States

Daniela Giardina
Disaster Risk Reduction and Resilience Advisor, Humanitarian Theme Team, Oxfam America, Boston, MA, United States

Constanza Gonzalez-Mathiesen
University of Melbourne; Bushfire and Natural Hazards Cooperative Research Centre, Melbourne, VIC, Australia; Universidad del Desarrollo, Concepción, Chile

Franciele Caroline Guerra
São Paulo State University (UNESP), Institute of Geosciences and Exact Sciences, Rio Claro, São Paulo, Brazil

Larissa Guschl
WeLoveTheCity, Rotterdam, The Netherlands

Richard Haigh
University of Huddersfield, Global Disaster Resilience Centre, Huddersfield, United Kingdom

Małgorzata Hanzl
Lodz University of Technology, Lodz, Poland

Karl Michael Höferl
Department of Geography, University of Innsbruck, Innsbruck, Austria

Marcela Hurtado
Department of Architecture, Technical University Federico Santa María, Valparaiso, Chile

Euan James Innes
Institute for Infrastructure and Environment, Heriot-Watt University, Edinburgh, United Kingdom

Astrid Catharina Mangnus
Copernicus Institute of Sustainable Development, Utrecht University, Utrecht, The Netherlands

Alan March
University of Melbourne; Bushfire and Natural Hazards Cooperative Research Centre, Melbourne, VIC, Australia

Kerri McClymont
School of Energy, Geoscience, Infrastructure and Society, Heriot-Watt University, Edinburgh, United Kingdom

David Morrison
School of Energy, Geoscience, Infrastructure and Society, Heriot-Watt University, Edinburgh, United Kingdom

Leonardo Nogueira de Moraes
University of Melbourne; Bushfire and Natural Hazards Cooperative Research Centre, Melbourne, VIC, Australia

Carolina Giraldo Nohra
Department Architecture and Design, Politecnico di Torino, Turin, Italy

Sérgio Cruz Oliveira
Centre for Geographical Studies, Institute of Geography and Spatial Planning, University of Lisbon, Lisbon, Portugal

Daniel E. Orenstein
Faculty of Architecture and Town Planning, Technion—Israel Institute of Technology, Haifa, Israel

Susana Pereira
Centre for Geographical Studies, Institute of Geography and Spatial Planning, University of Lisbon, Lisbon, Portugal

Evangelia Petridou
Risk and Crisis Research Centre, Mid Sweden University, Östersund, Sweden

Kari Pihl
Risk and Crisis Research Centre, Mid Sweden University, Östersund, Sweden

Eva Louise Posch
Department of Geography, University of Innsbruck, Innsbruck, Austria

Jose E. Ramirez-Marquez
Stevens Institute of Technology, Hoboken, NJ, United States

Eusébio Reis
Centre for Geographical Studies, Institute of Geography and Spatial Planning, University of Lisbon, Lisbon, Portugal

Robert Šakić Trogrlić
School of Energy, Geoscience, Infrastructure and Society, Heriot-Watt University, Edinburgh, United Kingdom

Pedro Pinto Santos
Centre for Geographical Studies, Institute of Geography and Spatial Planning, University of Lisbon, Lisbon, Portugal

Jörgen Sparf
Risk and Crisis Research Centre, Mid Sweden University, Östersund, Sweden

Robert Steiger
Department of Public Finance, University of Innsbruck, Innsbruck, Austria

Alexandre Oliveira Tavares
Earth Sciences Department and Center for Social Studies, University of Coimbra, Coimbra, Portugal

Guy Walker
School of Energy, Geoscience, Infrastructure and Society, Heriot-Watt University, Edinburgh, United Kingdom

José Luís Zêzere
Centre for Geographical Studies, Institute of Geography and Spatial Planning, University of Lisbon, Lisbon, Portugal

Bruno Zucherato
Human and Social Sciences Institute, Federal University of Mato Grosso/Araguaia Campus, Cuiabá, Brazil

Preface

Over the past 40 years, there has been a significant increase in the number of people affected by disasters globally. According to the Red Cross, an average of 354 disasters occurred throughout the world from 1991 to 1999, but the annual figure keeps increasing. In 2018 315 disasters led to 11,804 deaths, over 68 million people affected, and US$131.7 billion in economic losses around the world. The burden has not been shared equally: Asia suffered the highest impact, accounting for 45% of disaster events, 80% of deaths, and 76% of people affected (CRED, 2019). And the burden falls primarily upon those most impacted by systemic inequalities and injustices in society—often referred to as the "vulnerable."

Reducing disaster risk—and implicitly human vulnerability—is foundational to any semblance of sustainable development. Post-2015, if we are to aspire to the targets of the sustainable development goals (SDGs), we must first have a clear understanding of the differential and discriminatory impacts of disasters—and how risk is being created through "status quo" development in the first place. Understanding the processes that turn hazards into disasters leading to loss and damage is an initial step. In light of climate change and potential for modification of extreme weather events, this is critical.

For a long time a vision that valued the understanding of natural processes and technocratic approach to "taming" these processes prevailed; disasters have been widely called "natural" and framed as being caused by nature. One significant implication of this is that if nature caused the loss of life and property, no person is to blame. This allows a discussion of disasters removed from social and political critique, which some may revel in. However, there has been a gradual—and welcomed—shift that emphasizes that disaster risk emerges from society itself. We (humans) create risk. Disasters are a product of economic, social, cultural, and political processes, and while a hazard cannot be prevented, disasters can be (Chmutina and von Meding, 2019; Chmutina et al., 2019).

The Sendai Framework for Disaster Risk Reduction 2015–2030 (SFDRR) reinforces this argument by proposing four global targets that would contribute toward the reduction in mortality, number of affected people, direct economic loss in relation to GDP, and damage to critical infrastructure and disruption of basic services (UNDRR, 2015). There are also three targets contributing toward the increase in the number of countries with national and local DRR strategies, international cooperation between developed and developing countries, and the availability of multihazard early warning systems—to be achieved by 2030. The SFDRR asserts that effectiveness in disaster risk reduction (DRR) strategies requires an understanding of risk scenarios—on the hazard, exposure, vulnerability, and capacity dimensions—to adequately allow for informed decision-making. The focus is not only on postdisaster actions but also, and more importantly, on leveraging the actions in predisaster risk assessments, for prevention, mitigation, preparedness, and effective response to disasters. This is only possible when acknowledging that "disaster risk reduction is everyone's business" (Raju and Costa, 2018). National DRR strategies, whether or not consciously developed under a risk governance perspective (Aven and Renn, 2010), face the risk of assessing resilience simply as one more metric, adding to probabilistic studies on hazard and loss estimation. An approach that understands building resilience as a process provides more valuable insights in understanding and reducing disaster risk (Chmutina and von Meding, 2020).

Sequentially the first one is the knowledge they directly bring to the risk and concerns' assessment, crucial for the finding of the technical and nontechnical solutions and, more importantly, for the judgment and political commitment with disaster risk reduction, to be expressed in the risk management instruments. Assessments that improve our knowledge about disaster risk are directly connected to the relevance with which society, as a whole, faces this collective challenge. The effectiveness of risk planning, prevention, mitigation strategies, and emergency responses depends on that valuing, judgment, and commitment. The second practical utility is related to monitoring and accountability of disaster risk reduction strategies, so often intangible and hard to measure. Both qualitative and quantitative methods need to be applied to provide decision-makers with the most pertinent and conclusive indicators. It is however important to bear in mind "who" is measuring "what" and "why." It is critical that decision-makers challenge the normative approaches to disaster risk reduction and instead come up with the solutions that address the root causes and underlying issues that create vulnerability—and thus lead to disasters—in the first place.

This book "Understanding Disaster Risk: A multidimensional Approach" features 16 chapters authored by a mix of academic researchers and practitioners and comprises insights from cases around the world that address the first of the SFDRR's priorities—understanding disaster risk. The research presented here reflects the diversity of issues addressed by SFDRR Priority One: Understanding Disaster Risk. The chapters explore both the policy and the practice of risk assessment and management, ranging from the national to the street block and individual level. The chapters present innovative tools and theoretical frameworks that address the multidimensional character of disaster risk and contribute to its understanding and mitigation.

There are several prominent themes in the book. Many authors grapple with challenging concepts of resilience and vulnerability and their role in "building back better." Various approaches to vulnerability assessments comprise a significant part. Giampieri provides a comprehensive review of flood vulnerability indices in coastal cities. Innes and colleagues develop a comprehensive social vulnerability index to drought based on field-collected data. Guerra and Zucherato develop a local level vulnerability index in Brazil. Santos and colleagues correlate social vulnerability data with a flood susceptibility index. Finally, Espinoza-Valenzuela and Hurtado explore the cultural, institutional, and social roots of vulnerability to forest fires in Rapa Nui, Eastern Island, in support of a new risk governance paradigm.

Resilience and its role in Priority One are also addressed with different perspectives, scales, and methodologies. De Wit and colleagues present a thorough reflection on the uncertainty and complexity that transformative resilience faces, based on local participatory processes, systemic design, and literature review. Hanzl and colleagues develop the concept of urban resilience under the perspectives of urban metabolism and circular economy, through the optimization of land consumption. At the city scale, McClymont and colleagues make use of the abstraction hierarchy approach to capture the sociotechnical interactions between the tangible and intangible aspects that influence the capacity to absorb and recover from hazards' negative effects. Petridou and colleagues take the case of the Swedish municipal context and identify the gaps in integrating the resilience-related policies along with the sustainability-related ones. Posh and colleagues apply the agency toward resilience model in Nepal, a context in which the constraints posed by the structural conditions and the access to assets can be overcome by considering the individuals' goals and trade-offs.

Throughout the book, various hazards are explored; however, there is a particular—and timely—focus on wild and forest fires. Depietri and Orenstein focus on the cultural forcers related to the wildland-urban interface, while Gonzalez-Mathiesen and March address the same fire context but in regard to informal settlements. Both contribute to the design and planning of mitigation strategies and local-capacity strengthening. Finally, making use of a diverse set of knowledge sources, Nogueira de Moraes and March conduct a combined analysis of social and natural systems, aiming at the improvement of emergency management services and the resilience of local communities.

The interaction between the individual and collective actions is also an important part of "building back better." Giardina and Fullwood-Thomas apply Oxfam's approach in acquiring vulnerability and resilience-related knowledge and how it is used in providing adaptive interventions that improve people's capacities. Garcia-Tapia and Ramirez-Marquez bring the dimension of social media in linking the virtual communities they foster, with root-based disaster relief initiatives. Finally, Albert, Amaratunga, and Haigh study the threats posed by the impact of oil spills to the livelihood alternatives in the Niger Delta and how this process affects individuals' resilience.

This concept of resilience is prominent in all the contributions included in this book. Resilience is a frequently misused concept, but nevertheless, it plays a fundamental role in understanding disaster risk. Resilience has a multitude of meanings for the different stakeholders, academics, and decision-makers involved in DRR (Wang et al., 2020; Chmutina et al., 2016; Reghezza-Zitt et al., 2012; Levine, 2014; Alexander, 2013). It moved from being an explanation of how people act and cope with hazards, risks, and disasters to become a normative approached that is being "implemented" and "measured." As the concept of resilience is becoming more and more malleable, some argue that it faces the risk of becoming less useful in understanding the causes and drivers of disasters. Others however believe that it is through this malleability that resilience is able to act a boundary object that provides common ground for discussions. Additionally—assuming that it is possible to effectively measure resilience—there is still the theoretical challenge of deciding what should be measured and by whom. Quite often the metrics that are sought are the ones quantifying the capacity of communities to protect themselves against nature, which largely reduces the value of the concept and is counterproductive in reaching the roots of disasters. Other discussions revolve around community and individual resilience: does labeling someone "resilient" means that they do not need support? These and other discussions are important—however, it is thus important to remind ourselves of three underlying questions: resilience of what, resilience to what, and resilience by who? (Carpenter et al., 2001). It is only by considering these questions that we are able to unpack resilience contextually.

Five years into the 2015–2030 Sendai Framework period, this book helps us to reflect on the successes that have been achieved in relation to the Priority One—and on challenges that are yet to be tackled if we are to mainstream the idea of disaster risk governance in the next decade. Understanding disaster risk is multifaceted: it requires understanding of hazards, exposure, vulnerability, resilience, and capacities—and most importantly the actions (or the lack of such) that create the root causes of risk and vulnerability in the first place. As we are writing this editorial, COVID-19 pandemic is spreading around the world, once again highlighting that disasters are not natural and that resilience cannot be built on inequality and marginalization. Achieving the SFDRR goals is only possible if we act together in tackling the underlying social issues; as technocratic top-down approaches are failing, it is time to embrace successes and failure and learn from each other that resilience is not a sliver bullet.

References

Alexander, D.E., 2013. Resilience and disaster risk reduction: an etymological journey. Nat. Hazards Earth Syst. Sci. 13 (11), 2707–2716. https://doi.org/10.5194/nhess-13-2707-2013.

Aven, Terje, Renn, Ortwin, 2010. Risk Management and Governance: Concepts, Guidelines and Applications. Springer-Verlag, Berlin, Germany.

Carpenter, S., et al., 2001. From metaphor to measurement: resilience of what to what? Ecosystems 4 (8), 765–781. https://doi.org/10.1007/s10021-001-0045-9.

Chmutina, K., von Meding, J., 2019. A dilemma of language: 'natural disasters' in academic literature. Int. J. Disaster Risk Sci. 10, 283–292. https://doi.org/10.1007/s13753-019-00232-2.

Chmutina, K., Lizarralde, G., Dainty, A., Bosher, L., 2016. Unpacking resilience policy discourse. Cities 58, 70–79. https://doi.org/10.1016/j.cities.2016.05.017.

Chmutina, K., von Meding, J., Bosher, L., 2019. Language matters: dangers of the "natural disaster" misnomer. Contributing paper to the Global Assessment Report (GAR) 2019. UNDRR. Available at: https://www.preventionweb.net/publications/view/65974.

CRED, 2019. Natural Disasters 2018. CRED, Brussels. Available at: https://emdat.be/sites/default/files/adsr_2018.pdf.

Levine, S., 2014. Assessing resilience: why quantification misses the point, HPG Working Paper. London. Available at: http://www.odi.org/sites/odi.org.uk/files/odi-assets/publications-opinion-files/9049.pdf.

Raju, E., Costa, K., 2018. Governance in the Sendai: a way ahead? Disaster Prevent. Manage. 27 (3), 278–291. https://doi.org/10.1108/DPM-08-2017-0190.

Reghezza-Zitt, M., et al., 2012. What resilience is not: uses and abuses. Cybergeo: Eur. J. Geogr. https://doi.org/10.4000/cybergo.25554.

UNDRR, 2015. Sendai Framework for Disaster Risk Reduction 2015-2030. United Nations Office for Disaster Risk Reduction, Sendai. Available at: https://www.undrr.org/publication/sendai-framework-disaster-risk-reduction-2015-2030.

Wang, Y., et al., 2020. Conceiving resilience: Lexical shifts and proximal meanings in the human-centered natural and built environment literature from 1990 to 2018. Dev. Built Environ. 1, 100003. https://doi.org/10.1016/j.dibe.2019.100003.

Further reading

Chmutina, K., von Meding, J., 2020. Resilience. DisasterDecon's podcast episode recorded on Feb 24th 2020. Avaliable at: https://disastersdecon.podbean.com/e/s2e8-resilience-audience-special/.

Resilience in the Anthropocene

0.1

Fronika de Wit[a], Astrid Catharina Mangnus[b], and Carolina Giraldo Nohra[c]

Institute of Social Science, University of Lisbon, Lisboa, Portugal[a] Copernicus Institute of Sustainable Development, Utrecht University, Utrecht, The Netherlands[b] Department Architecture and Design, Politecnico di Torino, Turin, Italy[c]

CHAPTER OUTLINE

0.1.1 Introduction

High anthropogenic pressures on the earth system are exceeding the planetary boundaries on various scales. We are hitting the planetary ceiling: research shows that if humanity continues living the way it is doing, human well-being is at risk (Rockstrom, 2009) and irreversible changes to the earth system are impending (Lenton et al., 2008; Schellnhuber, 2009). The Great Acceleration has led us to a new geological epoch, the Anthropocene (Crutzen and Stoermer, 2000), in which humans are the dominating force that hold the future in their hands.

The main complicating factors to living in the Anthropocene are that all of its issues are interconnected and that there is a high level of uncertainty, which demands planetary stewardship: an alteration of the relationship between people and planet (Steffen et al., 2011). With the Sustainable Development Goals, adopted by the UN in 2015, we have a universal plan for people and planet—a roadmap for planetary stewardship. However, in order to achieve the SDGs and become planetary stewards, we need a completely new way of thinking that fits the turbulent context of the Anthropocene. Current governance approaches are failing to promote resilience, because of their disciplinary and top-down

Understanding Disaster Risk. https://doi.org/10.1016/B978-0-12-819047-0.00000-7

approach, in which the decision makers are not aware of the complexity of systems and do not take stakeholders' views into account. Folke (2016), "father" of modern resilience thinking, highlights that being resilient means having strategies and policies in place to deal with the unknown, which is a promising answer to the Anthropocene's main complicating factors.

Resilience thinking emerged in the 1970s from two different traditions: child psychology and ecosystem ecology. It developed into research streams like community resilience, climate resilience, disaster resilience, and development resilience. Folke et al. (2010) and Folke (2016) describe three forms of resilience: (1) *Persistence*, or "bouncing back," which includes continual change and adaptation, but remaining on the same pathways; (2) *Adaptability*, which is about innovation and change, but also remaining on the same pathways; and (3) *Transformability*, which is about shifting pathways. Where disaster resilience was more about bouncing back and climate resilience was about adapting to change, the "new" resilience thinking is about transformation: a reconfiguration of systems, values, and beliefs.

An important component of resilience thinking is the features and dynamics of complex systems. Research (Simonsen et al., 2014; Biggs et al., 2015) has provided seven building blocks for applying resilience thinking in order to increase the capacity to deal with unexpected change in complex social-ecological systems. These building blocks are: (1) maintaining diversity and redundancy; (2) managing connectivity; (3) managing slow variables and feedback; (4) treating social-ecological systems as complex systems; (5) encouraging learning; (6) broadening participation; and (7) polycentric governance systems.

This chapter looks at the potential of resilience thinking for addressing the challenges of the Anthropocene by outlining and synthesizing lessons learned in three different academic disciplines: human geography, futures studies, and design studies. How can we combine the insights of these different disciplines and rethink the concept of resilience outside of the "disciplinary box"? By looking at resilience thinking as a dynamic concept, we bring a transdisciplinary and holistic approach for resilience in the Anthropocene and show how this can be applied in practice to decision-making processes. The next section of the chapter outlines the methodology for the multidisciplinary literature review. The third section gives an overview of the results for each of the three disciplines. The fourth section discusses these findings and their relevance for the challenges of the Anthropocene, after which the chapter closes with a conclusion.

0.1.2 Methodology

When discussing a methodology for research that connects the knowledge contained in multiple disciplines, it is of importance to distinguish between three approaches to research: (1) multidisciplinary research; (2) interdisciplinary research; and (3) transdisciplinary research. According to Ramadier (2004), both multidisciplinary and interdisciplinary research are still based on disciplinary thinking. However, where multidisciplinary research is based on the juxtaposition of different disciplines, interdisciplinary research constructs a common model for the disciplines involved. Transdisciplinary research, on the other hand, goes beyond disciplinary thinking and its objective is "to preserve the different realities and to confront them."(2004, p. 434). In the case of this chapter, our aim is to connect the academic disciplines of futures studies, human geography, and systemic design in a way that goes beyond drawing on the three disciplines separately. This raises the need to develop an interdisciplinary approach.

To start this analysis, it is crucial to first take stock of the various ways in which the three disciplines define resilience. To get a concise but good overview, the first step in our methodology is a quite straightforward literature review. This literature review consists of filtering the 10 most cited articles in the three fields that mention resilience as a core concept. We do this by focusing on two of the most prominent literature sources across all academic disciplines: Scopus and Web of Science. We used the search terms "futures studies" AND resilience; "human geography" AND resilience; "systemic design" AND resilience. While this may seem like a rather straightforward review process, previously executed by a range of authors in all three fields (REFS), the first interdisciplinary hurdles show up here. While human geography is an established field with its own category in the search engines, futures studies are rather less so, and span disciplines. Therefore, the futures studies literature was further filtered by searching for "social science" and manually filtering out papers that were far outside the scope of this current chapter, such as is in the field of healthcare where resilience is a popular term also but used with a different meaning (e.g., recovering after disease). For systemic design, the search terms were modified slightly to narrow down the results: "design theory" * resilience * "systems theory." The results were filtered by amount of citations and the top 10 papers were scanned for their definition of resilience.

For each of the disciplines a brief synthesis of the definitions within the discipline is made, outlining the main points of interest and the general way the discipline talks about resilience. This "taking-stock" of the various disciplines serves as input for the discussion section, in which we look for differences, similarities, and a way to synthesize the knowledge from the three disciplines. A literature review such as this one can always be more extensive, but due to time limitations, we have chosen to focus on the 10 most cited articles to get a general overview and test this interdisciplinary approach in order to draw lessons for resilience in the Anthropocene—the main aim of this chapter.

0.1.3 Results

0.1.3.1 Insights from human geography

The reviewed articles on resilience in the field of human geography (see Table 0.1.1) show the discipline's flexibility and suitability to address challenges presented by the Anthropocene, such as climate change. However, many articles do not specify their definition of resilience, which they assume to be known by the reader. The reviewed articles show four key dimensions of resilience in the field of human geography: (1) nature-society relation; (2) scale and governance; (3) knowledge and power; and (4) local, place-based perspectives.

The first dimension of resilience research in human geography discusses the nature-society or nature-culture relation and points to the often-undertheorized social dimensions of resilience. All reviewed articles point to a certain extent to the importance of the relation between nature and society. Human geographers have played an important role in challenging the predominance of ecology-centric approaches to resilience. Brown (2014) discusses the social turn in resilience applied to global environmental change and concludes that there are significant advances and a much greater engagement and reflection on social dimensions, manifested in growing literature and debates on social dynamics. An et al. (2016) highlight the role of coupled human and natural systems (CHANS) in resilience research.

Table 0.1.1 Literature review results for human geography.

No.	Author (year)	Citation count	Definition	Journal
1	Brown (2014)	189	Author provides an overview of the applications and core concepts of resilience in different fields Strong focus on social resilience, defined "as the ability of communities to withstand shocks to their social infrastructure"	Progress in Human Geography
2	Osbahr et al. (2008)	154	"… the capacity to renew and reorganise their livelihoods in the face of [such] disturbances"	Geoforum
3	Cumming et al. (2013)	98	Landscape resilience and landscape sustainability are closely related. Perhaps the most relevant distinction is that common usage of the concept of sustainability emphasizes a "business as usual" scenario that focuses on current rates of exploitation and growth, whereas the concept of resilience emphasizes more heavily the ability of the system to cope with perturbations (and the related topics of uncertainty, innovation, and adaptation). Resilience further highlights the relevance of system memory and cyclical change (the "panarchy" of the adaptive cycle) although the adaptive cycle should not be conflated with the concept of resilience	Landscape Ecology
4	Devine-Wright (2013)	87	No definition; assumed to be known	Global Environmental Change
5	Jonas (2011)	70	"… resiliency is an internal property of (regional) territory, which must always be juxtaposed to the external risks to places and regions posed by global flows of investment and environmental and economic crises"	Progress in Human Geography
6	Gruby and Basurto (2013)	43	No definition; assumed to be known	Environmental Science and Policy
7	Crabtree (2006)	33	"… system resilience is dependent on flexibility and adaptability, in turn dependent on high levels of diversity and an apparent redundancy of components and functions" Following Perrings (1998, p. 221): "… the sustainability of a social system depends on the resilience of that system, and that the resilience of social systems in turn depends on a range of institutional and other properties … The implications are far-reaching. If the argument is right then a strategy of sustainable development has much less to do with the satisfaction of various efficiency measures than with institutional design, property rights, communication and trust"	Geoforum

No.	Author (year)	Citation count	Definition	Journal
Table 0.1.1 Literature review results for human geography—cont'd				
8	An et al. (2016)	30	"… human–nature systems are capable of retaining similar structures and functioning after disturbances"	Annals of the Association of American Geographers
9	Tschakert et al. (2017)	17	No definition; assumed to be known	Wiley Interdisciplinary Reviews—Climate Change
10	Scott (2015)	10	"Ecological resilience … refers to an environmental system's ability to change but maintain its structure and function for future development without shifting to another state" "Social resilience … has emerged referring to the capacity for social systems and communities to adapt to environmental and economic change, but without disrupting ecological resilience"	Progress in Human Geography

A second dimension of geographical resilience research is its focus on planetary boundaries: physical limits to human activities. This dimension relates to the need for a multilevel and multiscalar approach to sustainability and resilience governance. Cumming et al. (2013) discuss the concept of landscape resilience and highlight the need for finding an appropriate match between the scales of demands on ecosystems by human societies and the scales at which ecosystems are capable of meeting these demands. Gruby and Basurto (2013) highlight that less distributed decision making in the overall nested governance system could threaten the sustainability and resilience of coral reefs in the long term by constraining institutional innovation and diversity. Their results demonstrate the potential for interdisciplinary dialog to advance the research frontier on multilevel governance for large common pool resources. Also related to scale, Jonas (2011) discusses the concept of regional resilience and highlights that for resilience: "the knowledge of the internal capacities of regions matters as much as their external relations."

A third dimension that human geography adds to resilience research is the interaction between knowledge and power. The geography literature challenges the often very Eurocentric and neoliberal perspectives on resilience by using case studies from the Global South and emphasizing the importance of different epistemologies and ontologies, such as the example of local livelihood adaptation strategies in a village in Mozambique by Osbahr et al. (2008). Brown (2014) points to how resilience is similar to sustainability, as both concepts can act as boundary objects or bridging concepts and may be co-opted by different interests. Crabtree (2006) highlights how the main challenges for resilience building come from trust and power sharing issues. The fourth dimension is related to the previous one and points to the importance of place-based approaches and the incorporation of local knowledge. Brown (2014) points to the lack of a local-level dimension of resilience and a focus on community resilience. She discusses the growing number of community resilience projects. Various reviewed articles discuss the importance of place-based approaches to understand what matters to people (Crabtree, 2006; Devine-Wright, 2013; Tschakert et al., 2017).

0.1.3.2 **Insights from systemic design**

The transition from linear to holistic thinking requires approaches that encourage people to "think outside the box" and generate disruption (Considine, 2012). The role of design can be decisive in this process, as it deals with complex scenarios used to anticipate future situations and generate innovative outcomes. Recent work on design for sustainable development strategies reveals a transformation from a restricted technical, product, and process perspective to the large-scale system approach (Adams et al., 2016). Systemic Design provides a tool to design the flow of material and energy from one element of the system to another. This way it transforms outputs of one process into input for another, achieves zero emissions and generates resilient territories (Bistagnino, 2011).

To build resilient territories, systemic design thinking generates a multidisciplinary synergy tangible for all stakeholders in a decision-making process (Bason, 2014). It generates new relations among territorial entities that visualize hidden potentialities and boost a proactive collaboration among local actors. The Systemic Design approach includes other methodologies such as design thinking, codesign, user-centered design, bottom-up design, and participatory design. These all share an active engagement of users in the design process, consequently turning the end user into the focus of the resilience strategy formulation system and generating an innovative decision-making process (Allio, 2014). The conventional focus of policies is not the most efficient since it entails a top-down approach that does not consider the final users: the citizens. Therefore, participatory processes are a key element to design effective policy strategies, applying a bottom-up approach for policy planning (Ibid).

The Systemic Design expertise is proposed as an anticipatory tool for decision making presenting a new starting point across the holistic diagnosis or system mapping (Battistoni and Giraldo Nohra, 2017). The overview of such complex scenarios provides tools to encourage the generation of new cooperation channels among different local actors. Moreover, promotes a multidisciplinary approach that invites participants from different sectors to cocreate within an interdisciplinary scenario, new policies that will connect governments, citizenship, and industry. It provides an instrument which benefits all parties leading them to paths where all can reach new sustainable scenarios of economic profit and cooperation (Barbero, 2017) (Table 0.1.2).

0.1.3.3 **Insights from futures studies**

When considering resilience, the future is inherent in the concept: a resilient system can handle various future disturbances, which can make it last for a long time in that future. The field of futures studies takes an active position toward the various possible futures that can be, by systematically assessing the probably, the possible, and the preferable (Bell, 2004). Through participatory interventions, futures studies can enter into a reciprocal cycle with the objects of their research intervention and thus also shape the future (Vervoort et al., 2015). While also using the classic geological definition of the Anthropocene, the futures field increasingly uses the concept as an opportunity to push for sustainability transformations. By inviting communities to futures workshops, participants gain agency over their futures and can develop scenarios rooted in their own existing practices. Engaging with futures work can in turn increase community resilience, which is conceptualized as adaptive capacity, or the potential to adjust to many different future contexts (Deacon et al., 2018). In these workshops, the

No.	Author (year)	Citation count	Definition	Journal
		Table 0.1.2 Literature review results for systemic design.		
1	Fiksel (2003)	319	"An alternative is to design systems with inherent 'resilience' by taking advantage of fundamental properties such as diversity, efficiency, adaptability, and cohesion"	Environmental Science and Technology
2	Schwaninger (2012)	14	No definition; assumed to be known	Kybernetes
3	Filatova et al. (2016)	34	"Specifically, in the resilience literature a 'regime shift' is a change from one system state to another, although this concept applies to cases where the transition occurs over any timescale, abrupt or otherwise"	Environmental Modelling and Software
4	Ghazvini et al. (2017)	6	"... defined as the continual process of enhancing well-being for all citizens while also striving to prepare for and enhance recovery from acute disruptive events"	Sustainability (Switzerland)
5	ten Napel et al. (2011)	17	"Resilience theory is the model for maintaining system features in the presence of perturbations in ecosystems and social systems"	Livestock Science
6	Dearing et al. (2011)	31	No definition; assumed to be known	Environmental Management
7	Morse et al. (2013)	5	No definition; assumed to be known	Urban Ecosystems
8	Birgé et al. (2016)	14	Resilience theory recognizes that a limited set of ecological processes in a given system regulate ecosystem services, yet our understanding of these processes is poorly understood	Journal of Environmental Management
9	Klein (2012)	4	No definition; assumed to be known	Kybernetes
10	Fazey (2010)	32	No definition; assumed to be known	Ecology and Society

aforementioned building blocks for resilience (Simonsen et al., 2014; Biggs et al., 2015) could provide very useful checks and balances to preserve the quality of the resulting futures and to preserve the element of resilience throughout.

A strong example is the work of Sircar et al. (2013), who constructed resilient energy futures for multistakeholder processes in the United Kingdom. They started with existing futures studies and made those assessments more robust by drawing in a variety of stakeholders to "reflect about resilience in a set of possible future worlds" (Ibid.) in a series of workshops. The process highlighted overlapping definitions of resilience in each future, outlined interdependencies, and allowed for a deeper and more collaborative reflection on the longer-term resilience implications. A process such as this connects all the building blocks for resilience: it maintains diversity and connectivity, allows for feedback, recognizes the complexity of the energy system at hand, encourages learning and participation and implements a type of multistakeholder, polycentric governance. A few elements are important to keep in mind with regard to the coupling of futures studies and community resilience. In all futures work, a systematic approach, just power balance and selection of the stakeholders present in the process are key for a successful outcome (Wiek and Iwaniec, 2014). The facilitation and debriefing is also vital to ensure a real impact beyond a futures brainstorm and to move the process into truly resilient community development (Pereira et al., 2018) (Table 0.1.3).

Table 0.1.3 Literature review results for futures studies.

No.	Author (year)	Citation count	Definition	Journal
1	Sotarauta and Srinivas (2006)	45	"Well balanced co-evolution of policy and development'"	Futures
2	Gidley et al. (2009)	36	No definition; assumed to be known	Environmental Policy and Governance
3	Rickards et al. (2014)	30	Defined by what it is not: "Such is the pervasive uptake of resilience—given its own inherent adaptability to different contexts, including neoliberalism (Evans and Reid, 2013)—that Stumpp (2013) suggests it has now replaced sustainability as a core guiding concept in contemporary society, despite significant tensions between the individualistic subjectivity and adaptation rationale of dominant interpretations of resilience and the collective subjectivity and precaution rationale of sustainability (at least in its original formulation)"	Environment and Planning C: Governance and Policy
4	Blanton (2010)	11	The heart of resilience theory is a modified perspective on ecological succession that proposes an adaptive cycle beginning with a growth (r) phase, in which natural capital is accumulated (such as density of biomass), followed by a conservation phase (K), in which production and storage of energy and material is maintained at high levels. The instability of the K phase (see below) brings eventual collapse resulting in a release of accumulated capital (W phase) and reorganization (a phase)	Cross-Cultural Research
5	Jaroszweski et al. (2014)	8	[...] there are different perspectives on resilience. Local infrastructure managers will be concerned about the resilience of their infrastructure usually in isolation, or at least in isolation in terms of the mode. Passengers, however, will be interested in whether transport can meet their needs, with each	Progress in Physical Geography

No.	Author (year)	Citation count	Definition	Journal
			having a threshold for failure depending on their activity. Policy-makers, on the other hand, will want to ensure that transport meets the needs of the wider economy, and will be less concerned about the specific locations of failures	
6	Sircar et al. (2013)	8	Analyses of resilience often break the concept down into three generations. The first generation is that of robustness, i.e., how resistant a system is to hazards. The second generation of resilience is response and recovery, which focuses on whether the system can "bounce back" in the short term. The third generation of resilience connotes adaptation and evolution, such that there is a progressive "bounce forward" after a hazardous event	Futures
7	Poli (2015)	4	The capacity of people, communities, and organizations to manage and take advantage of the stress and excitement generated by the only certainty that we know—constant change	On the Horizon
8	Pernaa (2017)	1	No definition; assumed to be known	European Journal of Futures Research
9	Jasman and McIlveen (2011)	1	No definition; assumed to be known	On the Horizon
10	Wilson et al. (2020)	0	"...the capacity of the social system to absorb shocks and stressors while retaining its essential structures and functions"	Nature Climate Change

Table 0.1.3 Literature review results for futures studies—cont'd

0.1.4 Discussion

When comparing the disciplinary definitions for the concept of resilience, we can see that some of the reviewed articles discuss resilience without providing a definition: three articles for human geography, five for systemic design, and three for futures studies are without a definition for resilience. The authors assume their readers know the definition of resilience and do not feel a need to specify. Differences in defining resilience between the disciplines are more related to their specific focus. Human geography is more focused on space and scale and talks about concepts like "regional resilience" (Jonas, 2011) and "landscape resilience" (Cumming et al., 2013). Systemic design has a strong focus on the system features of resilience (ten Napel et al., 2011). Futures studies focuses more on policies that better prepare communities for the future (Grant, 2011).

Overall similarities between the reviewed articles are the link between resilience and sustainability and how the concepts have become almost interchangeable (Cumming et al., 2013; Rickards et al., 2014). Also, the reviewed articles from the different disciplines highlight the importance of an interdisciplinary approach that connects the ecological and social dimension (An et al., 2016; Scott, 2015; ten Napel et al., 2011) and looks at resilience from a system perspective (Sircar et al., 2013; Filatova et al., 2016; Fiksel, 2003). The reviewed articles also discuss the progress of conceptualization and the different generations of definitions for resilience (Sircar et al., 2013; Brown, 2014) and the importance of incorporating different perspectives of resilience (Jaroszweski et al., 2014). Another similarity between the reviewed articles is that they all highlight the importance of building capacity for change, this being a regime shift (Filatova et al., 2016), a renewal or reorganization of livelihoods (Osbahr et al., 2008), or constant change in general (Poli, 2015).

There are many interlinkages between the disciplinary insights on the conceptualization and operationalization of resilience, which we have synthesized in Fig. 0.1.1. The circles represent the dimensions of the three analyzed disciplines in relation to the concept of resilience, in which the size of the

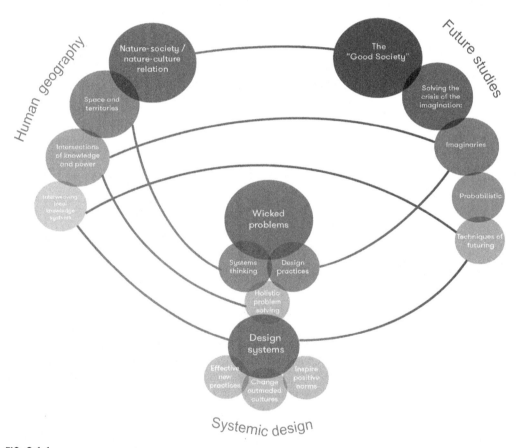

FIG. 0.1.1

The main disciplinary insights on resilience from human geography, futures studies, and systemic design and their interlinkages represented by the connecting lines.

circle represents the importance of that dimension for the discipline. The four key dimensions of resilience in the field of human geography, listed in the order of their importance, are: (1) *nature-society relation*: equal importance of both ecological and social dimensions; (2) *scale and governance*: multilevel and multiscalar governance; (3) *knowledge and power*: the need to go beyond a neoliberal perspective and western logics; and (4) local, place-based perspectives: making use of local knowledge and worldviews.

The five dimensions of futures studies research are: (1) *the Good Society*: the importance of values and normativity; (2) *solving the crisis of the imagination*: the immersive, experiential element of futures studies as an answer to problems with older versions of participation and multistakeholder governance; (3) *imaginaries*: collectively held, institutionally stabilized, and publicly performed visions of desirable futures; (4) *probabilistic* (future follows from past), *possibilistic* (gold standard—captures emergent futures) and *constructivist* ("anything can happen") *frameworks* for foresight; and (5) *techniques of futuring*: practices bringing together actors around one or more imagined futures and through which actors come to share particular orientations for action.

The dimensions in the field of systemic design are built around two main concepts: *wicked problems* and *design systems*. Wicked problems are a "class of social system problems which are ill-formulated, where the information is confusing, where there are many clients and decision makers with conflicting values, and where the ramifications in the whole system are thoroughly confusing" (Rittel and Webber, 1973). Addressing wicked problems has a theoretical and practical components, which are illustrated in Fig. 0.1.1 by the smaller circles System Thinking (theory) and Design Practices (practice). Both for the theoretical and practical component, there is a need for holistic problem solving, represented in the smaller circle in Fig. 0.1.1. The second main concept is Design Systems. Systems are designed in order to create effective new practices, change outmoded cultures are inspire positive norms.

The interlinkages between the disciplinary insights are represented by connecting lines in Fig. 0.1.1. A first interlinkage is between human geography's Nature-Society relationship and futures studies' Good Society: both are value-driven and normative dimensions. A second interlinkage is between human geography's dimension of scale and governance and systemic design's systems thinking, in which both pay attention to the multilevel, multiscalar, and multiactor dimension of the resilience challenge. A third interlinkage between the disciplines is the science-policy interface that they all address; interactions of knowledge and power (human geography), design practices and holistic problem solving (systemic design) and imaginaries (futures studies) are all related to a well-balanced coevolution of policy and development, as defined by Sotarauta and Srinivas (2006). The three disciplines also connect on their more methodological dimensions: incorporating local knowledge (human geography), design systems (systemic design), and techniques of futuring (futures studies). This shows the importance of the practical application of resilience research on an environmental, societal, and economic dimension.

These interlinkages between the disciplinary insights on resilience show us the common grounds and the need for a more collaborative approach and transdisciplinary policy pathways for building resilience in the Anthropocene. According to Nicolescu (2010) the level of reality is a key concept of transdisciplinarity. He emphasizes the importance of unifying levels of reality in building sustainable development and sustainable futures. He highlights the need for a new spirituality, and the need to unify the *homo religiosus* (spirituality) with the *homo economicus* (modern science) in order to get to the *homo sui transcendentalis* (Nicolescu, 1996). Ramadier (2004) describes two challenges of transdisciplinary thinking: (1) the *cognitive challenge*: the necessity to go beyond the notion of unity and to think in terms of articulation, by looking for the coherence of the multiple realities of the object under consideration; and (2) the *methodological challenge:* researchers are encouraged to unify their methodologies, to identify differences in level of reality. However, encouraging methodological unity prevents genuine transdisciplinarity and runs the risk to return to multidisciplinary thinking (domination of disciplines over the choice of methodology). The author therefore urges to pay attention to the articulation of methods.

In order to have a transdisciplinary perspective, it is of importance to further analyze the science-policy interface of resilience research. Lang et al. (2012) see transdisciplinarity as the interplay between societal and scientific practice and define it as: "a reflexive, integrative, method-driven scientific principle aiming at the solution or transition of societal problems and concurrently of related scientific problems by differentiating and integrating knowledge from various scientific and societal bodies of knowledge." They list design principles for using transdisciplinary sustainability research: (1) build a collaborative research team; (2) create joint understanding and definition of the sustainability problem to be addressed; (3) collaboratively define the boundary/research object, research objectives as well as specific research questions, and success criteria; (4) design a methodological framework for collaborative knowledge production and integration. Mauser et al. (2013) focus on transdisciplinary global change research and describe three steps: (1) codesign of the research; (2) coproduction of knowledge; (3) codissemination of the results.

0.1.5 **Final considerations**

The Anthropocene is a complex era and resilience thinking will need transformation to provide adequate responses to the challenges at hand. Our research on lessons learned for resilience in the Anthropocene from different perspectives, shows many similarities between the disciplines. Although they use different wordings and disciplinary jargon, Human Geography, Futures Studies, and Systemic Design all talk about the importance of stakeholder involvement and bottom-up approaches to support improved governance in times of complexity and uncertainty. However, the three disciplines can learn from each other in different ways. Anticipating the future and adapting to it is a key component of resilience. There is still room for further exploration and theorization of the relationship between anticipation and adaptation (Boyd et al., 2015). The coupling of the disciplines and their tools and methods with insights from other disciplines can provide a promising pathway to expand this knowledge further.

Grindsted (2018) emphasizes how science entangles itself in between keywords and buzzwords and asks for critical reflection on both how scientists make use of the power of reference between keywords and buzzwords and on the institutionalization of such concepts. Like with sustainability, we should be careful not to create a buzzword out of "resilience," turning it into an empty concept. In order to work in a transdisciplinary way, we need to use clear and practical descriptions of what we are working with. Future studies, with its futures workshops and Systemic Design expertise, with its design methods and systemic thinking, show the importance of hands-on methodologies to address policy making toward resilient territories and communities. Cultural geography adds the remark to be aware of our often very western and Eurocentric approach, and the need to go beyond western epistemologies and ontologies. Resilience is related to knowledge: collaboration between scientific knowledge and local knowledge gains from being based on equity and usefulness. Diverse knowledges can enhance the capacity to navigate complexity and increases equity and usefulness for all involved.

Complex systems in the Anthropocene need to be reframed in order to be solved. Resilience is all about hidden relationships and potentialities and how discovering new connections can enable. Making use of existing cultural potential and assets can foster social and economic benefits. To make communities more autopoietic and systems self-sustainable, local communities need economic and social empowerment. Therefore, we recommend a multidisciplinary approach for resilience thinking, which can support the creation of more efficient policies and decision making that can foster a better governance on sustainable development and disseminate innovative solutions to reinvent and shape more cohesive territories.

References

Adams, R., Jeanrenaud, S., Bessant, J., Denyer, D., Overy, P., 2016. Sustainability-oriented innovation: a systematic review. Int. J. Manage. Rev. 18 (2), 180–205. https://doi.org/10.1111/ijmr.12068.

Allio, L., 2014. Design Thinking for Public Service Excellence. vol. 1. Global Centre for Public Service Excellence, p. 28 (1).

An, L., Zvoleff, A., Liu, J., Axinn, W., 2016. Agent-based modeling in coupled human and natural systems (CHANS): lessons from a comparative analysis. In: Handbook of Applied System Science. 5608, Abingdon-on-Thames, Routledge, pp. 267–296. https://doi.org/10.4324/9781315748771.

Barbero, S., 2017. Systemic Design Method Guide for Policymaking: A Circular Europe on the Way. Vol. 1. Allemandi, Turin.

Bason, C., 2014. Design for Policy. Gower, Aldershot.

Battistoni, C., Giraldo Nohra, C., 2017. The RETRACE holistic diagnosis. In: Barbero, S. (Ed.), Systemic Design Method Guide for Policymaking: A Circular Europe on the Way. Allemandi, Turin, pp. 112–120.

Bell, W., 2004. Foundations of Futures Studies: Volume 2: Values, Objectivity, and the Good Society. Routledge, Abingdon-on-Thames.

Biggs, R.O., Peterson, G.D., Rocha, J.C., 2015. The regime shifts database: a framework for analyzing regime shifts in social-ecological systems. BioRxiv. https://doi.org/10.1101/018473.

Birgé, H.E., Bevans, R.A., Allen, C.R., Angeler, D.G., Baer, S.G., Wall, D.H., 2016. Adaptive management for soil ecosystem services. J. Environ. Manage. https://doi.org/10.1016/j.jenvman.2016.06.024.

Bistagnino, L., 2011. Systemic Design: Designing the Productive and Environmental Sustainability. Slow Food, Bra, Italy.

Blanton, R., 2010. Collective action and adaptive socioecological cycles in premodern states. Cross-Cult. Res. https://doi.org/10.1177/1069397109351684.

Boyd, E., Nykvist, B., Borgström, S., Stacewicz, I.A., 2015. Anticipatory governance for social-ecological resilience. Ambio 44 (S1), 149–161. https://doi.org/10.1007/s13280-014-0604-x.

Brown, K., 2014. Global environmental change I: a social turn for resilience? Prog. Hum. Geogr. 38 (1), 107–117. https://doi.org/10.1007/978-94-007-5784-4.

Considine, M., 2012. Thinking outside the box? Applying design theory to public policy. Politics Policy 40 (4), 704–724. https://doi.org/10.1111/j.1747-1346.2012.00372.x.

Crabtree, L., 2006. Sustainability begins at home? An ecological exploration of sub/urban Australian community-focused housing initiatives. Geoforum 37 (4), 519–535. https://doi.org/10.1016/j.geoforum.2005.04.002.

Crutzen, P.J., Stoermer, E.F., 2000. The Anthropocene. Global Change Newsletter 41 (41), 17–18. http://www.igbp.net/publications/globalchangemagazine/globalchangemagazine/globalchangenewslettersno4159.5.5831d9ad13275d51c098000309.html.

Cumming, G.S., Olsson, P., Chapin, F.S., Holling, C.S., 2013. Resilience, experimentation, and scale mismatches in social-ecological landscapes. Landsc. Ecol. 28 (6), 1139–1150. https://doi.org/10.1007/s10980-012-9725-4.

Deacon, L., Van Assche, K., Papineau, J., Gruezmacher, M., 2018. Speculation, planning, and resilience: case studies from resource-based communities in Western Canada. Futures 104, 37–46.

Dearing, J.A., Bullock, S., Costanza, R., Dawson, T.P., Edwards, M.E., Poppy, G.M., Smith, G.M., 2011. Navigating the perfect storm: research strategies for social ecological systems in a rapidly evolving world. Environ. Manage. https://doi.org/10.1007/s00267-012-9833-6.

Devine-Wright, P., 2013. Think global, act local? The relevance of place attachments and place identities in a climate changed world. Global Environ. Change 23 (1), 61–69. https://doi.org/10.1016/j.gloenvcha.2012.08.003.

Fazey, I., 2010. Resilience and higher order thinking. Ecol. Soc. https://doi.org/10.5751/ES-03434-150309.

Fiksel, J., 2003. Designing resilient, sustainable systems. Environ. Sci. Technol. https://doi.org/10.1021/es0344819.

Filatova, T., Gary Polhill, J., van Ewijk, S., 2016. Regime shifts in coupled socio-environmental systems: review of modelling challenges and approaches. Environ. Model. Softw. https://doi.org/10.1016/j.envsoft.2015.04.003.

Folke, C., 2016. Resilience (republished). Ecol. Soc. https://doi.org/10.5751/ES-09088-210444.

Folke, C., Carpenter, S.R., Walker, B., Scheffer, M., Chapin, T., Rockström, J., 2010. Resilience thinking: integrating resilience, adaptability and transformability. Ecol. Soc. 15(4). https://doi.org/10.1038/nnano.2011.191.

Ghazvini, M.S., Ghezavati, V., Raissi, S., Makui, A., 2017. An integrated efficiency-risk approach in sustainable project control. Sustainability (Switzerland). https://doi.org/10.3390/su9091575.

Gidley, J.M., Fien, J., Smith, J.A., Thomsen, D.C., Smith, T.F., 2009. Participatory futures methods: towards adaptability and resilience in climate-vulnerable communities. Environ. Policy Gov. https://doi.org/10.1002/eet.524.

Grindsted, T.S., 2018. Geoscience and sustainability – in between keywords and buzzwords. Geoforum 91 (May), 57–60. https://doi.org/10.1016/J.GEOFORUM.2018.02.029.

Gruby, R.L., Basurto, X., 2013. Multi-level governance for large marine commons: politics and polycentricity in Palau' s protected area network §. Environ. Sci. Policy 33, 260–272. https://doi.org/10.1016/j.envsci.2013.08.001.

Jaroszweski, D., Hooper, E., Chapman, L., 2014. The impact of climate change on urban transport resilience in a changing world. Prog. Phys. Geogr. https://doi.org/10.1177/0309133314538741.

Jasman, A., McIlveen, P., 2011. Educating for the future and complexity. On the Horizon. https://doi.org/10.1108/10748121111138317.

Jonas, A.E.G., 2011. Region and place: regionalism in question. Prog. Hum. Geogr. 36 (2), 263–272. https://doi.org/10.1177/0309132510394118.

Klein, L., 2012. The three inevitabilities of human being: a conceptual hierarchy model approaching social complexity. Kybernetes. https://doi.org/10.1108/03684921211257829.

Lang, D.J., Wiek, A., Bergmann, M., Stauffacher, M., Martens, P., Moll, P., Swilling, M., Thomas, C.J., 2012. Transdisciplinary research in sustainability science: practice, principles, and challenges. Sustain. Sci. https://doi.org/10.1007/s11625-011-0149-x.

Lenton, T.M., Held, H., Kriegler, E., Hall, J.W., Lucht, W., Rahmstorf, S., Schellnhuber, H.J., 2008. Tipping elements in the Earth's climate system. Proc. Natl. Acad. Sci. 105 (6), 1786–1793. https://doi.org/10.1073/pnas.0705414105.

Mauser, W., Klepper, G., Rice, M., Schmalzbauer, B.S., Hackmann, H., Leemans, R., Moore, H., 2013. Transdisciplinary global change research: the co-creation of knowledge for sustainability. Curr. Opin. Environ. Sustain. https://doi.org/10.1016/j.cosust.2013.07.001.

Morse, W.C., McLaughlin, W.J., Wulfhorst, J.D., Harvey, C., 2013. Social ecological complex adaptive systems: a framework for research on payments for ecosystem services. Urban Ecosyst. https://doi.org/10.1007/s11252-011-0178-3.

Nicolescu, B., 1996. "La Transdisciplinarité": Manifeste. Editions Du Rocher, Monaco. https://doi.org/10.5772/849.

Nicolescu, B., 2010. Methodology of transdisciplinarity – levels of reality, logic of the included middle and complexity. Transdiscip. J. Eng. Sci. https://doi.org/10.1080/02604027.2014.934631.

Osbahr, H., Twyman, C., Neil Adger, W., Thomas, D.S.G., 2008. Effective livelihood adaptation to climate change disturbance: scale dimensions of practice in Mozambique. Geoforum 39 (6), 1951–1964. https://doi.org/10.1016/j.geoforum.2008.07.010.

Pereira, L.M., Hichert, T., Hamann, M., Preiser, R., Biggs, R., 2018. Using futures methods to create transformative spaces: visions of a good Anthropocene in Southern Africa. Ecol. Soc. 23 (1). https://doi.org/10.5751/ES-09907-230119. art19.

Pernaa, H.K., 2017. Deliberative future visioning: utilizing the deliberative democracy theory and practice in futures research. Eur. J. Futures Res. https://doi.org/10.1007/s40309-017-0129-1.

Perrings, C., 1998. Introduction: resilience and sustainability. Environ. Dev. Econ. 3, 221–222.

Poli, R., 2015. Social foresight. On the Horizon. https://doi.org/10.1108/OTH-01-2015-0003.

Ramadier, T., 2004. Transdisciplinarity and its challenges: the case of urban studies. Futures. https://doi.org/10.1016/j.futures.2003.10.009.

Rickards, L., Ison, R., Fünfgeld, H., Wiseman, J., 2014. Opening and closing the future: climate change, adaptation, and scenario planning. Environ. Plann. C Gov. Policy. https://doi.org/10.1068/c3204ed.

Rittel, H.W.J., Webber, M.M., 1973. Dilemmas in a general theory of planning. Policy. Sci. https://doi.org/10.1007/BF01405730.

Rockstrom, J., 2009. A safe operating space for humanity. Nature. https://doi.org/10.1109/TIP.2011.2165549.

Schellnhuber, H.J., 2009. Tipping elements in the earth system. Proc. Natl. Acad. Sci. 106 (49), 20561–20563. https://doi.org/10.1073/pnas.0911106106.

Schwaninger, M., 2012. Making change happen: recollections of a systems professional. Kybernetes. https://doi.org/10.1108/03684921211229451.

Scott, M., 2015. Re-theorizing social network analysis and environmental governance: insights from human geography. Prog. Hum. Geogr. 39 (4), 449–463. https://doi.org/10.1177/0309132514554322.

Simonsen, S.H., Biggs, R., Schlüter, M., Schoon, M., Bohensky, E., Cundill, G., Dakos, V., et al., 2014. Applying Resilience Thinking: Seven Principles for Building Resilience in Social-Ecological Systems. Stockolm Resilience Centre.

Sircar, I., Sage, D., Goodier, C., Fussey, P., Dainty, A., 2013. Constructing resilient futures: integrating UK multi-stakeholder transport and energy resilience for 2050. Futures 49 (49), 49–63. https://doi.org/10.1016/j.futures.2013.04.003.

Sotarauta, M., Srinivas, S., 2006. Co-evolutionary policy processes: understanding innovative economies and future resilience. Futures. https://doi.org/10.1016/j.futures.2005.07.008.

Steffen, W., Persson, Å., Deutsch, L., Zalasiewicz, J., Williams, M., Richardson, K., Crumley, C., et al., 2011. The Anthropocene: from global change to planetary stewardship. Ambio. https://doi.org/10.1007/s13280-011-0185-x.

ten Napel, J., van der Veen, A.A., Oosting, S.J., Groot Koerkamp, P.W.G., 2011. A conceptual approach to design livestock production systems for robustness to enhance sustainability. Livest. Sci. https://doi.org/10.1016/j.livsci.2011.03.007.

Tschakert, P., Pandit, R., Lawrence, C., Barnett, J., Elrick-Barr, C., Tuana, N., New, M., Pannell, D., 2017. Climate change and loss, as if people mattered: values, places, and experiences. Wiley Interdiscip. Rev. Clim. Change. 8(5), e476. https://doi.org/10.1002/wcc.476.

Vervoort, J.M., Bendor, R., Kelliher, A., Strik, O., Helfgott, A.E.R., 2015. Scenarios and the art of worldmaking. Futures. https://doi.org/10.1016/j.futures.2015.08.009.

Wiek, A., Iwaniec, D., 2014. Quality criteria for visions and visioning in sustainability science. Sustain. Sci. 9 (4), 497–512. https://doi.org/10.1007/s11625-013-0208-6.

Wilson, R.S., Herziger, A., Hamilton, M., Brooks, J.S., 2020. From incremental to transformative adaptation in individual responses to climate-exacerbated hazards. Nat. Clim. Change 10 (3), 200–208.

RISK ASSESSMENT

Bridging two cultures of fire risk at the wildland-urban interface: The case of Haifa, Israel

1.1

Yaella Depietri and Daniel E. Orenstein

Faculty of Architecture and Town Planning, Technion—Israel Institute of Technology, Haifa, Israel

CHAPTER OUTLINE

1.1.1 Introduction

Fire regimes (or the long-term nature of fires in an ecosystem) (Brown, 2000) are commonly described based on the specific intensity, severity, seasonality, frequency, and pattern of the fire. These characteristics are themselves determined by weather (i.e., temperature, humidity, and wind), climate, vegetation, and topography, as well as by human factors, such as proximity to roads, trails, picnic areas, and habitations (Moreno et al., 2014). Different combinations of these factors give rise to different fire regimes. In Mediterranean ecosystems, fire regimes have changed in the past decades (Spyratos et al., 2007), particularly due to an increase in the severity of wildfires developing at the wildland-urban interface (WUI) (Calkin et al., 2014; Mell et al., 2010; Moreno et al., 2014; Radeloff et al., 2018).

The WUI is an area where built structures are in close proximity and intermingle with the peri-urban forest and other vegetated areas (Radeloff et al., 2005). The definition of its depth varies between 50 to 200 m around built-up areas and between 100 to 400 m around forested areas, depending on the system (Modugno et al., 2016). Some authors distinguish an "intermixed zone," where houses and wildland vegetation directly intermingle, and an "interface zone," where houses border wildland vegetation (Radeloff et al., 2018).

The term WUI is nearly exclusively used in the context of fire risk (i.e., a combination of the hazard features—fire in this case—and of the vulnerability of the system), this due to the combination of three elements: the human presence, wildland vegetation, and the narrow distance between them. Together these represent the potential for fire impacts (Stewart et al., 2007). At the WUI, the risk of fires is thus the result of two unmanaged growths: unrestricted and poorly planned urban expansion, and the unmanaged growth of the vegetation (Bar-Massada et al., 2014; Pyne, 2001). The main drivers behind the growth of the WUI are the desire of homeowners to be close to open spaces, in proximity to nature and recreational opportunities, and the search for affordable housing and/or privacy (Ewert, 1993; Hendricks and Mobley, 2018; Stewart et al., 2007). As a result, the spatial expansion of the WUI has been demonstrated to be the most important factor driving the increase in fire-suppression expenditures in the United States (Clark et al., 2016).

In order to take the necessary steps to identify enhanced adaptation strategies, it is fundamental to better understand these new sources of risk. In this chapter, we identify some of the specific elements of the system that co-determine fire risk (and then adaptation) at the WUI, and how these factors differ from those generally characterizing forest-fire risk in the open forest. With this aim in mind, we define risk as the combination of the hazard (wildfire in this case) and the vulnerability of the system. Vulnerability is itself generally described as a combination of: exposure (i.e., "the extent to which a unit of assessment falls within the geographical range of a hazard event"); susceptibility (i.e., "the predisposition of elements at risk to suffer harm"); and lack of coping capacity (i.e., the "limitations in terms of access to, and mobilization of, the resources of a community or a social-ecological system in responding to an identified hazard") (Birkmann et al., 2013, p. 200). As expressed by Priority 1 of the Sendai Framework (SF) for Disaster Risk (Understanding Disaster Risk), it is essential to characterize these components of risk to design tailored adaptation strategies (adaptation being the "longer-term and constantly unfolding process of learning, experimentation and change that feeds into vulnerability" to reduce risk) (Birkmann et al., 2013, p. 196). The chapter contributes to meeting Priority 1 of the SF by identifying the factors and specific characteristics of forest fire risk at the WUI and potential adaptation strategies, starting from the analysis of the available literature and following with an in-depth case study application focusing on the city of Haifa, in Israel. We thus target understanding disaster risk at the local level particularly contributing to the following subtask: "to apply risk information in all its dimensions of vulnerability, capacity and exposure of persons, communities, countries and assets, as well as hazard characteristics, to develop and implement disaster risk reduction policies" (Sendai Framework Priority 1, point 24, letter n).

In Section 1.1.1.1 we describe the social-ecological-technological systems (SETS) framework, which we apply toward a holistic understanding of fire risk at the WUI. We then analyze the literature on the topic to identify and discuss the main known factors contributing to this risk, especially in Mediterranean systems (Section 1.1.1.2). We then look at the available adaptation strategies to reduce sources of fire risk at the WUI, as identified in the relevant literature (Section 1.1.1.3). The subsequent sections focus on the characterization of fire risk at the WUI in the city of Haifa and on the identification and evaluation of potential adaptation strategies. Section 1.1.2 presents the methodology used for studying the target area, Section 1.1.3 the results, Section 1.1.4 the discussion, and Section 1.1.5 the conclusions.

1.1.1.1 The social-ecological-technological systems (SETS) framework

To analyze the literature on factors determining fire risk at the WUI, we adopt the social-ecological-technological systems (SETS) framework (Depietri and McPhearson, 2017; McPhearson et al., 2016a, 2016b). This integrated framework is specifically tailored to analyze environmental problems in the urban context where technological considerations assume an important role, one which is often neglected in the traditional conceptualization of coupled social-ecological systems (SES) (e.g., Folke, 2006; Walker et al., 2004). In the urban context, ecosystem services (e.g., water supply, water purification, water infiltration or urban cooling through the presence of vegetation) are often coproduced by ecosystem functions, human inputs, and technological solutions, as it is rarely possible to rely only on the services provided by unmanaged ecosystems (Depietri et al., 2016). In this way, the SETS framework adds previously overlooked dimensions to the understanding of urban ecosystems. The framework, presented in Fig. 1.1.1, is adopted here to analyze the challenge of fire risk at the WUI, which is indeed a multidimensional problem in which the social, ecological, and technological factors of risk are all important and often interrelated.

1.1.1.2 Forest-fire risk at the wildland urban interface in Mediterranean ecosystems

Mediterranean ecosystems, characterized by wet, mild winters, and dry, hot summers, are particularly fire prone (Pereira et al., 2017; Pyne, 2009). Expanding suburbanization and abandonment of traditional rural lifestyles, accompanied by a reduction in traditional agrarian activities, have further increased fire risk in these areas (Galiana-Martin et al., 2011). More recently, the WUI has become the focus of attention as an area particularly prone to fires, not least because of the high exposure of people, buildings, and infrastructures. Fire-related losses, land abandonment, and urban sprawl are highly interdependent in Mediterranean ecosystems (Darques, 2015), where the distribution of large burned surfaces are often found in close vicinity to cities (Modugno et al., 2016).

Based on a selection of studies from the literature addressing wildfire risk at the WUI, Table 1.1.1 lists the main factors contributing to wildfire risk in Mediterranean ecosystems as resulting from the combination of hazard features (a) and vulnerability of the system (itself a combination of exposure, susceptibility, and lack of coping capacity; b, c, and d, respectively). The components of risk are characterized according to the three dimensions of the SETS framework (i.e., social, ecological, and technological), depicting fire risk at the WUI as a distinctly multidimensional problem. Depending on the context, each of the listed variables and factors plays a unique role in creating fire risk. For instance, although ecological factors (e.g., type of trees, the continuity, density, and homogeneity of the forest) leading to fire regulating services or disservices (see Depietri and Orenstein, 2019) often play a major role in causing fire risk in large and sparsely inhabited forests, at the WUI the social (e.g., education and age of the population), economic (e.g., income and insurance), and technological

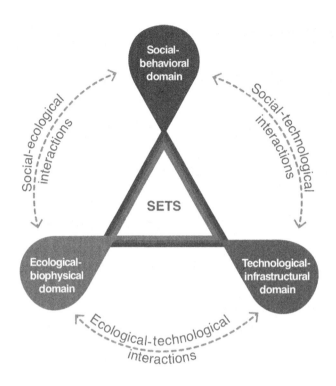

FIG. 1.1.1

The social-ecological-technological systems (SETS) framework.

Source: *Depietri, Y., McPhearson, T., 2017. Integrating the grey, green, and blue in cities: nature-based solutions for climate change adaptation and risk reduction. In: Kabisch, N., et al. (Eds)., Nature-Based Solutions to Climate Change Adaptation in Urban Areas, Theory and Practice of Urban Sustainability Transitions. Springer, Cham, pp. 91–109. doi:10.1007/978-3-319-56091-5_6.*

(e.g., flammability of the built-up area) components also significantly contribute in determining the vulnerability of the system, and consequently its risk. As depicted in Table 1.1.1, it emerges that in the urban context, the technological component is important to the understanding of fire risk, in contrast to the focus on the ecological dimension in the open forest.

1.1.1.3 Adapting to forest fires at the WUI: State of the art

Based on the characterization of fire risk at the WUI depicted in the previous section, adaptation strategies can be identified to address this new configuration of risk. These strategies are unique to the WUI context, rather than simply variations of those usually adopted to prevent and combat forest fires in the open forest. At the WUI, adaptation strategies for dealing with fire risk need to target all the social, ecological, and technological dimensions of the system at risk, as exemplified in Table 1.1.2.

While current approaches for managing wildfire risk focus primarily on forest management and fire suppression (Mahmoud and Chulahwat, 2018), at the WUI additional efforts are needed to define zoning guidelines for the management of vegetation (including buffer zones) and building codes (Pyne, 2001). According to the literature reviewed, strategies in these areas must consider that when people

Table 1.1.1 Factors determining vulnerability and fire risk at the WUI, classified according to the three SETS framework categories (i.e., the social, ecological, and technological) and to the components of risk: hazard (a) and vulnerability in its three components of: exposure (b), susceptibility (c), and coping capacity (d).

Social	Ecological	Technological
(a) Hazard		
Intentionally or accidentally set fires	Biomass (fuel) quantity and height, vegetation type, age and structure, presence of natural firebreaks (e.g., ponds, rivers, assemblages of little flammable vegetation)	Proximity to roads and picnic areas (i.e., ignition probability)
	Weather, climate	Flammable material outside of residential units (e.g., piles of wood, dead vegetation, garbage, construction material, flammable liquids or gas)
	Soil characteristics	
	Landscape diversity, spatial and structural patchiness	
	Topography, elevation	
(b) Exposure		
Density of the population	Biomass of the vegetation; concentration of animal species with restricted mobility or limited habitat	Density of built-up areas, number of dwellings; infrastructure of different types (e.g., roads, railroad tracks, energy infrastructure)
		Houses, buildings and infrastructure design and location with respect to topography (slope inclination)
		Recreational and aesthetic amenities
(c) Susceptibility		
Income, employment status	High concentration of high biomass, invasive and little adapted plant species; endangered animal species	Flammability properties of residential units (e.g., external and internal walls and roofing materials)
Mobility		Flammable material located inside residential units (e.g., nylon curtains, rugs, polyurethane furniture)
Homeowner behavior		
Legislation, law enforcement		
(d) Coping capacity		
Early warnings	Well adapted and local plant and animal species	Fire and smoke detectors
Availability of evacuation plans		Fire suppression and firefighting; availability of stock of water tanks, trucks, planes, fire brigades
		Availability of spark arresters

Table 1.1.2 Possible adaptation strategies for addressing fire risk at the WUI, compiled from the literature and classified according to the three components of the SETS framework.

Social	Ecological	Technological
Adaptation		
Preparing and administering education programming and raising awareness	Administering prescribed burns (in the more distant forested areas form the buildings)	Constructing or improving access roads for emergency vehicles
Preparing and distributing evacuation plans	Mechanical thinning, spacing, "limbing," and trimming, reducing stand and canopy density; maintenance of tall trees and big trees of fire-resistant species, while reducing and clearing young trees	Reducing flammability of structures by constructing with noncombustible materials
Provide individual landowners and residents who are at risk with knowledge of the regulations and best practices regarding building codes	Removing "ladder" fuels and surface vegetation	Removing flammable material (e.g., garbage or construction materials) from around buildings
Providing government incentives and regulations for better urban planning	Maintaining biodiversity and local species; removing invasive species	Implementing land-use plans for building density and location
Providing and maintaining insurance policies	Increasing landscape diversity and patchiness (e.g., through grazing or other agropastoral activities)	Providing fire hydrants, automatic sprinkler systems, cisterns, and other water sources
Reducing socioeconomic inequalities	Creating buffers (around built-up structures) and firebreaks	Securing propane tank storage
		Installing spark arresters and smoke and fire detectors
		Constructing watchtowers

build or purchase homes near forest areas, for the improved access to physical, mental, and aesthetic benefits, they also have expectations that their homes will be protected from fire. This situation sometimes precludes certain fire management strategies at the WUI, such as prescribed burns, due to concerns that the fire might escape control, leading to loss of life and property. On the other hand, residents might oppose strategies that are too invasive and might cause a significant change or an alteration of the peri-urban forest, affecting it's cultural value (Depietri and Orenstein, 2020). So, context-specific strategies are needed.

Reducing flammability of buildings is an important strategy for reducing fire risk at the WUI, and this is generally less relevant in open forests (Mell et al., 2010). Flying embers are the principal cause of house ignition at the WUI. Noncombustible materials are recommended to avoid ignition of buildings via embers. Other solutions, like removing wooden roofs, cleaning up garbage and construction materials around homes, building a defensible space, and facilitating access to firefighters, by widening the network of roads between buildings and the forested areas, are all important strategies for dealing with fires at the WUI (Pyne, 2001). According to Syphard et al. (2014), the most effective strategy is that of reducing woody cover by up to 40%, immediately adjacent to structures. While these and other context-specific strategies need to be envisaged in the WUI, the management of the forest should not be neglected in order to reduce fire risk in these areas. Yet, traditionally, the majority of wildfire suppression expenditures at the WUI has been directed toward fire suppression, and less has been invested in forest management policies (Schoennagel et al., 2009) or in technological solutions to reduce susceptibility.

Spatial patterns of urban development at the WUI affect fire risk. Higher-density urban development, for instance, reduces the cost of fire protection and overall fire risk, as most fire-related damage occurs in low-density housing areas (Radeloff et al., 2005). Clark et al. (2016, p. 656) found that "policies to control the spatial pattern of WUI development can be nearly as effective as policies that completely restrict WUI development." Isolated development, which increases fire risk, could be restricted by means of higher fees and taxes relative to those imposed on dense development, to account for higher fire-risk expenditures in these areas. Fire insurance policies could also have higher premiums for isolated development in high-risk areas (Clark et al., 2016).

We note, as Pyne (2018) suggests, that dealing with wildfires at the WUI requires merging two (now separate) approaches to deal with fire risk: one traditionally employed to deal with fires in open forests, and directed toward managing the vegetation, and another applied when dealing with fires in, and immediately outside, buildings. In the Haifa case study that follows, we investigate concretely how the risk from forest fires at the WUI is characterized and dealt with and how, to this end, combining two existing cultures of fire risk is needed to deal more effectively with this new threat.

1.1.2 Studying forest fires at the WUI in Haifa

1.1.2.1 Case study description

The city of Haifa sits on and develops around the northwest slopes of Mount Carmel (elevation 0–525.4 m above sea level) (Fig. 1.1.2). It has a population of about 281,000 inhabitants and is the third largest city in Israel. It is adjacent to the Carmel National Park, whose vegetation consists of different associations of *Pinus halepensis*, *Pistacia palestina*, *Cistus* sp., and *Quercus calliprinos* (Wittenberg and Malkinson, 2009). Similar vegetation is found in the city where the forest penetrates the built-up areas through a network of undeveloped wadis (or dry riverbeds). Most of the forests of the Carmel-Haifa region are the product of afforestation efforts carried out by the Jewish National Fund (or the Keren Kayemeth LeIsrael, KKL) since the beginning of the last century. In many areas, these planted forests are uniform, dense, monocultured, and even-aged, all characteristics that increase fire risk (Amir and Rechtman, 2006; Osem et al., 2008). The region has, in fact, been periodically affected by forest fires that have mostly occurred within the area of the Carmel National Park (such as the major fires in 1989, 1995, and 2010). A new risk of fires at the WUI has recently become evident due to a large urban fire in Haifa in November 2016. This fire was the first major fire in Israel exclusively urban (Tessler et al., 2019). The fire spread through the vegetated areas of the city, destroying 527 apartments in 77 buildings, leaving 1600 people homeless. The dry weather conditions facilitated the formation of a fast-spreading and intense fire, which proved difficult to extinguish. The event made clear to local authorities that the city needs to adapt to a new type of fire regime, not a fire spreading from the adjacent forest of the Carmel National Park, as had previously been considered, but one that originates

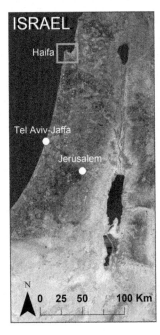

Source: Esri, DigitalGlobe, GeoEye, Earthstar Geographics, CNES/Airbus DS, USDA, USGS, AeroGRID, IGN, and the GIS User Community

FIG. 1.1.2

Location and satellite map of the city of Haifa.

within the city itself, at the WUI. This requires rethinking fire risk and the development of new strategies for the urban context of Haifa, different than those that had been designed to deal with fire risk in the open areas of the Carmel National Park.

1.1.2.2 **Methodology**

We began our study of the target area, Haifa, by reviewing professional committee reports produced following the 1989 and 2010 Carmel National Park fires, which provided recommendations for reducing fire risk in the park area (also see Pereira et al., 2017). Next, we conducted 13 in-depth interviews with experts in the field of fire ecology, management, and risk reduction, regarding the factors at play in the context of fire risk in Haifa. The experts included academics, independent experts, and local authorities, and were identified as the community of specialists most informed regarding the case of Haifa and fire risk. Interviewees were asked to identify areas particularly at risk from wildfire within the city and to indicate and comment on the factors of risk for those areas. Areas at risk (up to five for each respondent) were mapped using the online software Scribblemaps (https://www.scribblemaps.com/).

The software allowed respondents, together with the interviewer, to draw polygons on a satellite image of Haifa around areas of high risk. In the second part of the interview, the experts were asked to discuss the relevance to the urban and peri-urban context of five preselected fire-adaptation strategies that had been recommended by the postfire commissions to deal with fire risk in the Carmel National Park. These included: (1) thinning of the vegetation; (2) creating firebreaks and buffers; (3) introducing grazing; (4) removing pine trees; (5) diversifying the landscape and introducing patchiness through other activities, such as planting fruit trees. The experts were asked to assess these strategies to ascertain whether the recommendations given for dealing with the open forest fires in Carmel National Park were also applicable to the case of reducing fire risk at the WUI of Haifa, in the aftermath of the 2016 fire event. They were also asked to provide any additional comments, which often provided complementary and previously unconsidered information for the characterization of fire risk in the city. The interviews were carried out in English and lasted between 60 to 120 minutes. They were then transcribed and analyzed using Atlas.ti (https://atlasti.com/), a qualitative data analysis and research software for identifying common and recurring themes in the transcription of interviews.

We also supplemented the information derived from the interviews with transcripts from public lectures given at a postfire symposium held at the University of Haifa in November 2017, where local authorities, city fire department officials, military personnel, and academics gathered to discuss their institutions' perspectives on fire risk, fire prevention, and firefighting activities, focusing particularly on their experiences from the 2016 Haifa fire and its aftermath.

1.1.3 Results

1.1.3.1 Factors and configuration of fire risk in Haifa

In the initial part of the interviews, we focused on factors contributing to hazard intensity and exposure and less on susceptibility, which, as illustrated in Table 1.1.1c, is represented primarily by the social dimension. There was a consensus among the interviewees that, in the case of fires at the WUI in Haifa, the three main factors determining hazard intensity were: the type, location, height, and density of the vegetation; the orientation of the wadi or the green area with respect to wind direction; and the slope of the wadi and its inclination, especially with respect to the location of the built-up area (which increases exposure). Fig. 1.1.3 indicates the green areas of Haifa specified by the respondents as particularly at risk of fire due to different combinations of these three factors of risk. Most of these areas are concentrated in the southern parts of the city, where there is a prominence of wadis (ephemeral river beds) with an east-west orientation and dense tree growth, and where residential structures and infrastructure are built on the upper slopes of the wadis. Those areas that did not burn in the 2016 fire continue to be characterized by a high concentration of pine trees. Despite these specific areas, the respondents stressed that most of the city is, in fact, at risk from wildfire.

1.1.3.1.1 The hazard

The presence of dense forested areas and tall trees near infrastructures and buildings was considered to be a primary source of risk. Furthermore, while the factors were listed in a different order of relevance by the individual respondents, vegetation characteristics emerged as the primary factor for which intervention and manipulation is possible in order to reduce fire risk in the city.

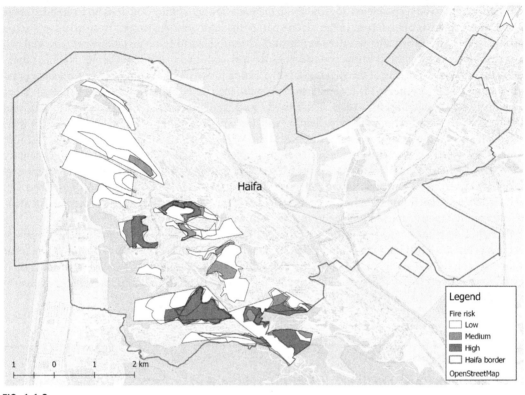

FIG. 1.1.3

Areas at high fire risk as indicated by fire experts.

Source: Depietri, Y., Orenstein, D.E., 2020. Managing fire risk at the wildland-urban interface requires reconciliation of tradeoffs between regulating and cultural ecosystem services. Ecosyst. Serv. 43 (in press).

Highly flammable Aleppo pines, especially those close to buildings, were of utmost concern to the experts. As mentioned, although occurring naturally in small numbers in the area, the widespread distribution of this species of pine in the region is mostly due to historical afforestation activities (Ne'eman et al., 1997). The species was selected for afforestation in the past because it grows quickly, requiring minimal care and water (Stemple, 1998). However, it proved to be invasive, spreading rapidly and contributing to fire risk. Most of the experts interviewed stressed that, the higher the density and the number of tall and mature Aleppo pines, the greater the risk. This species of pine is very flammable and produces fires of high temperatures. Burning cones can be propelled hundreds of meters, which further facilitates the spread of fire to nearby but disconnected forested areas. For some of the experts, the spread of Aleppo pines is the primary driver of more frequent catastrophic fires in Haifa and the Carmel. This represents one aspect of the ecological dimension of the hazard, when we apply the SETS framework to the case of Haifa.

Other biophysical and meteorological factors include the west-east orientation of the wadis, which increases fire risk by interacting with the dry and hot "Sharav" wind, a meteorological condition that greatly increases the occurrence and intensity of fires as it blows from the east. This wind can cause a rapid drop in humidity (to as low as 5%) and a rapid increase in temperature in a very short period of time (raising the temperature by up to 10°C in a span of a few hours).

Wildfires also burn faster and more intensely when climbing up a slope than when spreading along flat ground. According to the experts, a fire on a steep slope will burn with longer flame lengths. Furthermore, the wadis themselves produce a channel effect that leads to an even faster-moving fire. Some experts suggested that considering only the fuel load and the inclination of the slope would suffice in determining areas at high fire hazard. However, some important factors are less predictable, such as the precise wind direction, which can become chaotic once it reaches the urban fabric and the network of wadis.

Other important factors of risk to consider are the proximity of the forest to a road or to areas developed for outdoor activities. Areas of high interest for recreational activities (such as picnic areas) and areas within 50 m from a road were considered to be at higher risk due to the increased chance of inadvertent ignition. The presence of waste (which includes household trash, construction waste, vegetation clippings, and discarded furniture and tires) in the wadis and around buildings could increase fire intensity. These factors fall within the social-technological component of the SETS framework.

1.1.3.1.2 Exposure

If a built-up area, such as a residential area, is located upslope and the slope is steep and densely forested, the exposure and risk of damage from a wildfire increases. The widespread absence of buffer areas between the built-up areas and the forested areas further increases exposure of buildings and people to fires in Haifa. The steep slopes often hamper the construction of roads around the buildings, which, in the words of some experts, greatly increases the exposure of these areas as it makes them virtually indefensible once a fire erupts.

1.1.3.2 Approaches for adapting to fire risk at the WUI in Haifa

As mentioned, prior to the 2016 WUI fire in Haifa, policy makers and managers considered the risk of wildfires as coming primarily from the Carmel National Park, from the south and east of the city. This is in fact where most major fires had occurred in the past and they rarely reached the WUI of Haifa. However, the November 2016 fire, which started within the city and spread through its undeveloped vegetated corridors, made the local authorities and experts aware that the city should be prepared to deal with this new type of fire regime.

Experts were asked about the effectiveness of different ecosystem-based strategies to reduce wildfire intensity and exposure in Haifa among the five management options listed in the methodology section. Here we analyze how their recommendations also differed from those of the post-1989 and post-2010 fire commissions.

1.1.3.2.1 Thinning

Thinning was deemed an important option, but not enough if implemented in isolation. Thinning at the WUI, according to respondents, should be practiced especially to prevent fires from reaching tree canopies and for clearing the vegetation around the buildings. Canopy fires are indeed very difficult to stop, according to the experts. Further, on the steep slopes of Haifa's wadis, the distance between trees should be increased because the fire climbs faster there. Thinning is an effective practice to slow down fires in these cases. Low vegetation should be removed at the end of the summer to prevent ground fires from developing into canopy fires. This practice might detrimentally affect the landscape aesthetic value, according to the experts, but it should still be practiced in specific cases. Despite this widely shared opinion, thinning has not been systematically implemented in the city, even though the 2016 crisis helped create opportunities to advance intensive fire management practices in the context of restoration processes (see Tessler et al., 2019).

1.1.3.2.2 Firebreaks and buffers

In contrast to the recommendations for the Carmel Forest fire risk management, respondents in our research were wary regarding the effectiveness of firebreaks to reduce fire risk at the WUI of Haifa. In 2016, the fire jumped from one side of a road to the other and from one wadi to another. In these conditions, firebreaks would have been of little help. Also, considering that maintenance of firebreaks requires substantial and continued investment and that they affect the aesthetic, recreational, and ecological value of the green areas of Haifa (see also Depietri and Orenstein, 2020), respondents were understandably reticent to recommend them as an appropriate management strategy. Overall, firebreaks were considered the least cost-effective and desirable strategy among the five suggested to deal with wildfire risk in the city.

Buffers around the urban area, with roads to improve the access of firefighters at the precise interface between buildings and the forest, were deemed, in contrast, desirable. Buffers (providing a defensible space) were considered very important to prevent fires from reaching structures. However, some experts feared that a buffer could lead to an elimination of most of the wild vegetation within the city, particularly in the narrowest vegetated areas and wadis. Still, aiming for narrow, defensible buffers, such as the 30m depth suggested by FEMA (2008), may be proportional to and necessary for the conditions of Haifa. At the same time, defensible buffers accessible by roads for firefighting could provide new recreational opportunities, as the buffers would function as trails or paths for recreational activities. Orchards and gardens could supplant the dense vegetation in these buffer areas and could provide new recreational opportunities for inhabitants to enjoy nature at the WUI. Areas with typical, brush-like *maquis* vegetation would not need significant interventions because of their relatively low flammability. *Maquis* vegetation has high ecological and aesthetic value, it represents the characteristic vegetation of the area and is highly appreciated by city residents (Depietri and Orenstein, 2020).

1.1.3.2.3 Removing pine trees

Little doubt was expressed among the experts that management priority should be given to the removal, or significant reduction, of Aleppo pine trees from most of the green areas of Haifa, and particularly from areas close to buildings. This strategy was considered by some respondents as the most cost-effective ecosystem-management strategy among those suggested. Pine trees should be removed from within a minimum of 10 to 15m of houses and these areas should be planted, instead, with non- or little flammable, low vegetation (e.g., shrubs) and irrigated plants. Respondents specified that there should be at least 3m distance between each tree canopy in general, and even further apart in the case of pine trees. As Aleppo pines tend to resprout very easily, intensive management of vegetation would be required in the first few years right after the occurrence of a fire or after the mechanical removal of these trees.

Removing pine trees may face some community opposition, considering that Haifa residents feel that these trees are a natural part of the history and landscape of the city. Respondents recommend that pines be replaced with carob and oak trees, which are local species, and they are shorter and less flammable than pines. Oak trees located very close to buildings should nonetheless be thinned, although different, less strict standards should be applied in their case. Cypress trees are being considered for planting to reduce fire risk in Haifa, as they are less flammable than pines and can slow down the spread of fire.

1.1.3.2.4 Landscape patchiness
Orchards, olive trees, and fruit trees, experts agreed, could be planted in some areas of the wadis to increase landscape patchiness and diversity. These interventions could at the same time improve recreational opportunities (e.g., picnic areas) and increase the aesthetic value of the landscape. However, it may be a challenging practice to implement due to the topography of the city. It would require terracing wadi slopes and would entail additional investments in long-term management. Such strategies can nonetheless be implemented synergistically with the creation of buffer zones.

1.1.3.2.5 Grazing
Experts had some concerns regarding the introduction of domestic grazing in the urban context, mainly due to the need to pay shepherds and due to the potential inconveniences associated with raising herds of domesticated animals within the urban setting, including dangers to traffic and the smells that it might generate. In this regard, fences would need to be erected.

In any case, goats and sheep would be preferable to cattle, as cattle cause more disturbances to the soil, increase dung production, and create other nuisances. Cattle are still more suitable for removing herbaceous vegetation, while goats primarily feed on woody vegetation. The most efficient use of grazing would be to maintain potential firebreaks, but, as noted earlier, that would be of little relevance in Haifa, where firebreaks were considered by experts as ineffective. Local nature enthusiasts were positively predisposed to introducing grazing into Haifa's wadis, as reported in a companion study (Depietri and Orenstein, 2020).

1.1.3.2.6 Other strategies
Removing flammable garbage and construction material from close to buildings was considered as important for reducing fire risk in Haifa. Overgrowth of invasive species that colonize the disturbed environment around buildings, especially after a fire, should also be removed. In private gardens and lots, owners should be required, through regulations and incentives, to care for their lots and gardens and reduce the concentration of flammable vegetation and materials. Insurance schemes that consider owners' efforts to reduce fire risk on their property are deemed to be important mechanisms to lower aggregate fire risk in the city.

Besides these measures that focus on ecological aspects of the landscape, socio-economic and technological measures can be implemented to reduce risk. In Haifa and the Carmel National Park, fires do not start because of natural causes (e.g., lightning); they are only ignited by human activities. Most fires in the region start within 50 m of a road or close to army training camps. Therefore, a respondent suggested the importance of education and of awareness regarding fire risk as fundamental to insuring an adequate public response. People should be aware that they live in a fire-prone environment and should know how to prevent fires and how to react in the event of fire. Fire prevention and preparedness, through awareness campaigns, mechanisms for alerts, and evacuation plans, are important strategies for reducing losses.

Firewatchers and patrols should be organized to keep the city's most sensitive areas under control. Outdoor smoke and heat detectors placed in the forest at the WUI (a technological solution) are also potentially effective strategies to detect fire at its very onset, alerting firefighters at the very initial stages of fire propagation by sending real-time signals to the fire department. Firefighters would then have better chances to prevent the fire from spreading, especially in the case of favorable weather conditions. In urban areas, one cannot drop fire retardants or resort to aerial firefighting, as is done for fires in the open forest. The fire needs to be dealt with at much smaller spatial scales, sometimes from building to building. The time to react to fire spread is also much shorter when compared to that when dealing with fires in the open forest, and authorities need to act quickly to evacuate homes, schools, hospitals, and commercial areas.

According to the fire department of Haifa, in case of fire, there are more losses that occur in apartment buildings than in offices because fire regulations are currently too flexible for residential structures in the country. Typically, building codes only apply to new developments or to buildings undergoing comprehensive improvements or repairs, leaving existing buildings at risk. The maintenance of buildings is indeed poor regarding fire readiness. Technological solutions to diminish flammability of buildings are necessary. The fire department representative suggested during a lecture in 2017 that there are new technological developments that can be adopted to reduce fire risk in buildings, for example, windows that break at high temperatures so that smoke can flow out of the buildings. The use of flame-retardant building materials is also important in this context. Fireproof elevators could be made mandatory in high-rise buildings. Safe rooms with filters for smoke, sprinklers, smoke detectors, and escape routes should become an integral part of building construction. Also, new chemicals are available that can be added to water to make the liquid more effective in extinguishing fires. Finally, the fire department suggested the need to have smaller, but more widely distributed, fire stations across the city and to enlarge the firefighting crews (including volunteers). Evacuation centers should be clearly identified and made available during a fire event.

1.1.4 Discussion

The challenge of forest fires at the WUI brings together two cultures of fire: that of urban fires (which focuses on building types and structures) and that of fires in wildlands, as also suggested by Pyne (2018). In this research, it emerged clearly that the technological component gains a prominent role in determining the vulnerability of the system to forest fire at the WUI, while this is less relevant in the open forest, where prescribed burning, the diversification of the landscape, and the management of the vegetation are traditionally recommended to avoid catastrophic fires. At the WUI, the ecosystem management perspective for open spaces and the urban/indoor fire prevention and response perspective need to merge. This combined perspective became apparent from the analysis of the case of Haifa the drivers of fire risk in Haifa (e.g., location of buildings upslope and near a dense forest; the need to resort to technology to reduce risk).

Applying technological solutions may offer leeway regarding implementation of ecological management solutions, for example by allowing preserving some trees, since they are highly valued by local residents who feel that they contribute to the special forested character of the city (Depietri and Orenstein, 2020). Putting smoke and heat sensors in the open land around the urban areas, a strategy traditionally adopted in buildings, is now suggested as important for dealing with fire risk at the

WUI. Building materials need to be adapted to fire risk, as managing the vegetation alone might not be enough.

It emerged from the case study that some strategies generally considered as effective in the open forest might not be practicable at the WUI (e.g., firebreaks or grazing, which might generate ecosystem disservices). So, while current approaches to managing wildfires focus on fire suppression and managing fuel build-up (Mahmoud and Chulahwat, 2018), these might not be sufficient and more attention needs to be paid to technological and social aspects of the system to understand fire risk at the WUI, as also suggested by Mahmoud and Chulahwat (2018). As mentioned, these include using fireproof materials in buildings, installing automatic sprinkler systems and fire/smoke detectors, and managing vegetation in the ignition zone immediately adjacent to built-up areas.

Fig. 1.1.4 displays how understanding and adapting to fire risk at the WUI is specific to an area where the forest meets the city, and in which two different, traditionally distinct paradigms of risk need to be brought together at the theoretical and practical levels. In the case of Haifa, the fire department tends to approach this source of fire risk as they would with indoor fires, while ecologists and some fire experts tend to focus on ecosystem management or integrated approaches. Bringing together these two cultures of risk to assess and deal with fires at the WUI will require institutional reform and a dedicated cross-sectoral collaboration. An improved collaboration between the fire department and the environmental protection authorities is deemed fundamental to understanding and reducing wildfire risk at the WUI. To this end, an extended process of engaging with multiple stakeholders will be needed. Homeowners are also likely to be required to play a central role in lowering vulnerability to wildfires at the WUI, as also suggested by Mahmoud and Chulahwat (2018). Postfire restoration offers important opportunities to implement these principles. The municipality of Haifa, for instance, did implement some beneficial practices in the burnt areas after the 2016 fire, including reforesting with broadleaf trees, and building unpaved roads to improve access for firefighters (see Tessler et al., 2019).

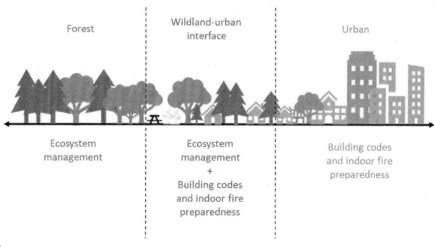

FIG. 1.1.4

Types of landscape and associated prevalent fire-risk reduction strategies. The figure shows that at the WUI two cultures of fire risk need to be brought together to adapt the system.

1.1.5 Conclusions

Understanding forest fires at the WUI requires a multidisciplinary and multisectoral approach in which different cultures of risk—those of urban, indoor fires and those of wildland fires—need to be brought together. By applying the SETS framework to the insights gathered through the literature and the Haifa case study, it becomes clear that the social, ecological, and technological components of risk and adaptation are all relevant when we are dealing with forest fires at the WUI. From our analysis it surfaced that the technological component of the system is fundamental, alongside ecological and social ones, to understand wildfire vulnerability and risk at the WUI. Forest fires at the WUI are more complex phenomena in which urban and forest elements of risk meet and intermingle. Innovative strategies and technological solutions are thus required to deal with this risk, in addition to the traditional management of the vegetation recommended to deal with fires in the open forest.

Acknowledgments

We thank Ronit Cohen for her help in designing Fig. 1.1.2 and for her assistance translating the expert lectures. This research was supported by a grant to Daniel E. Orenstein from the Israel Science Foundation (Grant No. 1835/16). Yaella Depietri was also partially funded by a Zeff Post-Doctoral Fellowship at the Technion - Israel Institute of Technology.

References

Amir, S., Rechtman, O., 2006. The development of forest policy in Israel in the 20th century: implications for the future. Forest Policy Econ. 8, 35–51. https://doi.org/10.1016/j.forpol.2004.05.003.

Bar-Massada, A., Radeloff, V.C., Stewart, S.I., 2014. Biotic and abiotic effects of human settlements in the wildland–urban interface. BioScience 64, 429–437. https://doi.org/10.1093/biosci/biu039.

Birkmann, J., Cardona, O.D., Carreño, M.L., Barbat, A.H., Pelling, M., Schneiderbauer, S., Kienberger, S., Keiler, M., Alexander, D., Zeil, P., Welle, T., 2013. Framing vulnerability, risk and societal responses: the MOVE framework. Nat. Hazards 67, 193–211. https://doi.org/10.1007/s11069-013-0558-5.

Brown, J.K., 2000. Introduction and fire regimes. In: Wildland Fire Ecosyst. Eff. Fire Flora. USDA For. Serv, pp. 1–8 (Gen. Tech. Rep. RMRS-GTR-42-Vol. 2).

Calkin, D.E., Cohen, J.D., Finney, M.A., Thompson, M.P., 2014. How risk management can prevent future wildfire disasters in the wildland-urban interface. Proc. Natl. Acad. Sci. U. S. A. 111, 746–751. https://doi.org/10.1073/pnas.1315088111.

Clark, A.M., Rashford, B.S., McLeod, D.M., Lieske, S.N., Coupal, R.H., Albeke, S.E., 2016. The impact of residential development pattern on wildland fire suppression expenditures. Land Econ. 92, 656–678.

Darques, R., 2015. Mediterranean cities under fire. A critical approach to the wildland–urban interface. Appl. Geogr. 59, 10–21. https://doi.org/10.1016/j.apgeog.2015.02.008.

Depietri, Y., McPhearson, T., 2017. Integrating the grey, green, and blue in cities: nature-based solutions for climate change adaptation and risk reduction. In: Kabisch, N. et al., (Eds), Nature-Based Solutions to Climate Change Adaptation in Urban Areas, Theory and Practice of Urban Sustainability Transitions. Springer, Cham, pp. 91–109. https://doi.org/10.1007/978-3-319-56091-5_6.

Depietri, Y., Orenstein, D.E., 2019. Fire-regulating services and disservices with an application to the Haifa-Carmel region in Israel. Front. Environ. Sci. 7. https://doi.org/10.3389/fenvs.2019.00107.

Depietri, Y., Orenstein, D.E., 2020. Managing fire risk at the wildland-urban interface requires reconciliation of tradeoffs between regulating and cultural ecosystem services. Ecosyst. Serv. 43. https://doi.org/10.1016/j.ecoser.2020.101108.

Depietri, Y., Kallis, G., Baró, F., Cattaneo, C., 2016. The urban political ecology of ecosystem services: the case of Barcelona. Ecol. Econ. 125, 83–100. https://doi.org/10.1016/j.ecolecon.2016.03.003.

Ewert, A.W., 1993. The wildland-urban interface: introduction and overview. J. Leis. Res. 25, 1–5. https://doi.org/10.1080/00222216.1993.11969905.

FEMA, 2008. Home Builder's Guide to Construction in Wildfire Zones; 2008. (Technical Fact Sheet Series).

Folke, C., 2006. Resilience: the emergence of a perspective for social–ecological systems analyses. Glob. Environ. Change 16, 253–267. https://doi.org/10.1016/j.gloenvcha.2006.04.002.

Galiana-Martin, L., Herrero, G., Solana, J., 2011. A wildland–urban interface typology for forest fire risk management in Mediterranean areas. Landsc. Res. 36, 151–171. https://doi.org/10.1080/01426397.2010.549218.

Hendricks, M.D., Mobley, W., 2018. To Avoid Future Catastrophes Like the California Fires, We Must Learn to Build Smarter. Huffington Post.

Mahmoud, H., Chulahwat, A., 2018. Unraveling the complexity of wildland urban interface fires. Sci. Rep. 8, 9315. https://doi.org/10.1038/s41598-018-27215-5.

McPhearson, T., Haase, D., Kabisch, N., Gren, Å., 2016a. Advancing understanding of the complex nature of urban systems. Ecol. Indic. https://doi.org/10.1016/j.ecolind.2016.03.054.

McPhearson, T., Pickett, S.T.A., Grimm, N.B., Niemelä, J., Alberti, M., Elmqvist, T., Weber, C., Haase, D., Breuste, J., Qureshi, S., 2016b. Advancing urban ecology toward a science of cities. BioScience 66, 198–212. https://doi.org/10.1093/biosci/biw002.

Mell, W.E., Manzello, S.L., Maranghides, A., Butry, D., Rehm, R.G., 2010. The wildland–urban interface fire problem—current approaches and research needs. Int. J. Wildland Fire 19, 238–251. https://doi.org/10.1071/WF07131.

Modugno, S., Balzter, H., Cole, B., Borrelli, P., 2016. Mapping regional patterns of large forest fires in wildland–urban interface areas in Europe. J. Environ. Manage. 172, 112–126. https://doi.org/10.1016/j.jenvman.2016.02.013.

Moreno, J., Arianoutsou, M., González-Cabán, A., Mouillot, F., Oechel, W., Spano, D., Thonicke, K., Vallejo, V., Vélez, R., 2014. Forest Fires Under Climate, Social and Economic Changes in Europe, The Mediterranean and Other Fire-Affected Areas of the World. FUME: Lessons Learned and Outlook.

Ne'eman, G., Perevolotsky, A., Schiller, G., 1997. The management implications of the Mt. Carmel research project. Int. J. Wildland Fire 7, 343–350. https://doi.org/10.1071/wf9970343.

Osem, Y., Ginsberg, P., Tauber, I., Atzmon, N., Perevolotsky, A., 2008. Sustainable management of Mediterranean planted coniferous forests: an Israeli definition. J. For. 106, 38–46. https://doi.org/10.1093/jof/106.1.38.

Pereira, M.G., Hayes, J.P., Miller, C., Orenstein, D.E., 2017. Fire on the hills: an environmental history of fires and fire policy in Mediterranean-type ecosystems. In: Vaz, E. et al., (Eds), Environmental History in the Making, Environmental History. Springer, Cham, pp. 145–169. https://doi.org/10.1007/978-3-319-41085-2_9.

Pyne, S.J., 2001. The fires this time, and next. Science 294, 1005–1006. https://doi.org/10.1126/science.1064989.

Pyne, S.J., 2009. Eternal flame: an introduction to the fire history of the Mediterranean. In: Chuvieco, E. (Ed.), Earth Observation of Wildland Fires in Mediterranean Ecosystems. Springer Berlin Heidelberg, Berlin, Heidelberg, pp. 11–26. https://doi.org/10.1007/978-3-642-01754-4_2.

Pyne, S., 2018. All wildfires are not alike, but the US is fighting them that way. The Conversation. http://theconversation.com/all-wildfires-are-not-alike-but-the-us-is-fighting-them-that-way-99251. Accessed 5 April 2020.

Radeloff, V.C., Hammer, R.B., Stewart, S.I., Fried, J.S., Holcomb, S.S., McKeefry, J.F., 2005. The wildland–urban interface in the United States. Ecol. Appl. 15, 799–805. https://doi.org/10.1890/04-1413.

Radeloff, V.C., Helmers, D.P., Kramer, H.A., Mockrin, M.H., Alexandre, P.M., Bar-Massada, A., Butsic, V., Hawbaker, T.J., Martinuzzi, S., Syphard, A.D., Stewart, S.I., 2018. Rapid growth of the US wildland-urban interface raises wildfire risk. Proc. Natl. Acad. Sci. U. S. A. 115, 3314–3319. https://doi.org/10.1073/pnas.1718850115.

Schoennagel, T., Nelson, C.R., Theobald, D.M., Carnwath, G.C., Chapman, T.B., 2009. Implementation of National Fire Plan treatments near the wildland-urban interface in the western United States. Proc. Natl. Acad. Sci. U. S. A. 106, 10706–10711. https://doi.org/10.1073/pnas.0900991106.

Spyratos, V., Bourgeron, P.S., Ghil, M., 2007. Development at the wildland–urban interface and the mitigation of forest-fire risk. Proc. Natl. Acad. Sci. U. S. A. 104, 14272–14276. https://doi.org/10.1073/pnas.0704488104.

Stemple, J., 1998. A brief review of afforestation efforts in Israel. Rangelands 20, 15–18.

Stewart, S.I., Radeloff, V.C., Hammer, R.B., Hawbaker, T.J., 2007. Defining the wildland–urban interface. J. For. 105, 201–207. https://doi.org/10.1093/jof/105.4.201.

Syphard, A.D., Brennan, T.J., Keeley, J.E., 2014. The role of defensible space for residential structure protection during wildfires. Int. J. Wildland Fire 23, 11. https://doi.org/10.1071/WF13158.

Tessler, N., Borger, H., Rave, E., Argaman, E., Kopel, D., Brook, A., Elkabets, E., Wittenberg, L., 2019. Haifa fire restoration project—urban forest management: a case study. Int. J. Wildland Fire 28, 485–494. https://doi.org/10.1071/WF18095.

Walker, B., Holling, C.S., Carpenter, S.R., Kinzig, A., 2004. Resilience, adaptability and transformability in social–ecological systems. Ecol. Soc. 9, 5.

Wittenberg, L., Malkinson, D., 2009. Spatio-temporal perspectives of forest fires regimes in a maturing Mediterranean mixed pine landscape. Eur. J. For. Res. 128, 297. https://doi.org/10.1007/s10342-009-0265-7.

A review of flood vulnerability indices for coastal cities

1.2

Mario A. Giampieri

Department of Urban Studies and Planning, Massachusetts Institute of Technology, Cambridge, MA, United States

Chapter outline

1.2.1 Introduction

The global population is becoming increasingly urban, with over 55% of people residing in urban areas as of 2018 and 68% projected to live in urban areas by 2050 (UN Department of Economic and Social Affairs, 2018). While much of this growth will occur across the urban gradient, urbanized coastal areas exhibit higher rates of population growth than inland areas and are more directly at risk from climate change-related sea level rise and coastal flood events than inland areas (Neumann et al., 2015). As of

2000, over 625 million people reside in low elevation coastal zones (LECZs) less than 10 m above sea level across the globe (Neumann et al., 2015). As coastal areas urbanize, the extent to which extreme weather events, augmented by changing climatic conditions, affect populated areas will continue to provide a challenge for policymakers and planners alike. Thirty-three percent of urban areas within LECZs are exposed to high-frequency floods as of 2000, and this number is expected to rise to 38% by 2030 (Güneralp et al., 2015).

The process of urbanization itself can increase vulnerability to flooding, as natural buffers such as wetlands or mangroves are replaced with hard structures (Alam and Collins, 2010; Papathoma and Dominey-Howes, 2003), and new residential areas are created adjacent to the coastline (Adelekan, 2010; Kebede and Nicholls, 2012), the effects of which will be exacerbated by sea level change (Muis et al., 2016; Wong et al., 2014). Vitousek et al. (2017) estimate that climate-change-induced sea-level rise in conjunction with storm surge will result in the doubling of coastal flooding frequency over the coming decades, particularly in areas near the equator. Yin et al. (2010) demonstrate that sea-level change can vary between and within oceanic basins, producing distinct and localized sea-level change conditions in coastal areas around the world. As urbanization continues, the variability and uncertainty associated with sea-level change predictions and extreme event frequency necessitate concerted effort to understand risk associated with flood events and to mitigate and adapt accordingly.

The Sendai Framework for Disaster Risk Reduction (SFDRR) outlines a strategy for mitigating the impact caused by various kinds of hazards based on four guiding priorities: understanding disaster risk; strengthening governance; investing in disaster risk reduction; and enhancing preparedness and adapting to changing global realities (UNISDR, 2015,[a] p. 14). Disaster risk is described as a probabilistic function of hazard, vulnerability, and the capacity to respond to or mediate damage from hazard events (UNGA, 2016; Field et al., 2012; Flanagan et al., 2011; McCarthy and Intergovernmental Panel on Climate Change, 2001). The SFDRR draws from the Hyogo Framework for Action in defining vulnerability as "the conditions determined by physical, social, economic and environmental factors or processes, which increase the susceptibility of a community to the impact of hazards" (UNISDR, 2005, 2015). The first priority outlined in the SFDRR is to understand disaster risk to be able to assess pre- and postdisaster conditions, as well as develop and implement disaster reduction policies and practices based on the most relevant quantitative and qualitative data available (UNISDR, 2015). This priority encompasses the imperative to identify relevant information and stakeholders; develop the capacity to collect and understand this information as it relates to disaster risk; and to effectively communicate this information to minimize risk by informing policy, preparation, and recovery actions.

Understanding potential impacts of hazards is key to minimizing loss of life and sustainable development. Issues related to resource availability, tenure, economic opportunity, and development goals in general are compounded by flood events and hazards in general (Pilon et al., 2004). Understanding disaster risk necessarily includes understanding the root socioeconomic, gender, political, and cultural inequalities that exist within and between places affected by hazards (UNISDR, 2015, p. 15). Vulnerability assessments can aid decision makers in understanding how hazards become disasters and inform how to mitigate future impacts in both ex-ante and ex-post situations (Mani Murali et al., 2013; Kaplan et al., 2009). This chapter provides a critical comparison of vulnerability assessment literature that focuses specifically on urbanized coastal areas with reference to goals of identifying relevant

---

[a]UNISDR is now known as UNDRR.

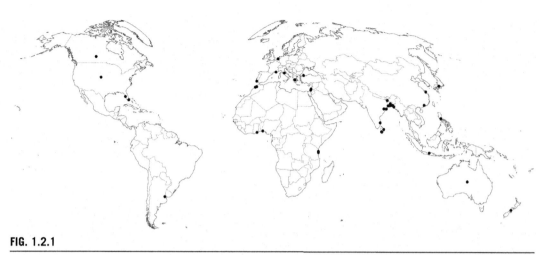

FIG. 1.2.1

Areas of interest of the vulnerability assessments included in this study.

information, generating baseline assessments of vulnerability, contributing to the larger body of knowledge on risk reduction, and effectively communicating results put forth in the first priority of the Sendai Framework for Disaster Risk Reduction.

A multitude of efforts to quantify vulnerability to different hazards have proliferated in recent years (Fig. 1.2.1) in effort to guide development and risk mitigation. However, different practitioners employ different conceptions of vulnerability when completing their assessments. A body of assessments focus exclusively on the biogeophysical components of a system, such as elevation, slope, geomorphology, and wave height to determine vulnerability (e.g., Gornitz, 1991). This approach has been labeled the "Risk-Hazard" model framework, in which exposure to stressors is the primary determinant of vulnerability, regardless of qualitative variation between personal, social, and political systems impacted (Turner et al., 2003). This model has been extended to capture the ways in which different nonhuman system elements may be impacted by hazards in different ways (e.g., Papathoma and Dominey-Howes, 2003). This conception of vulnerability will be referred to as an "external" approach in this study.

A second body of literature positions the concept of vulnerability in terms of the ways in which different persons, social groups, or political systems experience and are affected by hazard events differentially based on factors such as poverty, employment, education, race, and language proficiency (e.g., Wisner et al., 2004; Cutter et al., 2003; Flanagan et al., 2011). This body of work treats vulnerability as an internal or underlying condition that is exacerbated or made apparent during hazard events (Adger, 1996; Newton and Weichselgartner, 2014) and in the ensuing recovery period (Burton, 2015), and focuses less on exposure to hazards than the Risk-Hazard model. This conception of vulnerability will be referred to as an "internal" approach in this study.

A third body of literature expands the definition of vulnerability to also include some combination of resilience (expressed as coping mechanisms) and adaptive capacity relative to extreme events. Birkmann (2006, p. 19) cites Bohle (2001)'s establishment of the notion of the dual structure of vulnerability, which couples external factors (exposure) with internal factors (susceptibility) and coping capacity to describe this synthesis. In this way, elements of exposure and susceptibility are offset by qualities of different social and physical elements of a system (Kaplan et al., 2009; Balica et al., 2012;

Adger, 2006). This framework has been extended to capture the internal and external vulnerabilities of both human and natural system elements, resulting in an integrated, multiscalar, human-environment interaction-based understanding of vulnerability as it relates to hazard events (Turner et al., 2003; Birkmann, 2006; Birkmann and Fernando, 2008). This conception of vulnerability will be referred to as an "integrated" approach in this study.

1.2.1.1 Qualitative and quantitative measures of vulnerability

SFDRR illustrates the need to consider multiple, interacting elements of human-environment systems to be able to effectively respond to and offset the impacts of multihazard disaster risks (UNISDR, 2015, pp. 10–11). Quantitative measurement lends itself to hazard assessment by providing empirical and anecdotal observation of hazard events (Brakenridge et al., 2013), and is useful for vulnerability and resilience assessment by informing priority identification, monitoring progress over time, and comparing potential interventions (Committee on Increasing National Resilience, 2012). Despite the clear need for understanding of vulnerability across different parts of urban systems, gaps in data availability, resolution, and irregular or dated data vintage, as well as the ways in which different indicators are combined together to form composite measures make synthesizing this information a difficult task. Each of these issues also makes comparing across methodologies and geographies difficult. Combined or averaged assessment results can obfuscate extreme values that exist within subgeographies or relative to certain factors (Fekete, 2011; Alam and Collins, 2010; Adger, 2006). These challenges can limit the utility of results, or worse, provide a false sense of security or danger (Fekete, 2011). These misrepresentations may also affect the allotment of resources and policies aimed at reducing disaster risk.

Vulnerability is a complex and multifaceted topic. There have been numerous efforts across the globe made by governments, NGOs, and scholars operating at multiple scales to understand physical and nonphysical elements of vulnerability in order to guide risk reduction activities (see Table 1.2.1 for

ID	References	ID	References
1	Balica et al. (2012)	13	Atillah et al. (2011)
2	McInnes et al. (2013)	14	Kebede and Nicholls (2012)
3	Felsenstein and Lichter (2014)	15	Mustelin et al. (2010)
4	Bjarnadottir et al. (2011)	16	Appeaning Addo (2013)
5	Marfai and King (2008)	17	Li and Li (2011)
6	Newton and Weichselgartner (2014)	18	Das (2012)
7	Birkmann and Fernando (2008)	19	Martínez-Graña et al. (2016)
8	Torresan et al. (2012)	20	Kaplan et al. (2009)
9	Adelekan (2010)	21	Mani Murali et al. (2013)
10	Papathoma and Dominey-Howes (2003)	22	Kusenbach et al. (2010)
11	Alam and Collins (2010)	23	Barnard et al. (2015)
12	Mallick et al. (2011)	24	Kumar et al. (2010)

Table 1.2.1 Vulnerability assessments included in this study.

a sample from the academic literature). Based on different understandings and conceptions of vulnerability, these actors often rely on secondary data that describe the social, political, and physical elements of a system. Balica et al. (2012) observe that quantitative data can represent discrete and descriptive information that provide insight into the workings of the larger urban system. These data are often combined through a number of different statistical methods to develop a composite baseline assessment of vulnerability as a condition (Smit and Wandel, 2006). Others rely on structured or semi-structured interviews to determine the dynamic factors that lead to varying experiences with past hazard events, including the extent of damage experienced and components of the recovery process (e.g., Birkmann and Fernando, 2008; Adelekan, 2010; Alam and Collins, 2010).

Flooding poses context-specific challenges in different urbanized areas around the world, suggesting that place-based approaches to understanding flood risk and vulnerability are necessary (Cutter et al., 2008; Mallick et al., 2011). However, certain elements are shared across several of the observed studies for measuring vulnerability. There have been efforts to catalogue and compare these efforts previously. Khazai et al. (2014) created the VuWiki database, which provided an ontology-based assessment of 55 vulnerability-focused studies as the basis for organizing and disseminating knowledge and methods for vulnerability assessment. The VuWiki database includes a classification of studies in terms of the system being analyzed; the stressor(s) affecting that system; the reference framework used to define vulnerability; and the methodological approach for measuring vulnerability (Khazai et al., 2014, p. 59). As of this writing, the database created by Khazai et al. (2014) is no longer accessible at www.vuwiki.org, but cached versions of the website can be accessed through the Internet Archive (www.archive.org). This effort provides a valuable frame for understanding the subject and drivers for many extant vulnerability assessments. Beccari (2016) offers a comparison of 106 composite indices related to present-day (as opposed to emergent or future) multihazard risk, vulnerability, and resilience of governments and communities drawn from academic and gray literature and finds considerable variation in the included indicators, the statistical methods used to create composite scores, and the potential for communication and operationalization by decision makers. This study provides a basis for understanding the state of composite index creation for disaster risk reduction and the challenges therein, but is limited by the focus on baseline estimations as opposed to projective assessments, effectively excluding preemptive or projective vulnerability assessment, which is key from a planning and policy-making perspective to understanding future flooding impacts on coastal areas.

Fekete (2011) outlines some challenges posed by spatial vulnerability assessments, including lack of self-reflection or discussion of the limitations of studies, as well as limited understanding of how studies are received or used by intended audiences. The author notes that data availability (which varies by geographic scale and vintage), as well as the assumptions that are made in combining topically related (i.e., different attributes of social systems) or less related (i.e., combining social system attributes with physical system attributes) datasets that describe urban systems, is a major determining factor in a given study's output. Furthermore, the ways in which indicators are combined in terms of weighting and statistical algorithms is subjective, can have profound impacts on output results, and can even obfuscate which factors determine resulting vulnerability scores (Fekete, 2011 pp. 1165–1166). The idiosyncrasies of place and local populations can lead to or justify the subjective inclusion of indicator variables, but care must be taken when drawing conclusions about the validity of these factors for measuring vulnerability in other areas or comparing across geographies.

This chapter departs from these previous efforts by focusing on recent (i.e., within the last 20 years) efforts to calculate vulnerability to flood events for coastal areas and providing a critical assessment of the methods and scope of this body of work relative to the goals laid out in the SFDRR. The following section focuses on the selection method for sampled assessments and the criteria against which they are evaluated, followed by a discussion of the results. A conclusory section offers a summary assessment of the included studies and overarching trends in vulnerability measurement based on the sampled literature.

1.2.2 **Methods**

Candidate assessments of vulnerability were identified using the Web of Science Core Collection database (https://clarivate.libguides.com/webofscienceplatform/woscc). Studies were identified by topic searches for "flood vulnerability index" (634 results as of March 24, 2019) and "coastal vulnerability" (4087 results as of March 24, 2019), and by title searches for "flood vulnerability" (467 results as of March 24, 2019) and "coastal vulnerability" (515 results as of March 24, 2019). Duplicates were removed from these search results, yielding 4830 unique studies, which were further filtered by assessing relevance to human settlements versus nonhuman systems (e.g., wetlands) or specific pieces of infrastructure (e.g., dams). This ultimately yielded 150 unique vulnerability assessments of urbanized coastal areas. To keep the present study manageable, 24 studies were chosen based on perceived relevance for comparison, range of geographic focus (Fig. 1.2.2), and the number of times each has been cited in other work, which is taken as a proxy for their dissemination and usage by a wider audience. This selection does not provide an exhaustive review of the literature, but rather identifies a cross-section of existing vulnerability measurement efforts from which some observations and conclusions can be drawn (Figs. 1.2.2 and 1.2.3).

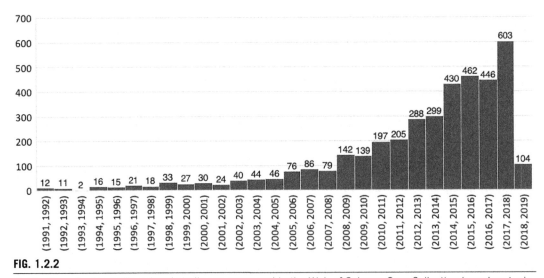

FIG. 1.2.2

Publication of vulnerability-related studies as measured in the Web of Science Core Collection based on topic searches as of March 24, 2019.

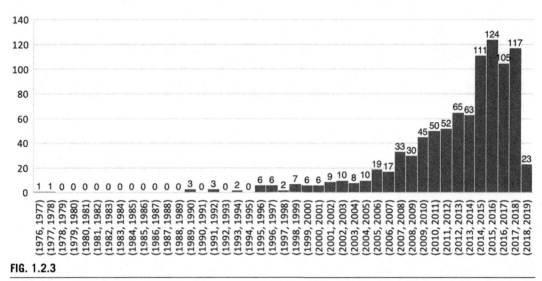

FIG. 1.2.3

Publication of vulnerability-related studies as measured in the Web of Science Core Collection based on title searches as of March 24, 2019.

Selected studies were entered into a matrix and categorized in terms of the given definitions of vulnerability, intended audience and goals, spatial scale of analysis, temporal focus (e.g., historical/contemporary/future), and the inclusion of various indicators across the categorical domains described in the SFDRR. Categorization was conducted, based on stated categorical grouping or the author's judgment based on context. Each of these dimensions are discussed in detail.

1.2.3 Results and discussion

1.2.3.1 Conceptions of vulnerability

Among the included literature, there exists clear separation of focus on different conceptions of vulnerability (e.g., internal, external, or integrated). Nine studies (37.5%) focused exclusively on exposure of urbanized areas to hazard, four studies (16.67%) focused solely on internal, latent measures of sensitivity or susceptibility, and eleven (45.8%) focused on a combination of internal and external factors resulting in an integrated approach. Table 1.2.2 lists the stated or implied sources of vulnerability from each assessment and the component parts of system vulnerability explored in each study. By definition, there is a clear distinction between each different category of assessment in terms of the elements and attributes of the systems that are included. External-focused assessments are concerned with the overall exposure of system elements to flooding, regardless of the implications of different population groups being affected by hazard. These assessments offer a valuable yet limited estimation of the number of system elements that may be affected by a flood event. Internal-focused assessments explore the sociodemographic, cultural, and political elements of systems, typically after an event has occurred, to understand why certain persons or communities experience flood damage differentially among affected system elements (e.g., Birkmann and Fernando, 2008; Alam and Collins, 2010).

Table 1.2.2 Positioning of vulnerability relative to urban system.

ID	Sources of vulnerability			Components of stated definition				
	Internal	External	Integrated	Hazard	Exposure	Susceptibility	Resilience	Adaptive capacity
1			+		+	+	+	+
2		+		+	+			
3	+				+	+	+	+
4			+	+	+	+	+	
5		+			+			
6			+			+	+	
7	+					+		+
8		+			+	+		
9			+		+	+	+	+
10		+			+	+		
11	+					+	+	+
12	+			+		+	+	
13			+		+	+		
14			+		+			
15			+		+	+		+
16		+			+			
17		+			+	+		
18			+	+	+	+		+
19		+			+			
20			+		+	+	+	
21			+		+	+	+	+
22			+		+	+	+	+
23		+		+	+			
24		+		+	+	+	+	

Integrated assessments consider both the extent of exposure and the susceptibility and coping/adaptive capacity of system elements relative to a flood. This type of assessment comprises the largest proportion of the sample, suggesting that there is both recognition of the importance of this approach, as well as the availability of necessary datasets to perform such an assessment.

1.2.3.2 Audience and stated goals

The majority of the sampled studies explicitly frame their assessments as efforts to inform government action at some scale. Sixteen studies explicitly gear their research toward a specific level of government (ranging from national to local), with local government being the most common stated audience ($n = 9$), followed by national government ($n = 5$), regional government ($n = 3$), and state or province government ($n = 2$). This stated focus on local governments suggests that vulnerability assessors consider local decision makers capable of minimizing vulnerability and/or guiding disaster response efforts during and immediately after hazard events, but it remains unclear whether this scale of government has the resources or capacity needed to act on vulnerability assessments to minimize damage or recover effectively (Næss et al., 2005). For example, Das (2012) notes that evacuation protocols are determined at the village level, while vulnerability assessments are conducted at the broader district level, and that this scalar mismatch has led to inefficiencies in operationalizing disaster risk reduction in the case of cyclone-prone districts in India.

An additional five studies list some unspecified level of government or coastal managers as the intended audience. Papathoma and Dominey-Howes (2003) and Atillah et al. (2011) focus more specifically on external vulnerability and loss potential and list insurance and re-insurance agencies as potential end users for their analytic frameworks as well. Mustelin et al. (2010)'s relatively localized study highlights the lack of focus on the particular conditions faced on the island of Zanzibar compared with mainland Tanzania and highlights the need to broaden the national discourse in risk management; they use their study to demonstrate local idiosyncrasies in efforts to influence national policy.

Interestingly, only Felsenstein and Lichter (2014) explicitly mention "the public" as an intended audience for their assessment—they note that the results of their study have been made available in a publicly available web map interface. This suggests that they consider the public to be interested in and capable of engaging with the results of their assessment and operationalizing this information in the decision-making process. This corresponds with the social justice perspective put forth in Morrow (2008), which argues that individuals and community groups should be considered active agents in determining and minimizing vulnerability. By comparison, each other study positions professional decision-makers as the intended audience for their assessments. This shifts the onus of communicating risk and vulnerability on governments and practitioners and away from the developers and aggregators of these frameworks. The degree to which these intended audiences actively engage with local communities in communicating and minimizing vulnerability varies on a case-by-case basis, but the disconnect between assessment and communication of results should be noted by planners and policymakers alike (Table 1.2.3).

1.2.3.3 Temporal scope

The surveyed literature focuses heavily on providing baseline assessments of extant vulnerability to flooding ($n = 19$; 79%). These initial efforts can provide a point of comparison for future researchers and stakeholders to evaluate adaptation and disaster risk reduction interventions and can help frame the conversation well into the future (for better or worse). A number of these studies also speculate on how the baseline vulnerability assessment may fluctuate based on changing climatic conditions, integrating explicit projections of sea-level change and inundation scenarios ($n = 14$; 58%). Although the methods and data used to estimate future conditions varies across the sampled literature, focus on vulnerability relative to dynamic processes marks a positive shift toward understanding the relationships between urbanized coastlines and climatic processes. Five studies (21%) focus on providing postevent explanations of experienced damage after extreme events in efforts to guide future preparation, as well as provide some insight on the overall functioning of local physical, economic, and political systems that contribute to increased vulnerability in areas of interest.

Table 1.2.3 Stated audiences, geographic scale, temporal scale, year of initial publication, source data type, and data vintage of included studies.

ID	Audience	Geographic Scale	Temporal Scale	Study Year	Data Source	Data Vintage
1	NG/A/I	City	C/F	2012	S	1993-2008
2	SG	10km² cells	C/F	2011	S	n/a
3	P/PP	Municipalities	C/F	2014	S	1997-2008
4	LG	County	C/F	2011	S	2009
5	RG	Region	C/F	2008	S	1992-2003
6	CM	Multiple	C/F	2014	S	n/a
7	NG/IA	Municipality	H	2008	P/S	2006
8	RG	Region	C/F	2012	S	2000-2006, n/a
9	LG/PP	Community	C/F	2010	P/S	2006, n/a
10	LG/I	Region	C	2003	P/S	2001
11	G/RM	Region	H	2010	P	post-2007
12	RM	Township	H/F	2011	P/S	2008
13	NG/LG/RM/I	City	C	2011	P/S	2007-2010
14	NG/RM	City	C/F	2012	S	2002-2005
15	NG	Island	H/C	2010	P/S	1953*-2005
16	CM	Municipality	C	2013	S	1980-2009
17	G	Region	C	2011	S	n/a
18	LG/RM	Village	H	2012	S	1999*
19	RM	Region	C/F	2016	S	1992*-2015
20	A/LG	Municipality	H	2009	P/S	2004-2008
21	SG/LG	Region	C/F	2013	S	1977*-2012
22	LG/RG	Municipality	C/F	2010	P	2000-2007
23	A/PP	Region	C/F	2015	S	1979*-2012
24	SG/LG	Region	C/F	2010	S	1972*-2006, n/a

Audience: P, *public;* IA, *international aid agencies;* G, *government (no scale specified);* NG, *national government;* RG, *regional government;* SG, *state government;* LG, *local government;* A, *academic institutions;* I, *re/insurance companies;* PG, *professional and NGO practitioners;* RM, *risk managers;* CM, *coastal managers. Geographic scale:* C, *city;* D, *subcity district;* G, *grid cell;* R, *region;* CZ, *coastal zone. Temporal scale:* H, *historical;* C, *contemporary;* F, *future. Data source:* P, *primary data;* S, *secondary data.* Red rows (dark gray *in print version*) indicate assessment positioning of vulnerability as external; blue (gray *in print version*) as internal; and purple (light gray *in print version*) as integrated. Note: The asterisk in the data vintage column denotes cases where historical data were sought out, and do not necessarily reflect lack of contemporary data sources.

1.2.3.4 **Geographic scale**

Nearly half ($n = 10$) of the surveyed studies focused on municipalities (ranging from villages to larger metropolises) as the scale of interest in their assessments. This may explain the degree of focus on local emergency managers and governments as the audience for the assessments discussed previously. Although climate change impacts and storm events do not comply with administrative boundaries, the municipal scale may be appropriate for certain flood preparation and response mechanisms. However, this scale of government is influenced both by state and national agendas as well as individual and community action, which can impede or facilitate adaptation measures taken at the municipal level (Adger et al., 2005; Næss et al., 2005). The scale of the coastal region was the second most common in the sample ($n = 9$). Identifying vulnerability to flooding at this scale highlights the need for inter-municipal (or regional/national) coordination to sufficiently address flood risks. Of the nine studies focused on the regional scale, seven deal strictly with external vulnerabilities, and one each for internal and integrated assessments. This is perhaps due to the regional scale of exposures (e.g., tsunami or storm surge), as opposed to the local-level coordination of reacting to extreme events.

1.2.3.5 **Unit of analysis**

The spatial resolution employed in each case has implications for how information is interpreted, and can reveal or obfuscate the different levels of vulnerability that exist at different scales depending on how results are aggregated (Jelinski and Wu, 1996). Confining an analysis to a discrete urban area has the benefit of measuring vulnerability at a familiar scale (such as a municipality), but often neglects to account for exogenous (but highly influential) factors such as interscalar economic dependence, cross-boundary reliance on infrastructure, or government resources at varying administrative levels, as well as drivers of vulnerability at the individual or community scale and hazard events themselves (Berkes and Ross, 2016).

1.2.3.6 **Variable inclusion**

Across the 24 surveyed vulnerability assessments, there were 86 unique indicators spanning social, economic, institutional, natural and built environment, and hazard categories (Fig. 1.2.4). Rationale for variable inclusion across the sampled literature is largely dependent on data availability (as discussed in Fekete, 2011 and Kebede and Nicholls, 2012), expert opinion in which domain experts are convened to select variables and weighting schemes, and reference to existing assessment methodologies. Variable categories largely adhere to the focus on internal or external vulnerabilities, and can be categorized as some combination of social, economic, institutional, natural environment, built environment, or hazard-related (Fig. 1.2.4). Variable selection is of the utmost importance in vulnerability assessments, as deciding what is being measured predetermines the range of possible outcomes that may be observed. A list of observed variables is shown in Fig. 1.2.5. Across the sampled literature, a number of statistical methods were used to combine variables, including principal component analysis, factor analysis, and arithmetic mean, but discussion of the merit of each method is outside the scope of this chapter.

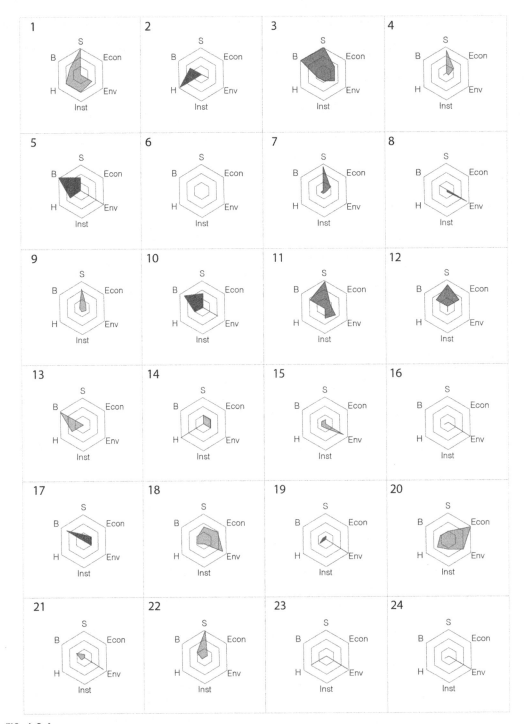

FIG. 1.2.4

Wind plots showing magnitude and distribution of factors relative to components of SFDRR. *Red plots (dark gray in print version)* indicate that the assessment focuses on external vulnerability, *blue (gray in print version)* on internal vulnerability, and *purple (light gray in print version)* on an integrated approach. *S*, social; *Econ*, economic; *Env*, environment; *Inst*, institutional; *H*, hazard; *B*, built environment.

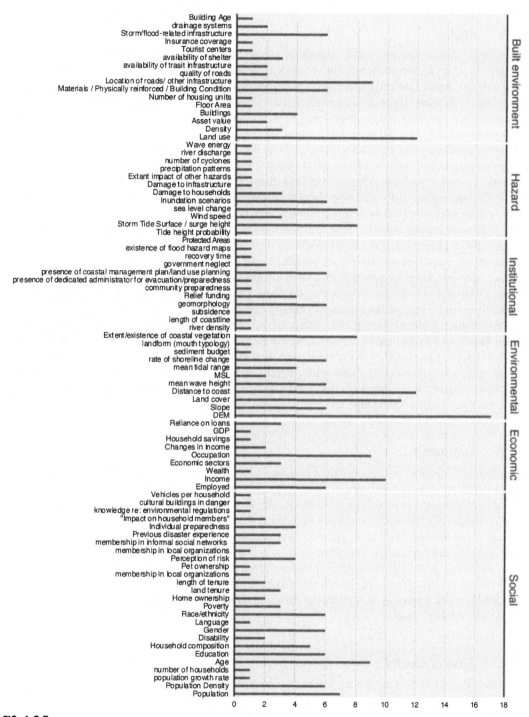

FIG. 1.2.5

Indicator occurrence by category.

1.2.3.7 Social

Observed variables related to social vulnerability center on individual or community characteristics that impede adequate preparation for or recovery from a flood event. Age (including the very young or the elderly), population and population density, education attainment and literacy, and gender were among the most commonly observed social indicators. Across the surveyed assessments, social-related indicators were the second-most prevalent (26 indicators occurring 82 times across 17/24 studies). Flanagan et al. (2011) posit that the relatively recent disciplinary focus on social vulnerability within the multihazard assessment discipline came as a reaction to traditional disaster management emphasis on the vulnerability of large infrastructural systems. Cutter et al. (2008) argue that vulnerability is fundamentally a social condition; after all, reducing disaster risk is focused on improving the human condition. The assumption that vulnerable populations stand to lose the most in extreme events makes sense, considering other factors equal. Newton and Weichselgartner (2014) reference Field et al. (2012)'s assessment that the main driver for future climate-related losses will be socioeconomic in nature as opposed to exposure-related, underscoring the criticality of improving social well-being as part of disaster risk reduction and adaptation planning.

The observed indicators generally described measures of susceptibility, or the degree to which affected persons may experience harm during a flood event due in large part to their (in)ability to evacuate safely or extant inequalities that impact quality of life in general (see e.g., Martin, 2015). There is a secondary category of sociodemographic indicators related to community function, including membership in formal or informal social networks, which are generally positioned relative to an individual's access to information regarding flood events and their ability to recover from floods. The disciplinary shift toward the study of social vulnerability signals the recognition that populations experience and respond to stressors in varying ways and that to adequately prepare for future extreme events, adaptation strategies must acknowledge this.

1.2.3.8 Economic

Economic indicators were less prevalent compared to other domains across the sampled literature (nine indicators occurring 36 times across 11/24 studies). Economic indicators in the sample either focus on qualities of the economic system (i.e., the diversity of economic sectors in an area) or individual financial assets such as household income. Both categories focus on the magnitude of loss from an extreme event, as well as the potential hardships incurred in recovering from an extreme event. Income and occupation were the two most prevalent economic indicators across the sampled literature. These indicators are measures of dependency between elements of the system as well as with external systems. The explicit role of multiscalar interactions was infrequently observed in the sampled literature (typically as it relates to institutional capacity) but should factor more heavily in future vulnerability studies.

1.2.3.9 Environmental

Indicators describing the nonbuilt environment were the most prevalent across the sample (15 indicators occurring 83 times across 20/24 studies). Furthermore, some measure of elevation was the most prevalent indicator across all categories (occurring in 15/24 studies). This frequency may be expected given the fundamentally volumetric nature of flooding. Observed environmental indicators include both measures of the relationship between urban elements and natural elements (e.g., proximity of an urbanized area to water) as well as characteristics of the land itself (e.g., elevation, slope, and land cover). In an urban context, the former category should be framed as the result of spatial planning and development decisions: they do not describe characteristics of the natural environment as such but rather the ways in which urban areas have enmeshed themselves within the environment. Explicitly acknowledging the relationship between environmental conditions and spatial planning and management policies can allow for more proactive, holistic adaptation moving forward.

1.2.3.10 Built environment

Observed indicators related to the built environment focus on either the kind of assets that stand to be damaged during a flood event or the existence of flood prevention infrastructure, including barriers and evacuation infrastructure. Built environment indicators were the third most prevalent category across the sample (15 indicators occurring 56 times in 17/24 studies). Land use (observed in a third of sampled assessments) is a critically important indicator as it denotes whether or not an area is residential or contains vital infrastructure and should be included in any measurement of flood vulnerability. Building type similarly denotes use, but provides more detail and can indicate a structure's susceptibility to floods based on condition or construction style. As coastal areas around the world continue to urbanize, the insights derived from flood vulnerability analyses should be utilized to improve the siting, construction, and monitoring of future built areas, beyond solely diagnosing current conditions.

1.2.3.11 Institutional

Institutional-related indicators were less diverse and generally less frequent than indicators in other categories (eight indicators occurring 17 times in 10/24 studies). The observed institutional capacities are more difficult to quantify than indicators in other categories: government neglect, flood management policies, and consideration for coastal land use planning are qualitative and can be difficult to assess before, during, or after a flood event. However, institutional aptitude and planning for flood events is critical to minimize human casualty and to ensure persistence of urban system functionality. Ten studies included measures of institutional indicators, with the presence of coastal planning and relief funding being the most frequently observed measures. The assessments that include these measures mostly gear their analysis toward risk managers and government at multiple scales. Mallick et al. (2011) make the important point that relief funding, without local capacity building, can promote cycles of poverty, elevated levels of risk, and perpetual dependence on government after flood events, as opposed to building local capacity and minimizing vulnerability before an event occurs. Only three studies explicitly include government capacity as indicators in their studies, suggesting the need for continued engagement between risk management professionals, academics, and the intended government audiences.

1.2.3.12 Hazard

The majority of studies included indicators related to historical or projected flood events, although hazard-related indicators were the second-least frequent across the literature (12 indicators occurring 35 times across 19/24 studies). The majority of hazard-related indicators are used to establish scenarios under which system vulnerability may be exposed (e.g., storm surge height/inundation scenarios and sea-level change projections). In some cases, these scenarios are based on historical events, as in Papathoma and Dominey-Howes (2003) and Atillah et al. (2011). These studies benefit from the empirical measures of past flood events to inform current and future hazard risk, although this practice can mislead decision makers to believe that past events, even if extreme, represent the maximum potential extent of flood hazard. This could ultimately lead to a false sense of security outside of the historical inundation zone; however, given the difficulty and uncertainties inherent in modeling possible storms, historical data can provide a useful baseline. Other researchers employ projections established by national and international bodies related to future sea-level change to establish projective future inundation, although the majority of the studies that include measures of sea-level change typically focused on periodic flooding as opposed to stochastic events like tsunamis (four of seven). Still others employ a hybrid technique, coupling future sea level and precipitation projections with extant storm surge projections (e.g., McInnes et al., 2013; Kebede and Nicholls, 2012; Balica et al., 2012).

1.2.3.13 Data sources

As vulnerability is a multifaceted condition spanning physical and nonphysical elements of urban systems, researchers must draw from a range of data sources to perform an evaluation. In the observed literature, 15 of 24 studies relied solely on secondary data sources, seven on mixed primary and secondary sources, and two on primary survey and interview data only. Often times, researchers draw from national, state-sponsored data collection efforts such as censuses. However, these datasets are generated at relatively large intervals (e.g., every 10 years), which can lead to drastic mismatches between the condition being represented and the year the data reflect. Furthermore, researchers are often required to combine data from a number of different sources and geographic and temporal scales when conducting vulnerability assessments (see Table 1.2.3). Input datasets are collected for a variety of reasons (many of which do not include flood vulnerability assessment) for various stakeholder groups, leaving vulnerability assessors with few options for data selection. As hazard vulnerability, and particularly flooding, continues to draw the attention of planners and policymakers in coastal cities, greater attention should be devoted to collecting, organizing, and disseminating data in a way that facilitates appropriate data combination across temporal and spatial units (Fig. 1.2.6).

1.2.4 Conclusion

This chapter relies on an illustrative sample of 24 vulnerability assessment studies to explore the current status of vulnerability measurement and calculation for flooding in coastal cities. Each study contributes to the SFDRR's priority to understand disaster risk by identifying relevant scientific information related to vulnerability, drawing from prevailing conceptions of vulnerability as it relates to biogeophysical and social, economic, and institutional factors, and generating an assessment of vulnerability, which can then be used to inform policy and prepare for future hazard events. Each study was categorized relative to conceptions of vulnerability, audience and stated goals, temporal and spatial scale of interest, and the categories and frequencies of input indicators.

    Across the literature, the prevalent categories of vulnerability indicators are social, economic, environmental, institutional, infrastructural, and noninfrastructural built environment, which correspond

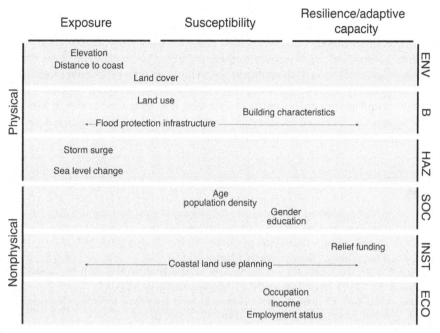

FIG. 1.2.6

Matrix of commonly observed indicators segmented by categorical domain, conceptual relation to vulnerability, and physical/nonphysical manifestation. *ENV*, nonbuilt environment; *B*, built environment; *HAZ*, hazard; *SOC*, social; *INST*, institutional; *ECO*, economic.

to the dimensions outlined in the Sendai Framework for Disaster Risk Reduction. Social indicators acknowledge that vulnerability affects different populations differently. Economic indicators begin to unravel the interrelationships between individual susceptibility and system-level robustness. Environmental indicators focus both on shifting climatological events, as well as the implications of urbanization relative to the natural environment in coastal settings. Institutional indicators are difficult to measure, but underscore the need for increased capacity of governments, as well as collaboration between governments at multiple scales. Built environment indicators pertain to traditional elements of flood management systems, which remain essential parts of a complex sociospatial urban system, as well as and the distribution of urban elements in space.

Although the SFDRR has identified the need for integrated, multidomain (and multihazard) risk assessments to guide development and disaster risk reduction, there remains a divide between assessments that focus solely on biogeophysical elements of vulnerability, and those that include social, economic, and institutional measures of vulnerability as it relates to flooding. While both physical and nonphysical facets of vulnerability are important for planners and policymakers at every stage of a disaster (before, during, and after the event), understanding the intersection between physical and social components of an urban system is essential not just to reduce disaster risk, but to improve quality of life and overall wellbeing. The implications of either physical or nonphysical attributes of a system, relative to a flood event, can shift dramatically when viewed in isolation or in combination with other factors. For instance, relatively low levels of physical exposure, when coupled with high levels of susceptibility, have consequences and require interventions different than either of those conditions alone. Furthermore, interventions addressing either physical or nonphysical elements of vulnerability can result in both positive and negative consequences on other system elements. Less than half of the observed studies assume this integrated approach; the majority contend with either internal or external elements of the system. As we proceed into an era of steady urbanization and climatic uncertainty, the need to produce comprehensive vulnerability assessments that incorporate internal and external elements of urban systems is clear if we are to understand disaster risk and ultimately reduce it as outlined in the Sendai Framework for Disaster Risk Reduction.

References

Adelekan, I.O., 2010. Vulnerability of poor urban coastal communities to flooding in Lagos, Nigeria. Environ. Urban. 22, 433–450. https://doi.org/10.1177/0956247810380141.

Adger, W.N., 1996. Approaches to vulnerability to climate change. Change 2006, 268–281.

Adger, W.N., 2006. Vulnerability. Global Environ. Change 16, 268–281.

Adger, W.N., Arnell, N.W., Tompkins, E.L., 2005. Successful adaptation to climate change across scales. Global Environ. Change 15, 77–86. https://doi.org/10.1016/j.gloenvcha.2004.12.005.

Alam, E., Collins, A.E., 2010. Cyclone disaster vulnerability and response experiences in coastal Bangladesh. Disasters 34, 931–954. https://doi.org/10.1111/j.1467-7717.2010.01176.x.

Appeaning Addo, K., 2013. Assessing coastal vulnerability index to climate change: the case of Accra – Ghana. J. Coast. Res., 2, 1892–1897.

Atillah, A., El Hadani, D., Moudni, H., Lesne, O., Renou, C., Mangin, A., Rouffi, F., 2011. Tsunami vulnerability and damage assessment in the coastal area of Rabat and Salé, Morocco. Nat. Hazards Earth Syst. Sci. 11, 3397–3414. https://doi.org/10.5194/nhess-11-3397-2011.

Balica, S.F., Wright, N.G., van der Meulen, F., 2012. A flood vulnerability index for coastal cities and its use in assessing climate change impacts. Nat. Hazards 64, 73–105.

Barnard, P.L., Short, A.D., Harley, M.D., Splinter, K.D., Vitousek, S., Turner, I.L., Allan, J., Banno, M., Bryan, K.R., Doria, A., Hansen, J.E., Kato, S., Kuriyama, Y., Randall-Goodwin, E., Ruggiero, P., Walker, I.J., Heathfield, D.K., 2015. Coastal vulnerability across the Pacific dominated by El Niño/Southern Oscillation. Nat. Geosci. 8, 801–807. https://doi.org/10.1038/ngeo2539.

Beccari, B., 2016. A comparative analysis of disaster risk, vulnerability and resilience composite indicators. PLoS Curr. 8. https://doi.org/10.1371/currents.dis.453df025e34b682e9737f95070f9b970.

Berkes, F., Ross, H., 2016. Panarchy and community resilience: sustainability science and policy implications. Environ. Sci. Pol. 61, 185–193.

Birkmann, J., 2006. Measuring vulnerability to promote disaster-resilient societies: conceptual frameworks and definitions. In: Birkmann, J. (Ed.), Measuring Vulnerability to Natural Hazards—Towards Disaster Resilient Societies. United Nations University Press, Tokyo, Japan, pp. 9–54.

Birkmann, J., Fernando, N., 2008. Measuring revealed and emergent vulnerabilities of coastal communities to tsunami in Sri Lanka. Disasters 32, 82–105. https://doi.org/10.1111/j.1467-7717.2007.01028.x.

Bjarnadottir, S., Li, Y., Stewart, M.G., 2011. Social vulnerability index for coastal communities at risk to hurricane hazard and a changing climate. Nat. Hazards 59, 1055–1075. https://doi.org/10.1007/s11069-011-9817-5.

Bohle, H.-G., 2001. Vulnerability and criticality: perspective from social geography. IHDP Update 2, 1–7.

Brakenridge, G.R., Syvitski, J.P.M., Overeem, I., Higgins, S.A., Kettner, A.J., Stewart-Moore, J.A., Westerhoff, R., 2013. Global mapping of storm surges and the assessment of coastal vulnerability. Nat. Hazards 66, 1295–1312. https://doi.org/10.1007/s11069-012-0317-z.

Burton, C.G., 2015. A validation of metrics for community resilience to natural hazards and disasters using the recovery from Hurricane Katrina as a case study. Ann. Assoc. Am. Geogr. 105, 67–86.

Committee on Increasing National Resilience to Hazards and Disasters, Committee on Science, Engineering, and Public Policy, 2012. Disaster Resilience: A National Imperative. National Academies Press.

Cutter, S.L., Boruff, B.J., Shirley, W.L., 2003. Social vulnerability to environmental hazards. Soc. Sci. Q. 84, 242–261.

Cutter, S.L., Barnes, L., Berry, M., Burton, C., Evans, E., Tate, E., Webb, J., 2008. A place-based model for understanding community resilience to natural disasters. Global Environ. Change 18, 598–606.

Das, S., 2012. The role of natural ecosystems and socio-economic factors in the vulnerability of coastal villages to cyclone and storm surge. Nat. Hazards 64, 531–546. https://doi.org/10.1007/s11069-012-0255-9.

Fekete, A., 2011. Spatial disaster vulnerability and risk assessments: challenges in their quality and acceptance. Nat. Hazards 61, 1161–1178. https://doi.org/10.1007/s11069-011-9973-7.

Felsenstein, D., Lichter, M., 2014. Social and economic vulnerability of coastal communities to sea-level rise and extreme flooding. Nat. Hazards 71, 463–491. https://doi.org/10.1007/s11069-013-0929-y.

Field, C.B., Barros, V., Stocker, T.F., Dahe, Q. (Eds.), 2012. Managing the Risks of Extreme Events and Disasters to Advance Climate Change Adaptation: Special Report of the Intergovernmental Panel on Climate Change. Cambridge University Press, Cambridge. https://doi.org/10.1017/CBO9781139177245.

Flanagan, B.E., Gregory, E.W., Hallisey, E.J., Heitgerd, J.L., Lewis, B., 2011. A social vulnerability index for disaster management. J. Homel. Secur. Emerg. Manage. 8 (1). Article 3.

Gornitz, V., 1991. Global coastal hazards from future sea level rise. Palaeogeogr. Palaeoclimatol. Palaeoecol. (Global Planet. Change Sect.) 89, 379–398.

Güneralp, B., Güneralp, İ., Liu, Y., 2015. Changing global patterns of urban exposure to flood and drought hazards. Glob. Environ. Change 31, 217–225. https://doi.org/10.1016/j.gloenvcha.2015.01.002.

Jelinski, D.E., Wu, J., 1996. The modifiable areal unit problem and implications for landscape ecology. Landsc. Ecol. 11, 129–140.

Kaplan, M., Renaud, F.G., Lüchters, G., 2009. Vulnerability assessment and protective effects of coastal vegetation during the 2004 Tsunami in Sri Lanka. Nat.l Hazards Earth Syst. Sci. 9, 1479–1494. https://doi.org/10.5194/nhess-9-1479-2009.

Kebede, A.S., Nicholls, R.J., 2012. Exposure and vulnerability to climate extremes: population and asset exposure to coastal flooding in Dar es Salaam, Tanzania. Reg. Environ. Change 12, 81–94. https://doi.org/10.1007/s10113-011-0239-4.

Khazai, B., Kunz-Plapp, T., Büscher, C., Wegner, A., 2014. VuWiki: an ontology-based semantic wiki for vulnerability assessments. Int. J. Disast. Risk Sci. 5, 55–73.

Kumar, T.S., Mahendra, R.S., Nayak, S., Radhakrishnan, K., Sahu, K.C., 2010. Coastal vulnerability assessment for Orissa State, East Coast of India. J. Coast. Res. 26, 523–534.

Kusenbach, M., Simms, J.L., Tobin, G.A., 2010. Disaster vulnerability and evacuation readiness: coastal mobile home residents in Florida. Nat. Hazards 52, 79–95. https://doi.org/10.1007/s11069-009-9358-3.

Li, K., Li, G.S., 2011. Vulnerability assessment of storm surges in the coastal area of Guangdong Province. Nat. Hazards Earth Syst. Sci. 11, 2003–2010. https://doi.org/10.5194/nhess-11-2003-2011.

Mallick, B., Rahaman, K.R., Vogt, J., 2011. Social vulnerability analysis for sustainable disaster mitigation planning in coastal Bangladesh. Disaster Prev. Manage. 20, 220–237. https://doi.org/10.1108/09653561111141682.

Mani Murali, R., Ankita, M., Amrita, S., Vethamony, P., 2013. Coastal vulnerability assessment of Puducherry coast, India, using the analytical hierarchical process. Nat. Hazards Earth Syst. Sci. 13, 3291–3311. https://doi.org/10.5194/nhess-13-3291-2013.

Marfai, M.A., King, L., 2008. Potential vulnerability implications of coastal inundation due to sea level rise for the coastal zone of Semarang city, Indonesia. Environ. Geol. 54, 1235–1245. https://doi.org/10.1007/s00254-007-0906-4.

Martin, S.A., 2015. A framework to understand the relationship between social factors that reduce resilience in cities: application to the City of Boston. Int. J. Disast. Risk Reduct. 12, 53–80. https://doi.org/10.1016/j.ijdrr.2014.12.001.

Martínez-Graña, A.M., Boski, T., Goy, J.L., Zazo, C., Dabrio, C.J., 2016. Coastal-flood risk management in central Algarve: vulnerability and flood risk indices (South Portugal). Ecol. Indic. 71, 302–316. https://doi.org/10.1016/j.ecolind.2016.07.021.

McCarthy, J.J., Intergovernmental Panel on Climate Change (Eds.), 2001. Climate Change 2001: Impacts, Adaptation, and Vulnerability: Contribution of Working Group II to the Third Assessment Report of the Intergovernmental Panel on Climate Change. Cambridge University Press, Cambridge, UK; New York.

McInnes, K.L., Macadam, I., Hubbert, G., O'Grady, J., 2013. An assessment of current and future vulnerability to coastal inundation due to sea-level extremes in Victoria, southeast Australia. Int. J. Climatol. 33, 33–47. https://doi.org/10.1002/joc.3405.

Morrow, B.H., 2008. Community Resilience: A Social Justice Perspective (No. 4). CARRI Research Report, Oak Ridge National Laboratory, Miami, FL.

Muis, S., Verlaan, M., Winsemius, H.C., Aerts, J.C.J.H., Ward, P.J., 2016. A global reanalysis of storm surges and extreme sea levels. Nat. Commun. 7, 11969. https://doi.org/10.1038/ncomms11969.

Mustelin, J., Klein, R.G., Assaid, B., Sitari, T., Khamis, M., Mzee, A., Haji, T., 2010. Understanding current and future vulnerability in coastal settings: community perceptions and preferences for adaptation in Zanzibar, Tanzania. Popul. Environ. 31, 371–398.

Næss, L.O., Bang, G., Eriksen, S., Vevatne, J., 2005. Institutional adaptation to climate change: Flood responses at the municipal level in Norway. Global Environ. Change 15, 125–138. https://doi.org/10.1016/j.gloenvcha.2004.10.003.

Neumann, B., Vafeidis, A.T., Zimmermann, J., Nicholls, R.J., 2015. Future coastal population growth and exposure to sea-level rise and coastal flooding – a global assessment. PLOS One. 10, e0118571.

Newton, A., Weichselgartner, J., 2014. Hotspots of coastal vulnerability: a DPSIR analysis to find societal pathways and responses. Estuar. Coast. Shelf Sci. 140, 123–133. https://doi.org/10.1016/j.ecss.2013.10.010.

Papathoma, M., Dominey-Howes, D., 2003. Tsunami vulnerability assessment and its implications for coastal hazard analysis and disaster management planning, Gulf of Corinth, Greece. Nat. Hazards Earth Syst. Sci. 3, 733–747. https://doi.org/10.5194/nhess-3-733-2003.

Pilon, P.J., Dengo, M., Brewster, M., Peichert, H., Chaudhry, A., Sainio, M., Harding, J., Rencoret, N., Barrantes, M., Kluser, S., Jezeph, D., Barrett, C., Grabs, W., Davis, D.A., Halliday, R.A., Paulson, R., 2004. Guidelines for Reducing Flood Losses. United Nations Department of Economic and Social Affairs, Population Division.

Smit, B., Wandel, J., 2006. Adaptation, adaptive capacity and vulnerability. Global Environ. Change 16, 282–292.

Torresan, S., Critto, A., Rizzi, J., Marcomini, A., 2012. Assessment of coastal vulnerability to climate change hazards at the regional scale: the case study of the North Adriatic Sea. Nat. Hazards Earth Syst. Sci. 12, 2347–2368. https://doi.org/10.5194/nhess-12-2347-2012.

Turner, B.L., Kasperson, R.E., Matson, P.A., McCarthy, J.J., Corell, R.W., Christensen, L., Eckley, N., Kasperson, J.X., Luers, A., Martello, M.L., Polsky, C., Pulsipher, A., Schiller, A., 2003. A framework for vulnerability analysis in sustainability science. Proc. Natl. Acad. Sci. U. S. A. 100, 8074–8079. https://doi.org/10.1073/pnas.1231335100.

UN Department of Economic and Social Affairs, 2018. World Urbanization Prospects: The 2018 Revision (Key Facts). United Nations Department of Economic and Social Affairs, Population Division.

UNGA (United Nations General Assembly), 2016. Report of the open-ended intergovernmental expert working group on indicators and terminology relating to disaster risk reduction (No. A/71/644). United Nations.

UNISDR (United Nations International Strategy for Disaster Reduction), 2005. Hyogo Framework For Action 2005–2015. United Nations, Hyogo, Japan.

UNISDR (United Nations International Strategy for Disaster Reduction), 2015. Sendai Framework for Disaster Risk Reduction 2015–2030; 2015. United Nations, Geneva, Switzerland.

Vitousek, S., Barnard, P.L., Fletcher, C.H., Frazer, N., Erikson, L., Storlazzi, C.D., 2017. Doubling of coastal flooding frequency within decades due to sea-level rise. Sci. Rep. 7. https://doi.org/10.1038/s41598-017-01362-7.

Wisner, B., Blaikie, P., Cannon, T., Davis, I., 2004. At Risk: Natural Hazards, People's Vulnerability, and Disasters, second ed. Routledge, London; New York.

Wong, P.P., Losada, I.J., Gattuso, J.-P., Hinkel, J., Khattabi, A., McInnes, K.L., Saito, Y., Sallenger, A., 2014. 5. Coastal systems and low-lying areas. In: Climate Change 2014: Impacts, Adaptation, and Vulnerability. Part A: Global and Sectoral Aspects. Contribution of Working Group II to the Fifth Assessment Report of the Intergovernmental Panel on Climate Change. Cambridge University Press, Cambridge England; New York, pp. 361–409.

Yin, J., Griffies, S.M., Stouffer, R.J., 2010. Spatial variability of sea level rise in twenty-first century projections. J. Clim. 23, 4585–4607. https://doi.org/10.1175/2010JCLI3533.1.

Further reading

Biesbroek, G.R., Swart, R.J., Carter, T.R., Cowan, C., Henrichs, T., Mela, H., Morecroft, M.D., Rey, D., 2010. Europe adapts to climate change: Comparing National Adaptation Strategies. Global Environ. Change 20, 440–450.

Borden, K.A., Shmidtlein, M.C., Emrich, C.T., Piegorsch, W.W., Cutter, S.L., 2007. Vulnerability of U.S. cities to environmental hazards. J. Homel. Secur. Emerg. Manage. 4 (2). Article 5.

Cooper, J.A.G., McKenna, J., 2008. Social justice in coastal erosion management: the temporal and spatial dimensions. Geoforum 39, 294–306. https://doi.org/10.1016/j.geoforum.2007.06.007.

Cutter, S.L., 2016. The landscape of disaster resilience indicators in the USA. Nat. Hazards 80, 741–758.

Füssel, H.-M., Klein, R.J.T., 2006. Climate change vulnerability assessments: an evolution of conceptual thinking. Clim. Change 75, 301–329. https://doi.org/10.1007/s10584-006-0329-3.

Hewitt, K., 1983. The idea of calamity in a technocratic age. In: Hewitt, K. (Ed.), Interpretations of Calamity From the Viewpoint of Human Ecology. Allen and Unwin, Boston, pp. 1–32.

Kelly, P.M., Adger, W.N., 2000. Theory and practice in assessing vulnerability to climate change and facilitating adaptation. Clim. Change 47, 325–352.

Menoni, S., Molinari, D., Parker, D., Ballio, F., Tapsell, S., 2012. Assessing multifaceted vulnerability and resilience in order to design risk-mitigation strategies. Nat. Hazards 64, 2057–2082.

Oliver-Smith, A., Hoffman, S.M., 2002. Why anthropologists should study disasters. In: Hoffman, S.M., Oliver-Smith, A. (Eds.), Catastrophe & Culture: The Anthropology of Disaster. School of American Research Press, Santa Fe, NM, pp. 3–22.

Schneider, S., Sarukhan, J., 2001. Overview of Impacts, Adaptation, and Vulnerability to Climate Change (No. AD5.WG2.1), Working Group II: Impacts, Adaptation and Vulnerability. Intergovernmental Panel on Climate Change.

Simon, G., 2016. Cities, Nature and Development: The Politics and Production of Urban Vulnerabilities, first ed. Routledge. https://doi.org/10.4324/9781315572123.

Smit, B., Burton, I., Klein, R.J.T., Wandel, J., 2000. An anatomy of adaptation to climate change and variability. Clim. Change 45, 223–251.

UNESCO, n.d. Flood Vulnerability Indices (FVI) [WWW Document]. UNESCO-IHE. http://unihefvi.free.fr/indicators.php.

Damaging flood risk in the Portuguese municipalities

1.3

Pedro Pinto Santos[a], Susana Pereira[a], Eusébio Reis[a], Alexandre Oliveira Tavares[b], José Leandro Barros[c], José Luís Zêzere[a], Ricardo A.C. Garcia[a], and Sérgio Cruz Oliveira[a]

Centre for Geographical Studies, Institute of Geography and Spatial Planning, University of Lisbon, Lisbon, Portugal[a]
Earth Sciences Department and Center for Social Studies, University of Coimbra, Coimbra, Portugal[b] Centre for Social Studies, Universidade de Coimbra, Coimbra, Portugal[c]

CHAPTER OUTLINE

1.3.1 Introduction

Riverine floods are responsible for significant human and material losses. Climate change scenarios pose additional challenges to flood susceptibility and risk assessment (Kundzewicz et al., 2014; Winsemius et al., 2015). While in recent decades the increases in global flood damage have mainly been driven by population and economic activity growth in flood-prone areas, climate change projections launch a high level of uncertainty about future impacts, showing, however, that flood protection standards can play a relevant part in mitigating such effects (Winsemius et al., 2015). Assessing flood risk based on the physical dimensions of a flood hazard is an intrinsically complex endeavor (Santos et al., 2019), due to the difficulty in data collection and modeling extreme events (Kjeldsen et al., 2014; Benito et al., 2004). In addition, assessing flood exposure, losses, and vulnerability is also a challenging task because of the individual and social complexity of the processes that control both causes and effects (Tapsell et al., 2002; Rufat et al., 2015), making concepts such as sociohydrology relevant in understanding the physical-human relations related to flooding (Viglione et al., 2014; Baldassarre et al., 2017).

In terms of flood risk management at regional and national levels, there is a need to use administrative or physical terrain units to comprehensively resume the complexity of input data from distinct dimensions (Fekete, 2009). In Portugal, for example, water and flood-related management are structured in regional hydrographical units. Other authors use the municipal level to assess not only hazard but also vulnerability to floods, from which recommendations for decision-makers are drawn—e.g., the Social and Infrastructure Flood Vulnerability Index (Fekete, 2009)—arguing that social vulnerability assessments have the capacity of unveiling communities' fragilities and strengths in the face of specific stressors or hazards. A study of the linkages between flood susceptibility, vulnerability, and the underlying territorial resilience was previously conducted in mainland Portugal in the river mouth of the Douro river (Ferrari et al., 2019), producing local and valuable knowledge of flood-risk mitigation. A basin level (Santos and Tavares, 2015) and municipal level approach conducted in a set of 23 municipalities in inland Portugal (Santos et al., 2014) demonstrated the connection between flood impacts and the territorial processes, identifying priorities in terms of risk management strategies.

The generality of methodological approaches applied to assess flood susceptibility is not usually expressed at the municipal level. Flood susceptibility in mainland Portugal was recently assessed on a cell-by-cell basis using flow accumulation, slope, and relative permeability data (Santos et al., 2019), and posteriorly converted to the municipal level. The proposal of Sá and Vicêncio (2011) is another example of a municipal-level assessment, which considered five subindices, related not exclusively to susceptibility: historical losses, urban and peri-urban land use, length of the drainage network, hourly maximum rainfall, and population density.

Other examples of a cell-by-cell susceptibility analysis include the work of Cunha et al. (2017), who used morphologic analysis through GIS algorithms that identify wet (concave) and dry (convex) systems. Jacinto et al. (2015) combined the accumulated mean contribution of flow, slope (to derive a cost distance matrix), and flow curve number, using the official flood-prone area mapping and flood-loss records to calibrate the weight of each factor. Santos and Reis (2018) applied a methodology similar to Santos et al. (2019) to a medium-size watershed in Central Portugal. Finally, an interesting study following the quantitative principles of Patton and Baker (1976) assessed flood susceptibility having the hydrographical basin as the terrain unit to infer the basins' capacity to generate fluvial flooding (Leal and Ramos, 2013).

In the preceding perspective on flood risk characterization, the incorporation of disaster databases in the process of building knowledge is essential. Besides the strict description of losses, disaster databases can be used as forms of calibrating or validating their probability and severity (Barnolas and Llasat, 2007), and they might also be used—with caution, however—in validating flood susceptibility conditioning factors (Li et al., 2012).

This research aims to contribute to the understanding of the contexts of flood disasters at the municipal level in mainland Portugal, considering three levels of information: flood susceptibility, flood losses, and social vulnerability. It intends to differentiate municipalities by cross-analyzing these dimensions with the aim of producing strategic, national-level knowledge, with the capacity of supporting decision making in defining flood risk priorities. By doing so, the research attempts to comply with and to contribute to the achievement of the objectives of the Sendai Framework for Disaster Risk Reduction (UNISDR, 2015[a]) with regard to the building and application of disaster risk knowledge enunciated in its first priority.

1.3.2 Study area

Mainland Portugal is located in southern Europe, representing an area of 89,089 km^2 (Fig. 1.3.1). The main rivers that drain the mainland Portuguese territory are the Minho (300 km in length, 77 of which are in Portugal), Douro (927 km, 330 of which are in Portugal), Tagus (875 km, 225 of which are in Portugal) and Guadiana (711 km, 240 of which are in Portugal) rivers. These international basins cover more than 50% of the mainland Portuguese territory. The longest exclusively Portuguese rivers are the Mondego (basin area of 6.645 km^2 and 220 km of river length), the Sado (7.692 km^2 and 175 km, respectively) and the Vouga (3.635 km^2 and 143 km, respectively), whose river mouth is the city of Aveiro (DGT, 2018; APA, 2018) (Fig. 1.3.1).

Susceptibility to floods is particularly high in the floodplains of the largest basins' lower sectors, where fluvial geomorphology controls the extension of the flooded area. On the other hand, the basins' area, permeability, and total rainfall control peak flows and flood duration. Considering these factors, the basin of the Tagus River can be identified as the one in mainland Portugal where susceptibility to floods is highest.

A previous analysis of a flood loss database for mainland Portugal showed that the impacts have a different frequency and severity depending on the type of flood (Pereira et al., 2016). Thus according to the DISASTER database (Zêzere et al., 2014), while the vast majority of evacuated and displaced persons are associated with slow onset floods, the type of flood that—in proportion to the number of cases—aggregates the most fatalities and injuries is the flash flood.

Regarding social vulnerability (SV), a recent assessment identified the most vulnerable municipalities in the northern region, particularly in the Douro River valley (Tavares et al., 2018). The main drivers of SV are the population's fragile economic condition, unemployment, age, and state-assistance dependency. The more vulnerable territories present low economic dynamism and insufficient public and private infrastructure and services. Low SV is found generically in the main cities of Porto and Lisbon and in the Algarve region (southern Portugal).

1.3.3 Data and methods
1.3.3.1 Flood susceptibility

Susceptibility is the spatial incidence of hazard in a specific area. In this study, flood susceptibility expresses the propensity for an area to be affected by flooding. The method for assessing susceptibility prioritized the flood-prone areas that are affected by recurrent flooding events of fluvial origin.

---

[a]UNISDR is now known as UNDRR.

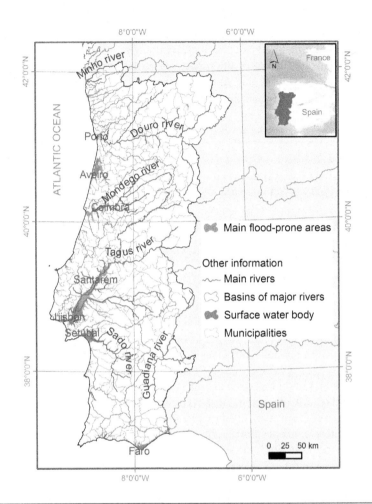

FIG. 1.3.1

Main flood-prone areas in mainland Portugal used in the flood susceptibility assessment at the municipal level.

Sources: APA, 2018. SNIAmb—Environment Information National System. https://sniamb.apambiente.pt/; DGT, 2017. Portuguese Administrative Official Chart. http://www.dgterritorio.pt/cartografia_e_geodesia/cartografia/carta_administrativa_oficial_ de_portugal_caop_/.

The rationale is to consider only the main flood-prone areas, assuring the maximum uniformity of the input data, considering that the methodology is required to be consistent for the entire mainland country.

The flood susceptibility assessment was based on the official flood-prone areas and on the official geologic map, 1:500,000 scale (Oliveira et al., 1992) (Fig. 1.3.1). The presence of alluvial deposits was considered to be the main testimony of flooding, and thus an indicator of flood susceptibility. Additionally, flood hazard mapping for the 1000-year return period, produced on the framework of the Floods Directive, and a previous 100-year flood hazard study authored by the National Laboratory of Civil Engineering (LNEC) were also considered for the municipal assessment of flood susceptibility.

These datasets are available on the Portuguese SNIAmb webpage (APA, 2018). Municipal flood susceptibility was computed as the percentage of the municipal area covered by the overlay of the previously mentioned three data layers that compose the main flood-prone areas, classified in 10 quantiles. The surface water bodies that include estuarine areas, coastal lagoons, and intertidal zones were excluded from the computation. The municipal boundaries of the 278 municipalities were obtained from the Portuguese Administrative Official Chart (DGT, 2017).

1.3.3.2 **Historical record of flood losses**

Damaging floods (excluding undefined-type, urban, and flash flood-related disasters) registered in the DISASTER database (Zêzere et al., 2014) for the period 1865–2015 were used. The database includes only flood disasters in which at least one dead, injured, missing, evacuated, or displaced person is recorded. This means that flood occurrences with uniquely material consequences are not included. Although geographical contexts change in terms of land use and exposure, given the exceptionality of flood severity associated with this type of flood, it was considered statistically more correct to include all the losses registered on the database since 1865. Four variables were defined for the 278 municipalities, expressing flood losses defined by their density: density of flood disaster cases, casualties, displaced and evacuated persons, expressed by their number per square kilometer. The use of density allows for a direct comparison of severity between municipalities. The flood disaster records were posteriorly cross-analyzed with municipal flood susceptibility and municipal social vulnerability.

1.3.3.3 **Social vulnerability**

Social vulnerability was calculated as the product of criticality and support capability, as described in Eq. (1.3.1) (Mendes et al., 2010; Tavares et al., 2018):

$$\text{Social Vulnerability} = \text{Criticality} \times (1 - \text{Support Capability}) \qquad (1.3.1)$$

Criticality is defined by the set of characteristics and behaviors of individuals that determine the degree of resistance and resilience to respond or deal with disasters and catastrophic events. Support capacity is defined by the coverage and diversity of infrastructure and equipment that drives the response of communities to disasters and catastrophes. Most of the input data was derived or extracted directly from the 2011 population census (INE, 2011) and from a web-based database with territorial and social indicators (FFMS, 2017). For each factor of the equation, a principal component analysis (PCA) was performed (Mendes et al., 2010). The statistical procedure consists of normalizing values to the z-score, eliminating redundant variables, applying a PCA, and interpreting the cardinality of the explicative variables inside each principal component (PC).

For criticality, a final set of 22 variables was retained and self-grouped into six principal components (PCs) (in parenthesis, the percentage of explained variance): PC1—Risk groups (30% of explained variance), PC2—Economic conditions (13%), PC3—Disadvantaged population (12%), PC4—Level of income (7%), PC5—Employment (6%) and PC6—Dependent population (5%). For support capability, a final set of 12 variables was used, aggregated into three PCs: PC1—Civil protection response (30%), PC2—Economic and environmental dynamism (22%), and PC3—Logistics and services capacity (12%).

The final score of criticality and support capability is the result of the sum of the scores of all the PCs, without weighting. Both scores are normalized to the interval [0, 1]. Applying Eq. (1.3.1), the final SV score was obtained and classified into five classes, according to the standard deviation (SD): very low: <-1 SD; low: $[-1, -0.5$ SD]; moderate: $[-0.5, +0.5$ SD]; high: $[0.5, 1$ SD]; very high: ≥ 1 SD.

1.3.3.4 Multivariate analysis

1.3.3.4.1 Cluster analysis

The 278 municipalities were classified according to cluster membership, performed with the following six variables, normalized to the z-scores: flood susceptibility (percentage of the municipality area covered by the main flood-prone areas); density of flood-related disaster cases and density of number of casualties, displaced, and evacuated persons (number of cases and number of persons by square kilometer); and final scores of social vulnerability (nondimensional).

In an initial stage, the R packages factoextra and NbClust were used in order to calculate the average silhouette width, a parameter the maximum inflection of which indicates the optimal number of clusters to be used in k-means clustering (Rousseeuw, 1987). The graphic for the normalized data was not clearly conclusive (Fig. 1.3.2), indicating an optimal number of 4 clusters: 1 cluster alone grouped 274 municipalities, leaving 4 municipalities for the remaining 3 clusters. This clustering was considered insufficient, although it served as a reference to test a higher number of clusters. After several experiments using SPSS, the final number of 10 clusters was considered balanced in terms of (i) providing homogeneous groups of municipalities and (ii) representing the diversity among clusters, while presenting the second highest average silhouette width (≈ 0.35). The 10 clusters allow a reasonable and

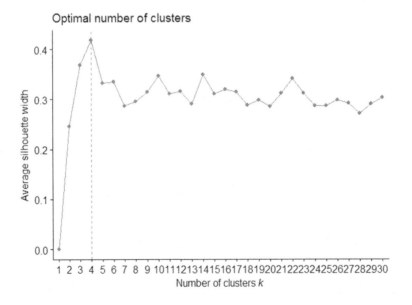

FIG. 1.3.2

Average silhouette width of k-means cluster analysis using normalized data.

meaningful interpretation of the intrinsic differences among clusters. The analysis of the mean value of the variables in each cluster—an output presented by the final cluster centers—was particularly useful in understanding the characteristics of each cluster.

1.3.3.4.2 Principal components analysis

The PCA was applied to the 278 municipalities in order to better understand the social and territorial contexts in which disasters occur, with which severity and under which susceptibility conditions. Two PCAs were conducted using normalized data: one for criticality and another for support capability.

The PCA for criticality used the z-scores of the following 11 variables: flood susceptibility (i.e., percent of the municipality area covered by the main flood-prone areas), density of flood-related disaster cases, density of number of casualties, density of displaced, and density of evacuated persons; and the scores of the six PCs obtained in the SV assessment. The PCA performed for support capability was quite similar to this but, instead of the scores of criticality, the scores of the three PCs of support capability were used, consisting of a total of eight variables. The resulting scores in each PC were mapped, providing an interesting insight into the spheres of flood risk governance in which a concerned analysis and allocation of resources may be required.

1.3.4 Results

1.3.4.1 Components of the municipal flood disasters

1.3.4.1.1 Flood susceptibility

Flood susceptibility is absent in 140 of the 278 municipalities, where a null percentage of main flood-prone areas is found (Fig. 1.3.3A). The distribution of flood susceptibility is highly asymmetric within the study area, with the lower sections of the major rivers concentrating the highest susceptibility. According to this methodology, only 35 municipalities (12.6%) have more than 10% of the area as flood-prone. There are 13 municipalities with more than 25% of the area with susceptibility to flooding, while 2 municipalities—Golegã and V.F. Xira, both located in the lower Tagus floodplain—present more than 50% of their area as flood-prone. In summary, high flood susceptibility is found in the municipalities that cover the lower sectors of the Minho, Mondego, Tagus, and Sado basins, as observed for disaster cases (Fig. 1.3.3A). From the 35 municipalities with more than 10% of the respective area as prone to major floods, 31 are located on the lower sections of those basins.

1.3.4.1.2 Historical flood losses

Damaging floods include 811 cases that resulted in 1 or more casualties, missing, injured, evacuated, or displaced persons in mainland Portugal. The majority of cases are located along the Tagus, Mondego, Douro, and Minho rivers (Fig. 1.3.3A). The municipality with the highest number of cases is Coimbra (87 cases), located on the Mondego River basin, in the transition of the hilly and low-permeability terrains of the Hesperian Massif and the low-lying morphology of the Meso-Cenozoic basin. Moreover, the river crosses Coimbra, a medium-sized city (Figs. 1.3.1 and 1.3.3) with historically recurrent flood disaster events in both margins.

In mainland Portugal, the 138 municipalities that feature flood susceptibility (i.e., a nonnull percentage of main flood areas) represent 92.4% of the total evacuated persons (7958 out of 8617) and 95.8% of the total displaced persons (32,326 out of 33,718) recorded in the DISASTER database. Such

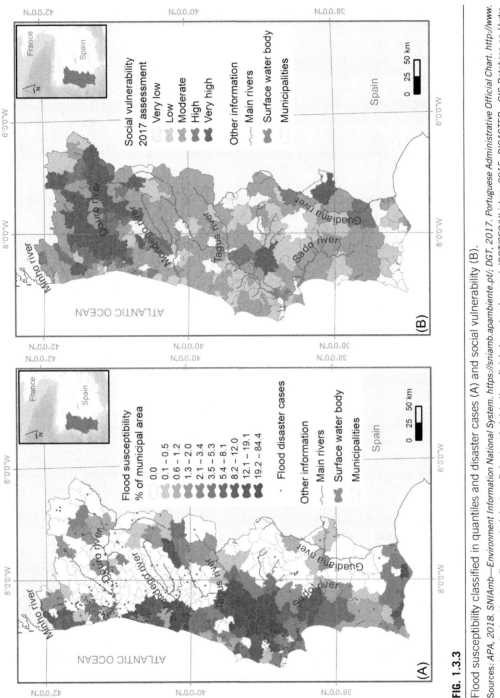

FIG. 1.3.3

Flood susceptibility classified in quantiles and disaster cases (A) and social vulnerability (B).

Sources: APA, 2018. SNIAmb—Environment Information National System. https://sniamb.apambiente.pt/; DGT, 2017. Portuguese Administrative Official Chart. http://www. dgterritorio.pt/cartografia_e_geodesia/cartografia/carta_administrativa_oficial_de_portugal_caop_/; IGOT/CEG/ULisboa, 2015. DISASTER—GIS Database on Hydro-Geomorphological Disasters in Portugal: A Tool for Environmental Management and Emergency Planning. http://riskam.ul.pt/disaster/.

figures express a high agreement between the flood loss records and the results of the flood suscepti-bility assessment. On the other hand, these same 138 municipalities only account for 65 of the 251 casualties (25.9%).

The 140 municipalities classified with the absence of main flood-prone areas represent 17.4% of the total number of disaster cases (141 in 811 cases). This is related to the fact that the susceptibility input data represents the most relevant contexts of flooding in mainland Portugal (specifically, the alluvial deposits and the Floods Directive hazard map) which, however, does not cover all the damaging floods that may be associated with local drivers of flood disasters. On the other hand, the flood losses database based on newspaper descriptions—although robust in terms of damage description and spatial and temporal coverage—is not exempt from inaccuracies in terms of the classification of the flood type.

1.3.4.1.3 Social vulnerability

The most relevant drivers of criticality are related to risk groups, disadvantaged and dependent pop-ulation (PCs 1, 3, and 6), as well as to the economic condition and employment (PCs 2 and 5). The geographical expression of higher criticality highlights the northern and inland municipalities of the Douro basin and some sparsely distributed municipalities on inland central and southern regions. The surrounding municipalities of the cities of Lisbon and Porto present low scores of support capa-bility due to the imbalance between the population and the existing public and private equipment and services that are relevant to the communities' disaster risk reduction and recovery. A detailed analysis of these results in comparison with a previous SV assessment is available in Tavares et al. (2018).

Relating both concepts—criticality and support capability—the results of the SV assessment at the municipal level identify a dichotomous scenario, as in general terms, the northern region is socially more vulnerable than the south (Fig. 1.3.3B).

1.3.4.2 Understanding the contexts of municipal flood disasters

Cross-analyzing the data collected regarding flood susceptibility, flood losses, and the criticality and support capability components of social vulnerability, it is possible to achieve new outputs that im-prove the knowledge of flood disaster contexts at the municipal level.

1.3.4.2.1 Cross-tabulation

There is a clear inverse correlation between the percentage of floodable area and the number of mu-nicipalities classified according to the number of flood disaster cases (Table 1.3.1). In fact, the 130 municipalities that present absence of disaster cases correspond to a low mean percentage of municipal territory with susceptibility (2.4%). To this same group of 130 municipalities, a mean criticality of 0.6 is observed, a value above the average of mainland Portugal (Table 1.3.1).

The 6 municipalities with the highest number of flood-related disaster cases (cf. the very high class in Table 1.3.1), also embrace more than half of the 8617 evacuated persons, more than one-third of the displaced persons and 35% of casualties (88 in 251 persons). The same municipalities present a mean floodable area of 18.1% of the municipality area. These observations support the conclusion that there exists a high disparity of flood impacts associated with the main flood-prone areas in mainland Por-tugal. The severest flood susceptibility contexts correspond to situations of low SV (mean of 0.180) and low criticality (mean of 0.360 in the range 0–1).

Table 1.3.1 Comparison between the damaging floods, susceptibility, and social vulnerability.

No. of flood disaster cases	No. of municipalities	% Floodable area	No. of casualties	No. of displaced persons	No. of evacuated persons	Criticality	Support capability	Social vulnerability
	Sum	Mean	Sum	Sum	Sum	Mean	Mean	Mean
Very low (0)	130	2.4	0	0	0	0.6	0.5	0.3
Low (1–5)	112	3.4	104	4805	970	0.5	0.5	0.3
Moderate (6–10)	19	9.1	46	1532	919	0.5	0.4	0.3
High (11–25)	11	20.5	13	14,337	2120	0.4	0.4	0.2
Very high (26–87)	6	18.1	88	13,044	4608	0.4	0.5	0.2
Mainland Portugal	**278**	**4.3**	**251**	**33,718**	**8617**	**0.5**	**0.5**	**0.3**

Flood susceptibility is classified in five quantiles and compared with the criticality, support capability, and social vulnerability in Fig. 1.3.4. The use of a logarithmic scale on the y-axis is justified by the fact that the very low susceptibility class (with 0% of area covered by flood-prone areas) aggregates a high number of municipalities (140) on the 1st quantile. The very high susceptibility class does not occur in municipalities with very high criticality (Fig. 1.3.4A). However, there are 4 and 13 municipalities classified, respectively, with very low and low support capability located in areas with very high susceptibility (Fig. 1.3.4B). The first group is located on the lower Tagus and the Mondego basins, while the second group is more sparsely distributed along the Sado, Vouga, Tagus, and Mondego basins, as well as in the small watersheds that drain the southern district of Faro.

1.3.4.2.2 Cluster classification

The cluster analysis identified 10 profiles of municipalities grouped by the final cluster centers. The interpretation of the final cluster centers allowed characterization of the municipalities that belong to each cluster (Fig. 1.3.5), expressing the distinct combinations of characteristics in terms of susceptibility, flood cases and losses, and social vulnerability.

Cluster 1 includes two municipalities, characterized by a relatively high density of displaced persons despite a very low susceptible area (Fig. 1.3.4). One of the municipalities is located at the mouth of the Douro River and the other in the upper sector of the floodplain of the Tagus River, i.e., both are located at the end of narrow fluvial sectors of the respective valleys.

Cluster 2 is composed of one municipality (Santarém, in the Tagus lower basin) mainly characterized by the relatively high density of evacuated persons and relatively low z-scores for the other variables.

The variable with the highest score in Cluster 3 is the density of casualties (Fig. 1.3.5), although, in the context of mainland Portugal, that value is not among the most concerning ones, which are found in clusters 5, 6, and 10.

Cluster 4 gathers 12 municipalities, most of them in the Tagus basin, but also one in the Mondego and three in the Vouga basin (river mouth in Aveiro) (c.f. Fig. 1.3.1). The cluster is characterized by a comparatively high portion of municipal areas with flood-prone areas but low numbers of losses (Fig. 1.3.5).

Clusters 5 and 6 include only one municipality each, both covering vast areas of the Tagus floodplain. Mortality is higher in cluster 5, while temporary evacuation is the most significant feature in cluster 6 (the cluster with the highest mean of evacuated persons).

Clusters 7 and 9 are similar in all variables except social vulnerability. The 75 municipalities that belong to cluster 7 are classified with the highest SV in mainland Portugal, while losses and susceptibility are the lowest of all the 10 clusters. In cluster 9 the SV z-score is closest to zero (Fig. 1.3.5).

Cluster 8 is composed of a single municipality, located in the intermediate course of the Douro River in mainland Portugal (Peso da Régua) that is characterized by a high density of displaced persons and cases, but low percentage of inundated area. This combination points out the existence of a well-identified hotspot of population exposure overlaid with small sectors of flood susceptibility.

Cluster 10 includes the municipality of Porto, the most remarkable of mainland Portugal in terms of density of flood disaster cases and number of casualties by square kilometer (Figs. 1.3.5 and 1.3.6).

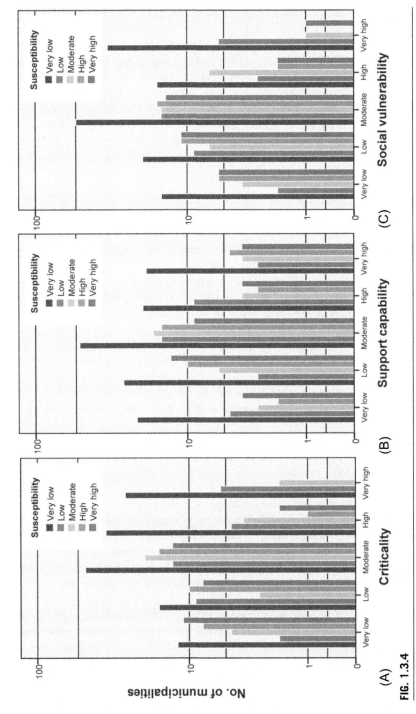

FIG. 1.3.4

Municipality classification of flood susceptibility by class of criticality (A), support capability (B), and social vulnerability (C).

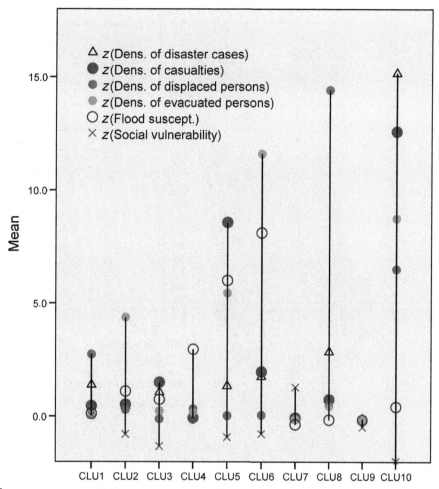

FIG. 1.3.5

Mean normalized values (z-scores) of the six variables used in cluster analysis, in each cluster.

1.3.4.2.3 Principal component analysis

PCA performed with variables describing flood susceptibility, flood losses, and the principal components (PCs) of the 2017 criticality assessment generated the association of variables presented on the rotated component matrix (Table 1.3.2). The associations self-generated by PCA and observable in the matrix do not express a cause-effect relationship between the explicative variables inside each PC, as flood losses are not exclusively explained alone, neither by the physical context of susceptibility, which is more static in time, nor by the socioeconomic context described by criticality and support capability. It should also be considered that the conditions that led to the reported historical flood losses in the past are not the current conditions of exposure and vulnerability. Nevertheless, some relationships, useful for risk management, can be drawn from the resulting data.

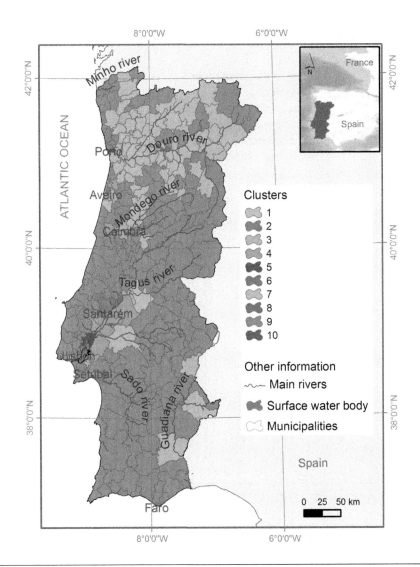

FIG. 1.3.6

Clusters characterizing flood susceptibility, flood cases and losses from DISASTER database, and social vulnerability.

Sources: *APA, 2018. SNIAmb—Environment Information National System. https://sniamb.apambiente.pt/; DGT, 2017. Portuguese Administrative Official Chart. http://www.dgterritorio.pt/cartografia_e_geodesia/cartografia/carta_administrativa_oficial_de_portugal_caop_/.*

PC1 does not associate flood susceptibility (percentage of floodable area) with variables related to density of losses, but shows a negative association with level of income (PC4 of criticality), meaning that municipalities with higher flood losses coincide with municipalities with higher levels of income, since in the point of view of criticality the higher scores in PC4 represent low levels of income—and the cardinality of this variable in the principal component 1 is negative (−0.541 of loading), while the

Table 1.3.2 Rotated component matrix of variables describing flood susceptibility, flood losses, and the principal components of criticality.

Variables	Components				
	PC1	PC2	PC3	PC4	PC5
Density of flood disaster cases	**0.911**	0.109	0.111	0.033	−0.009
Density of casualties	**0.844**	0.259	−0.032	0.042	−0.007
Density of evacuated persons	**0.680**	**0.482**	−0.014	0.231	−0.035
Density of displaced persons	**0.637**	−0.078	0.310	−0.030	0.111
Criticality PC4—Level of income	**−0.541**	0.395	**0.401**	0.183	0.088
% of floodable area	0.208	**0.744**	−0.217	0.280	−0.002
Criticality PC2—Economic conditions	−0.005	**−0.545**	−0.107	0.159	0.094
Criticality PC1—Risk groups	−0.155	**−0.400**	0.110	0.281	−0.165
Criticality PC3—Disadvantaged population	0.186	−0.070	**0.856**	−0.025	−0.029
Criticality PC5—Employment	−0.088	0.039	0.022	**−0.893**	−0.024
Criticality PC6—Dependent population	0.008	−0.033	−0.004	0.005	**0.973**

Extraction method: PCA. Rotation method: Varimax. Loading values ≤ -0.4 and ≥ 0.4 are shown in bold.

loss-related variables present positive loadings. Plotting the scores, the two higher classes (above 0.5 SD) clearly outline the more impacted areas, while the municipalities with scores below −0.5 SD never registered a flood disaster-type loss in the 1865–2015 period (Fig. 1.3.7, on the left).

PC2 does slightly associate high susceptibility with a high record of evacuated persons and, negatively, with economic conditions, which agrees with PC1 in regard to the expression of level of income. The association of high flood susceptibility with evacuated persons is symptomatic of the type of damage caused by flooding on large basins and floodplains, a characteristic in particular related to slow-onset floods: predictable onset of peak flows and temporary disruption, but rarely causing casualties or permanent displacement.

The other PCs (3–5) do not present as explicative the variables related to flood susceptibility or losses, making them less relevant for the understanding of flood risk. Despite that, they provide an understanding of the drivers of criticality—employment and dependent population—which can be comparable among municipalities. As an example of a possible interpretation, the scores of PC3 are plotted (Fig. 1.3.7, on the right): the most concerning contexts of flood impacts are found in the areas that simultaneously present high scores in PC2 and PC3, because PC3 expresses to a certain extent the local communities' recovery capacity.

Regarding the relation between the PCs of the 2017 support capability assessment, flood susceptibility, and flood losses, the rotated component matrix (Table 1.3.3) associates positively the contexts of high losses with high scores in support capability PC3 (coverage of logistics and services). The percentage of flood susceptibility per municipality is strongly associated with the density of evacuated persons shown before in Table 1.3.2, but negatively associated with the support provided by civil protection infrastructures (support capability PC1), which may indicate the need for a reinforcement of the installed capacity, at least and in particular, regarding flood risk. The PC3 main explicative variable is only the support capability PC related with economic and environmental dynamism, not allowing for relational analysis with the other variables.

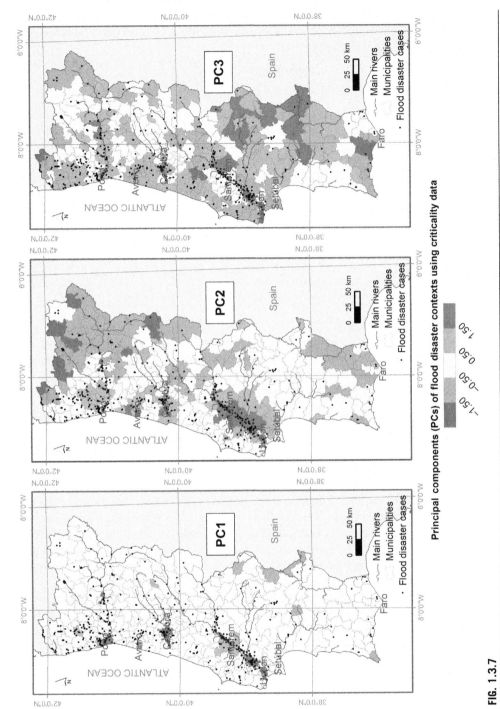

Principal components (PCs) of flood disaster contexts using criticality data

FIG. 1.3.7

PC1, PC2, and PC3 of the flood disaster contexts using criticality data. Explicative variables are listed in Table 1.3.2. Classification is done using the standard deviation.

Sources: APA, 2018. SNIAmb—Environment Information National System. https://sniamb.apambiente.pt/; DGT, 2017. Portuguese Administrative Official Chart. http://www.dgterritorio.pt/cartografia_e_geodesia/cartografia/carta_administrativa_oficial_de_portugal_caop_/; IGOT/CEG/ULisboa, 2015. DISASTER—GIS Database on Hydro-Geomorphological Disasters in Portugal: A Tool for Environmental Management and Emergency Planning. http://riskam.ul.pt/disaster/.

Table 1.3.3 Rotated component matrix of variables describing flood susceptibility, flood losses, and the principal components of support capability.

Variables	Components		
	PC1	PC2	PC3
Density of flood disaster cases	**0.895**	0.237	−0.003
Density of casualties	**0.800**	0.378	0.016
Density of displaced persons	**0.665**	−0.011	−0.095
Support Capability PC3—Logistics and services capacity	**0.571**	−0.247	0.061
% of floodable area	0.066	**0.841**	0.160
Density of evacuated persons	**0.599**	**0.616**	0.154
Support Capability PC1—Civil protection response	0.019	**−0.536**	0.275
Support Capability PC2—Economic and environmental dynamism	−0.027	0.006	**0.936**

Extraction method: PCA. Rotation method: Varimax. Loading values ≤ -0.4 and ≥ 0.4 are shown in bold.

The overlay of low scores in PC3 and high scores in PC2 identify the areas in which flood impacts might be more severe in terms of the economic recovery (Fig. 1.3.8, center and right).

1.3.5 **Discussion**

The multivariate analysis techniques—cross-tabulation, cluster, and PCA analysis—applied to the "raw" results presented in Section 1.3.4.1 allow a deepening of our knowledge of the contexts and drivers of flood disasters.

The patterns of flood susceptibility and losses allow three main groups of municipalities to be distinguished. The first is the group including the municipalities with a percentage of flooded area higher than 12% (9th and 10th quantiles), and in which displacement and evacuation prevail over loss of life. The second group includes those municipalities with a small percentage of flooded area (below 1.3%, i.e., 2nd and 3rd quantiles), which registered several types of impacts, as exemplified by clusters 8 and 10. Finally, the last group corresponds to the municipalities with negligible flood susceptibility, which is confirmed by the low number and low density of reported cases and losses.

Two types of limitations can be identified in this study. One, regarding the historical database of flood losses that along the 150-year timeframe allows coverage of a wide range of flood hazard typologies and severities, covers territorial contexts in which losses have occurred in the past, but which may not correspond to the current contexts of hazard, exposure, and vulnerability. For example, the construction of dams along most hydrographic basins during the second half of the 20th century altered the frequency-magnitude relationship of floods in the largest river basins.

A second limitation is related to the SV assessment, in which criticality and support capability were assessed for the entire municipal territory, thus not expressing the specific territorial conditions where flooding occurs. Nevertheless, such assessment is relevant for understanding potential indirect impacts, as it also sets the context of the communities' proneness to be impacted as well as their ability to provide assistance and assure daily contingency during and after a flood event.

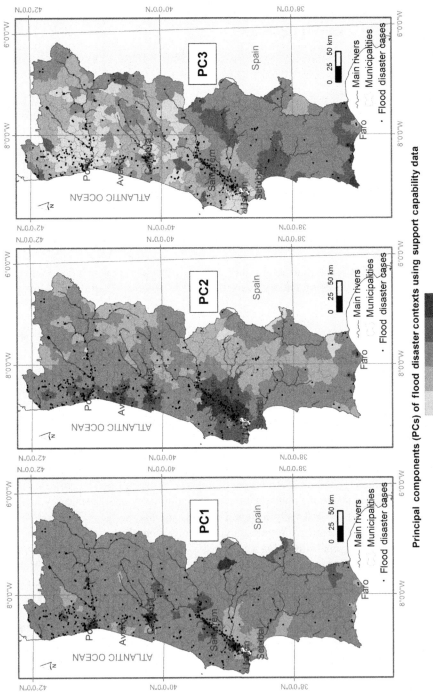

Principal components (PCs) of flood disaster contexts using support capability data

FIG. 1.3.8

PC1, PC2, and PC3 of the flood disaster contexts using support capability data. Explicative variables are listed in Table 1.3.3. Classification is done using the standard deviation.

Sources: APA, 2018. SNIAmb—Environment Information National System. https://sniamb.apambiente.pt/; DGT, 2017. Portuguese Administrative Official Chart. http://www.dgterritorio.pt/cartografia_e_geodesia/cartografia/carta_administrativa_oficial_de_portugal_caop_/; IGOT/CEG/ULisboa, 2015. DISASTER—GIS Database on Hydro-Geomorphological Disasters in Portugal: A Tool for Environmental Management and Emergency Planning. http://riskam.ul.pt/disaster/.

National disaster risk management must consider the spatial distribution of casualties in regard to the susceptibility mapping that identifies the major flood-prone areas. In fact, the data shows that flood-related mortality is historically related to other contexts of susceptibility—i.e., floods in small basins—and to particular contexts of exposure and individual risk behavior during the flood event (Pereira et al., 2016, 2017), highlighting the relevance of collecting and analyzing loss databases (Barnolas and Llasat, 2007). Both underconsidered drivers of casualties require the close attention of decision makers, not only those from the civil protection-related fields, but also the planners and practitioners working in mobility, spatial planning, meteorological forecasting, and early warning systems.

Flood consequences, translated into the number of occurrences and number of displaced, evacuated, and fatalities, present a strong asymmetric distribution among the Portuguese municipalities that, to a certain extent, is evidenced by the existence of clusters with a reduced number of individuals. However, not only the physical constraints of flood processes have to be considered to explain this pattern, as exposure and social vulnerability also play a relevant role in explaining the historical record of flood losses (e.g., Baldassarre et al., 2017). The same finding was observed through the application of PCA, showing that the self-aggregation of variables around each principal component is driven by a combination of sociological and hydrological factors. One of the PCs linked high flood susceptibility with high number of evacuated persons while, in opposition, a high number of fatalities are not directly associated with high susceptibility. Cost-benefit analysis, as recognized by the European Floods Directive, needs to integrate the multidimensional drivers of disasters and to weight the balanced efficiency of structural and nonstructural measures.

1.3.6 Conclusions

This research comes in line with the first priority of the Sendai Framework for Disaster Risk Reduction (UNISDR, 2015), understanding disaster risk, which stands for the need to develop policies and practices in risk management based on an understanding of all the dimensions of disaster risk: vulnerability, capacity, exposure, and hazard characteristics.

Despite the high uncertainty regarding the increase of flood susceptibility in mainland Portugal in a climate change scenario, considering the intrinsic high irregularity of extreme rainfall events and the role of existing and future dams, flood risk management must dedicate efforts toward correcting past errors in spatial planning through risk management strategies, water management and mobility, and civil protection actions. Moreover, future dynamics in land use change point toward an increase in artificialized areas (Meneses et al., 2018), posing the challenge of conforming this trend with the need to reduce exposure to flooding. In this regard, the research identified municipalities with imbalances in terms of percentage of floodable area and severity of losses, which present themselves as priorities for civil protection and spatial planning solutions.

A future research line consists of adding to the loss analysis the impacts of flooding on agriculture, as most of the flood-prone areas are subject to prolonged flooding. This economic impact is also related to the socioeconomical context reflected by vulnerability studies.

Acknowledgments

This work was financed by national funds through FCT—Portuguese Foundation for Science and Technology, I.P., under the framework of the project FORLAND—Hydro-geomorphologic risk in Portugal: driving forces and application for land use planning (PTDC/ATPGEO/1660/2014) and by the Research Unit UID/GEO/00295/2019. Pedro Pinto Santos is financed by FCT through the project with the reference CEEIND/00268/2017.

References

APA, 2018. SNIAmb—Environment Information National System. https://sniamb.apambiente.pt/.

Baldassarre, G.D., Saccà, S., Aronica, G.T., Grimaldi, S., Ciullo, A., Crisci, M., 2017. Human-flood interactions in rome over the past 150 years. Adv. Geosci. 44, 9–13. https://doi.org/10.5194/adgeo-44-9-2017.

Barnolas, M., Llasat, M.C., 2007. A flood geodatabase and its climatological applications: the case of catalonia for the last century. Nat. Hazards Earth Syst. Sci. 7 (2), 271–281. https://doi.org/10.5194/nhess-7-271-2007.

Benito, G., Lang, M., Barriendos, M., Llasat, C., Francés, F., Taha, O., Varyl, R., Enzel, Y., Bardossy, A., 2004. Use of systematic, palaeoflood and historical data for the improvement of flood risk estimation. Review of scientific methods. Nat. Hazards 31, 623–643. https://doi.org/10.1023/B:NHAZ.0000024895.48463.eb.

Cunha, N.S., Magalhães, M.R., Domingos, T., Abreu, M.M., Küpfer, C., 2017. The land morphology approach to flood risk mapping: an application to Portugal. J. Environ. Manage. 193, 172–187. https://doi.org/10.1016/j.jenvman.2017.01.077.

DGT, 2017. Portuguese Administrative Official Chart. http://www.dgterritorio.pt/cartografia_e_geodesia/cartografia/carta_administrativa_oficial_de_portugal_caop_/.

DGT, 2018. Atlas de Portugal. http://dev.igeo.pt/atlas/Cap1/Cap1.html.

Fekete, A., 2009. Assessment of Social Vulnerability for River-Floods; 2009. Landwirtschaftlichen Fakultät.

Ferrari, S., Oliveira, S., Pautasso, G., Zêzere, J.L., 2019. Territorial Resilience and Flood Vulnerability. Case Studies at Urban Scale in Torino (Italy) and Porto/Vila Nova de Gaia (Portugal). Springer, Cham, pp. 147–174. https://doi.org/10.1007/978-3-319-76944-8_10.

FFMS, 2017. PORDATA—Database of Contemporary Portugal. https://www.pordata.pt/.

INE, 2011. Population Census—2011. INE, Lisbon.

Jacinto, R., Grosso, N., Eusébio, R., Dias, L., Santos, F.D., Garrett, P., 2015. Continental Portuguese Territory Flood Susceptibility Index—contribution to a vulnerability index. Nat. Hazards Earth Syst. Sci. 15, 1907–1919. https://doi.org/10.5194/nhess-15-1907-2015.

Kjeldsen, T.R., Macdonald, N., Lang, M., Mediero, L., Albuquerque, T., Bogdanowicz, E., Brázdil, R., et al., 2014. Documentary evidence of past floods in Europe and their utility in flood frequency estimation. J. Hydrol. 517, 963–973. https://doi.org/10.1016/j.jhydrol.2014.06.038.

Kundzewicz, Z.W., Kanae, S., Seneviratne, S.I., Handmer, J., Nicholls, N., Peduzzi, P., Mechler, R., et al., 2014. Flood risk and climate change: global and regional perspectives. Hydrol. Sci. J. 59 (1), 1–28. https://doi.org/10.1080/02626667.2013.857411.

Leal, M., Ramos, C., 2013. Susceptibilidade Às Cheias Na Área Metropolitana de Lisboa Norte. Factores de Predisposição e Impactes Das Mudanças de Uso Do Solo. Finisterra 95, 17–41.

Li, K., Wu, S., Dai, E., Zhongchun, X., 2012. Flood loss analysis and quantitative risk assessment in China. Nat. Hazards 63 (2), 737–760. https://doi.org/10.1007/s11069-012-0180-y.

Mendes, J.M., Tavares, A.O., Freiria, S., Cunha, L., 2010. Social vulnerability to natural and technological hazards: the relevance of scale. In: Bris, B., Soares, C.G., Martorell, M. (Eds.), Reliability, Risk and Safety: Theory and Applications. ESRA and Taylor & Francis Group, pp. 445–451.

Meneses, B.M., Reis, E., Vale, M.J., Reis, R., 2018. Modelling the land use and land cover changes in Portugal: a multi-scale and multi-temporal approach. Finisterra 53 (107), 3–26. https://doi.org/10.18055/Finis12258.

Oliveira, J.T., Pereira, E., Ramalho, M., Antunes, M.T., Monteiro, J.H., 1992. Geological Map of Portugal, 1:500.000, fifth ed. Portuguese Geological Services, Lisbon, Portugal.

Patton, P.C., Baker, V.R., 1976. Morphometry and floods in small drainage basins subject to diverse hydrogeo-morphic controls. Water Resour. Res. 12 (5), 941–952. https://doi.org/10.1029/WR012i005p00941.

Pereira, S., Zêzere, J.L., Quaresma, I., Santos, P.P., Santos, M., 2016. Mortality patterns of hydro-geomorphologic disasters. Risk Anal. 36 (6), 1188–1210. https://doi.org/10.1111/risa.12516.

Pereira, S., Diakakis, M., Deligiannakis, G., Zêzere, J.L., 2017. Comparing flood mortality in Portugal and Greece (Western and Eastern Mediterranean). Int. J. Disaster Risk Reduct. 22, 147–157. https://doi.org/10.1016/j.ijdrr.2017.03.007.

Rousseeuw, P.J., 1987. Silhouettes: a graphical aid to the interpretation and validation of cluster analysis. J. Comput. Appl. Math. 20 (C), 53–65. https://doi.org/10.1016/0377-0427(87)90125-7.

Rufat, S., Tate, E., Burton, C.G., Maroof, A.S., 2015. Social vulnerability to floods: review of case studies and implications for measurement. Int. J. Disaster Risk Reduct. 14, 470–486. https://doi.org/10.1016/j.ijdrr.2015.09.013.

Sá, L., Vicêncio, H., 2011. Risco de Inundações—Uma Metodologia Para a Sua Cartografia. Territorium 15, 227–230.

Santos, P.P., Reis, E., 2018. Assessment of stream flood susceptibility: a cross-analysis between model results and flood losses. J. Flood Risk Manage. 11(S2). https://doi.org/10.1111/jfr3.12290.

Santos, P.P., Tavares, A.O., 2015. Basin flood risk management: a territorial data-driven approach to support decision-making. Water (Switzerland) 7 (2), 480–502. https://doi.org/10.3390/w7020480.

Santos, P.P., Tavares, A.O., Zêzere, J.L., 2014. Risk analysis for local management from hydro-geomorphologic disaster databases. Environ. Sci. Policy 40, 85–100. https://doi.org/10.1016/j.envsci.2013.12.007.

Santos, P.P., Reis, E., Pereira, S., Santos, M., 2019. A flood susceptibility model at the national scale based on multicriteria analysis. Sci. Total Environ. 667, 325–337. https://doi.org/10.1016/j.scitotenv.2019.02.328.

Tapsell, S.M., Penning-Rowsell, E.C., Tunstall, S.M., Wilson, T.L., 2002. Vulnerability to flooding: health and social dimensions. Philos. Trans. R. Soc. A: Math. Phys. Eng. Sci. 360 (1796), 1511–1525. https://doi.org/10.1098/rsta.2002.1013.

Tavares, A.O., Barros, J.L., Mendes, J.M., Santos, P.P., Pereira, S., 2018. Decennial comparison of changes in social vulnerability: a municipal analysis in support of risk management. Int. J. Disaster Risk Reduct. 31, 679–690. https://doi.org/10.1016/j.ijdrr.2018.07.009.

UNISDR, 2015. Sendai framework for disaster risk reduction 2015-2030. In: Third World Conference on Disaster Risk Reduction, Sendai, Japan, 14–18 March https://doi.org/A/CONF.224/CRP.1.

Viglione, A., Baldassarre, G.D., Brandimarte, L., Kuil, L., Carr, G., Salinas, J.L., Scolobig, A., Blöschl, G., 2014. Insights from socio-hydrology modelling on dealing with flood risk—roles of collective memory, risk-taking attitude and trust. J. Hydrol. 1–12. https://doi.org/10.1016/j.jhydrol.2014.01.018.

Winsemius, H.C., Aerts, J.C.J.H., van Beek, L.P.H., Bierkens, M.F.P., Bouwman, A., Jongman, B., Kwadijk, J.C.J., et al., 2015. Global drivers of future river flood risk. Nat. Clim. Change, 1–5. https://doi.org/10.1038/nclimate2893.

Zêzere, J.L., Pereira, S., Tavares, A.O., Bateira, C., Trigo, R.M., Quaresma, I., Santos, P.P., Santos, M., Verde, J., 2014. DISASTER: A GIS database on hydro-geomorphologic disasters in Portugal. Nat. Hazards 72 (2), 503–532. https://doi.org/10.1007/s11069-013-1018-y.

Social vulnerability to drought in rural Malawi

1.4

Euan James Innes[a], Robert Šakić Trogrlić[b], and Lindsay C. Beevers[a]

Institute for Infrastructure and Environment, Heriot-Watt University, Edinburgh, United Kingdom[a] School of Energy, Geoscience, Infrastructure and Society, Heriot-Watt University, Edinburgh, United Kingdom[b]

CHAPTER OUTLINE

Understanding Disaster Risk. https://doi.org/10.1016/B978-0-12-819047-0.00006-8

1.4.1 Introduction

Drought is a hazard that is particularly complex and difficult to model. Due to its slow nature and often incremental development, it is challenging to define the start of a drought event, its severity, and when it has ended. Drought hazard interacts in complex and subtle ways with the environmental, social, and economic systems of an affected area, resulting in significant and often widespread impacts (Mishra and Singh, 2010). A recent example of such impacts is the drought emergency in Kenya in 2018, leaving 2.7 million people food insecure (Guimarães Nobre et al., 2019). Critical to understanding this complex relationship between the exposed population and hazard is the concept of vulnerability. When two communities are exposed to the same hazard, why may one come to more harm than the other? What factors make that community more susceptible to harm? What makes a community more resilient to harm, and how can communities increase their resilience? These are important questions for policy makers, governments, and aid agencies dealing with drought around the world; they must be answered to ensure that interventions and assistance are targeted in the most effective manner.

Unfortunately, vulnerability is inherently unmeasurable (Cutter et al., 2003). To overcome this difficulty, indicator studies have been developed over the past 20 years as a method of quantifying vulnerability in such a way that communities can be compared and assessed in a consistent manner. Most often they are applied to kinetic hazards such as storms, earthquakes, and floods, with drought vulnerability a notably understudied area. This study is an attempt to apply a community-based rapid assessment tool for measurement and mapping of social vulnerability to drought in a subsistence agriculture context. The tool has been applied to the Nsanje district of rural Malawi. Nsanje was chosen as the study area due to its history of vulnerability and marginal subsistence agriculture economy. Thus the aim of this study is to outline and field test a theoretical model for rapid assessment of social vulnerability to drought.

1.4.1.1 Drought as a concept, hazard, and disaster

Drought as a hazard can be defined as an extreme shortfall in the availability of water, either through precipitation (meteorological drought), soil moisture (agricultural drought), or river flows (hydrological drought). Drought as an aberration defines it separately from aridity, which is the long-running climatic condition of low precipitation, and from seasonal aridity, the presence of a normal climatic dry season in the year (Mishra and Singh, 2010). However, all hydrological systems experience highs and lows, so at what point does normal seasonal variation become drought? Unlike flooding, the exact boundaries of drought can be hard to define. Drought is the most complex of all natural hazards (Iglesias et al., 2009). With kinetic hazards such as flooding, the frequency and magnitude of the events are the key to the understanding of risk (UNDP, 2012). With drought, it is more often the duration of the event that causes harm (UNDP, 2012).

Drought is characterized by its slow onset, with precipitation failure causing the stock levels of surface water bodies, aquifers, and soil moisture to fall. It is difficult to know when a drought has begun or indeed ended; in fact, it may only be possible to define retrospectively, once a deficit has been established. Drought tends to be regional in its spatial scale, and similarly it may be hard to define the exact boundaries of the drought zone, as moisture levels may transition gradually to normal. Due to these uncertainties, there may be hundreds of definitions of drought. Most of these drought definitions are based on hydro-meteorological calculations [e.g., Standard Precipitation Index (SPI) and the

Palmer Drought Severity Index (PDSI)] (Paulo and Pereira, 2006). These are assessments of meteo-rological drought, expressing an unusual deficit in precipitation. While these may be useful analyses, they do not measure any impact on the systems receiving the rainfall. It is common to develop this approach to examine agricultural drought, whereby the characteristics of the agricultural system are analyzed for signs of harm. A further definition is hydrological drought, where surface water bodies and stream flows are observed to fall below normal levels. As essentially a runoff process, often the conditions of hydrological drought will occur after those of meteorological and agricultural droughts (at least in the context of rain-fed agriculture), as the former deals purely with precipitation and the latter with in-catchment processes. The impacts on hydrological systems will often occur later (NDMC, 2017).

While these drought definitions relate to explicitly measurable metrics, the true complexity of drought as a hazard, and as a disaster, is in its interactions with the impacts on the underlying socio-economic system, or socioeconomic drought. This type of drought can be defined as a shortfall in the supply of economic goods caused by a deficit in precipitation. In many cases this will overlap with agricultural drought in a shortfall of production, but may also affect public water supply, and hydro-power. Harm from drought is not direct but is a result of complex interactions and pressures within the economy and social structure of the exposed population (UNDP, 2012).

1.4.1.2 Risk, hazard, and vulnerability

Social vulnerability, exposure, hazard, and risk are terms with different meanings depending on the source used. For the purposes of this chapter, the terms are defined as follows:

Hazard—A natural event or phenomenon causing or potentially causing harm, i.e., flood, drought, earthquake, etc. Here a hazard shall refer to agricultural and resource drought, i.e., the failure of the environmental system to provide enough water to produce agricultural goods and provision of drinking water (UNISDR, 2019[a]).

Exposure—The extent of the receptor under threat from the hazard. In this study, this is defined as the area of land under threat and the population affected (UNISDR, 2019).

Vulnerability (or susceptibility)—The characteristics of the receiving population which determine the magnitude of harm. The United Nations/International Strategy for Disaster Reduction (UNISDR, 2004) defines vulnerability as "conditions determined by physical, social, economic and environmental factors or processes which increase the susceptibility of a community to the impact of hazards."

Social vulnerability—The subject of this chapter; as defined by Cutter et al. (2003), it is the com-binations of factors that influence the susceptibility of groups to harm, such as age, race, income, etc., but they also extend the definition to "place inequalities," or the characteristics of the communities and their build environment including "urbanization, growth rates and economic vitality."

Risk—The combined effect and interaction of hazard, exposure, and vulnerability (Iglesias et al., 2009). Risk is the true measure of the occurrence of drought as a disaster: that the hazard has impacted the vulnerable receptor and caused harm (Birkmann, 2007). Part of the concept of vulnerability is the concept of resilience or adaptive capacity. Adaptive (or coping) capacity is a measure of the extent to

---

[a]UNISDR is now known as UNDRR.

which an organization or community has the resources to positively respond to hazard to reduce risk (UNISDR, 2019).

1.4.1.3 The measurement of vulnerability and indicator studies

There are many different approaches and methodologies to assessing vulnerability to natural hazards. These have been reviewed by, for instance, Cutter et al. (2009), Birkmann (2007), Zarafshani et al. (2016), Mwale (2014), Mwale et al. (2015), Adger (2006), and Schmidtlein et al. (2008). These studies range widely in their scale and approach. In keeping with the approach developed by Cutter et al. (2003), many are based on existing data. For example, Antwi-Agyei et al. (2012) used crop production and rainfall data to model vulnerability to drought in Ghana; Kim et al. (2015) used similar county-level data to produce a vulnerability-to-drought map of South Korea, while Yaduvanshi et al. (2015) combined this approach with remote imaging to produce drought vulnerability mapping for drought risk over tropical India. Similarly, Malcomb et al. (2014) developed a model to map the relative levels of vulnerability to climate change across Malawi, and Ayantunde et al. (2015) uses participatory interviews to map the vulnerability to drought in communities in the West African Sahel. What is clear, however, is that there is no clear consensus or one defined method to capture this information.

Measurement of vulnerability is inherently difficult; while it may be possible to say what factors make a person vulnerable, it is not possible to measure vulnerability directly (Cutter et al., 2003). This poses a difficulty for policy makers to target interventions effectively to the most vulnerable communities. Over time the concept of vulnerability indicators has developed, most notably in Cutter and colleagues 2003 paper "The Social Vulnerability Index." An indicator is a direct and quantifiable value that stands as a proxy for vulnerability. The factors affecting vulnerability to drought in general vary widely, particularly at different scales of discretization (Jülich, 2014). Vulnerability varies between individuals and within households; thus scaling from household level to community level may average information and lower the resolution of the data. In selecting indicators, it is important to take note of scale and locality (Birkmann, 2007) to ensure their relevance for the specific context. Social vulnerability is a complex concept, consisting of multiple different facets that may impact on an individual's or community's ability to respond to or deal with the effects of a hazard. There may be many factors that govern these vulnerabilities in a population. In general, vulnerability mapping is less developed than hazard mapping (Birkmann, 2007).

A notable methodology in the context of community disaster management is the Community Based Disaster Risk Index (CBDRI), developed by Bollin et al. (2003). The CBDRI is an attempt to create a baseline assessment of the relative risk of communities. It was developed using a review of previous studies, primarily focused on South America. Indicators were selected for:

Validity—Accurate measure of vulnerability
Reliability—The extent to which it could be remeasured
Sensitivity—How the measure will change with a developing situation
Availability—Whether data is easily obtained
Objectivity—Essential reproducibility

The CBRDI intention is that the information can be collected by questionnaire, augmented by preexisting data where available. This approach allows for much finer resolution than relying solely on national or regional scale indicators, and it has the advantage of receiving input directly from the studied community. These indicators are only intended to be part of the picture and are combined with qualitative and participatory methods. To turn the indicators into a true index scheme, they need to be scaled and weighted to ensure that the relative importance of each measure is accounted for appropriately.

1.4.2 **Hydrohazards in Malawi**

Malawi is one of the most densely populated and least-developed countries in the world (CIA, 2017). It is a demographically young country, with 46.5% of the population in the 0–14 age bracket. While Malawi's population growth rate has slowed slightly in recent decades, it is still the second highest in the world, with an annual growth of 3.32%. Literacy is low at 65.8% of the population (CIA, 2017). The population is highly dispersed, with 83.7% living outside the cities. It is a poorly developed nation and is primarily dependent on agriculture, which makes up 90% of its exports, with tobacco, tea, and sugar as the leading products. Agriculture accounts for 40% of the GDP, and this dependence makes Malawi highly susceptible to economic damage from extreme weather events that impact on crop growth (World Bank, 2010). Malawi and the wider Sub-Saharan Africa region are increasingly being affected by climate change, with more unstable rainfall causing floods and droughts (Malcomb et al., 2014). The rapid rate of population growth is putting further pressure on the economy and environment of Nsanje through deforestation, soil loss, and increased competition for resources (Mwale et al., 2015).

The majority of Malawi falls within the Shire subcatchment of the greater Zambezi basin and Lake Malawi, which dominate the water system of the country. The climate of the Shire subbasin follows a cyclical wet/dry season, typical of the tropics. The wet season runs from November/December to March/April, corresponding to the Southern hemisphere summer. The dry season runs from May to October, accounting for the southern winter. Temperatures are typically cooler in the winter, with an average of 16°C rising to an average of 22°C in the summer (Euroconsult, 2007). Annual flooding in swamps, marshlands, and floodplains is part of the natural cycle of the Zambezi basin; local cultures have adapted to this pattern and account for it in their lives. The flooding pattern is crucial to supporting the fisheries and floodplain agriculture, through sediment deposition and nutrient transfer. However, the Shire basin is also prone to less frequent, extreme floods and droughts, for which infrastructure and populations are less prepared, leading to loss of assets and lives. The region is also often prone to sudden flash floods (World Bank, 2010). The climate of the Zambezi basin is highly sensitive to the El Niño Southern Oscillation (FAO, 2018). Situated on the west of the Indian Ocean, the Zambezi is subject to the inverse conditions of El Niño, generally causing lower rainfall and higher temperatures and correlating to drought occurrences in the basin. An example of this is the failure of the 2015 rainy season. Conversely the "cold phase" of the ENSO (La Niña) causes greater than average rainfall in the Zambezi. During the 2015–16 La Niña event, Malawi was hit by the highest rainfall on record, causing extreme flooding, displacing hundreds of thousands of people, damaging thousands of hectares of land, and causing over 100 deaths (Malawian Government, 2016). Similar flooding occurred during Hurricane Idai in 2019, with 1.1 million people projected to have insufficient food supplies during the September 2019 to March 2020 period (UNICEF, 2019).

Flooding in Malawi accounts for ∼40% of all natural hazards (Nillson et al., 2010), and although annual floods can support agrobased livelihoods, given the appropriate timing and magnitude, flooding also seriously affects income generated from agriculture, together with damaging infrastructure and impacting local communities' social, economic, cultural, and psychological values (Lumumba Mijoni and Izadkhah, 2009; Šakić Trogrlić et al., 2017). In a detailed study of flood vulnerability in the Southern Malawi, Mwale et al. (2015) found that vulnerability is shaped predominantly by the socioeconomic and environmental susceptibility of the underlying communities. Acknowledging the high levels of vulnerability, Chawawa (2018) argues that local communities are not helpless victims, because they have developed a set of adaptive strategies based on their long-lasting experience with flooding. However, a recent study by Šakić Trogrlić et al. (2019) indicates that local people are limited in the extent to which they can apply their local adaptation, due to processes that generate vulnerabilities in the first place being outside their direct area of influence [e.g., poverty, environmental degradation, disaster risk reduction (DRR) planning].

Droughts have similarly had a severe impact on the communities and economy of Malawi. Malawi is particularly sensitive to drought, with its economy reliant on agriculture and Shire River hydropower for electricity. During droughts, low flows in the Shire result in blackouts. Since 1982 there have been eight major droughts in Malawi. It is estimated that Malawi has lost 1% of its GDP per year to drought during that period (World Bank, 2016). The most recent severe drought was during the 2015–16 growing season. The Malawian Government estimated the failure of the rains left 6.5 million people, 39% of the population, food insecure during 2016–17 (Malawi Gov., 2016). The impacts were particularly severe in the Southern Region, with a state of emergency being declared in Nsanje District. The data collection for this study presented in this chapter was completed during late 2016, while the impact of the drought was underway.

With a greater focus being placed historically on flooding, drought is understudied in Malawi. The objective of this chapter is to therefore contribute to filling this knowledge gap by laying out a location-specific assessment tool for the local variation in vulnerability to drought.

1.4.3 **Methods**

1.4.3.1 **Case study area in Nsanje**

Malawi is organized into three administrative regions: North, Central, and South. Each region of Malawi is subdivided into a number of districts, and districts are further subdivided into Traditional Authority (TA) areas, as shown in Fig. 1.4.1. Nsanje is a district in the South administrative region.

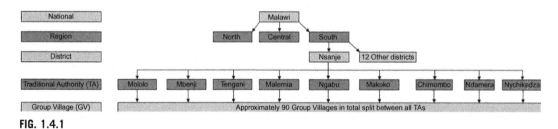

FIG. 1.4.1

Administrative divisions of Malawi, with a detailed overview of Nsanje.

Beneath the TA there are approximately 82 Group Villages constituting the lowest level of government organization in Malawi. Group Villages (GVs) vary in land area and population. The GVs are represented by Village Development Committees (VDCs), part of which are the Village Civil Protection Committees (VCPCs). Nsanje is predominantly situated on the Shire's west bank, with the exception of Mololo, located on the east bank, as shown in Fig. 1.4.2.

At the confluence of the Shire and Ruo rivers, there is a wide marsh area known as the Elephant Marsh. A smaller marshland area, named the Ndindi Marsh, is situated to the south of the district. From the Shire, the middle third of Nsanje is a flat plain through which the main road (M1) between Mozambique and the North of Malawi runs. The western third of the district rises to a mountain range approximately 500 m above the plain, as shown in Fig. 1.4.3.

In 2018, the population of Nsanje was reported to be 299,168, an increase of 24% since the previous survey in 2010 (MNSO, 2019). The Lower Shire is considered to be the most vulnerable region of Malawi, and 45% of the population is food insecure in each year (Malcomb et al., 2014). Nsanje has a dependency ratio of 1.1, meaning that approximately half of the population is out of the economically active age range of 15–65. The average household size for Nsanje is 4.7, which is one of the highest in the rural parts of the country (MNSO, 2019). Ninety percent of the population of the Shire Valley is made up of subsistence farmers, generally working less than 1 ha of land. This land is often unproductive, with maize crops often returning only 30% of their theoretical maximum yield (Mwale, 2014). Livelihoods are supplemented by fishing, animal rearing, and casual agricultural labor. Livelihoods in Nsanje are precarious, as 90% of the population is exposed to flooding (Atkins, 2012), which particularly threatens the fertile marsh area surrounding the Shire River.

1.4.3.2 Theoretical framework

The initial theoretical framework for use in the study is the CBDRI, discussed in Section 1.4.1, but modified in this study for application to droughts. The method is a community level framework that can be scaled to the appropriate spatial resolution. Further, the CBDR can be applied to vulnerability in the context of any hazard by using a weighting metric to make the context specific. The weighting should be specific to the hazard and situation required. For instance, community building codes may be a significant factor in earthquake safety, but less so in relation to droughts. Weighting is inherently difficult, as it relies on the subjective assessment of the practitioner. Bollin et al. (2003) list this as a specific weakness of their approach.

The intention of the CBDRI is that, with effective weighting, it should be applicable to any hazard. However, many of the indicators used may not be relevant to the nonkinetic characteristics of drought: for example, building codes, emergency response drills, etc. Also, the impacts of drought are primarily felt through the socioeconomic loss of livelihood and food supply. Thus this part of the CBDRI required modification for this application.

Zarafshani et al. (2016) attempted to apply a community-based indicator model to drought in agricultural communities. Their approach to the risk equation was akin to the CBDRI; however, the framework was adapted by focusing on indicators such as suitability of crop, nonfarm income, and irrigation, and included individual psychology as a significant factor in the resilience to hazard (i.e., the attitude of a farmer when faced with a problem and how they respond under stress). They placed the formulating of indicators within the wider context of bridging the divide between theory and government intervention, and plotted a continual improvement loop, as shown in Fig. 1.4.4.

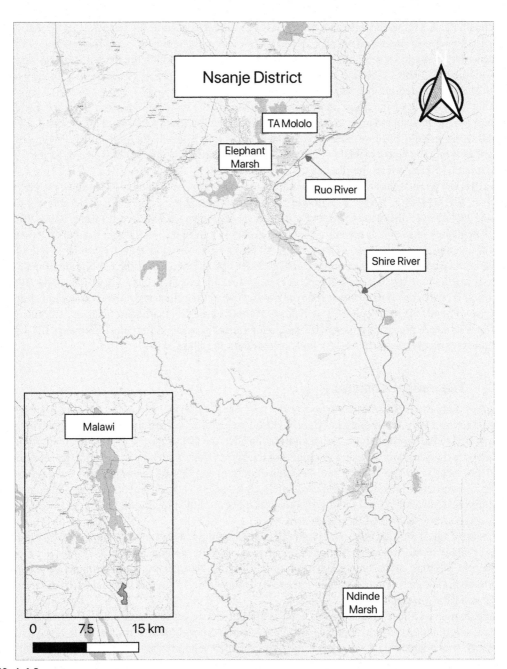

FIG. 1.4.2

Nsanje District, inset Malawi with Nsanje highlighted.

FIG. 1.4.3

Nsanje central plain foreground with western hills in background.

Image author's own.

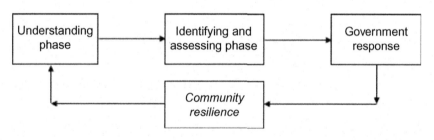

FIG. 1.4.4

Phases of drought risk management.

Adapted from Zarafshani, K., Sharafi, L., Azadi, H., Van Passel, S., 2016. Vulnerability assessment models to drought: toward a conceptual framework. Sustainability 8(6), 588.

The objective of this study is to assess the spatial variation of social vulnerability to drought in Nsanje, which fits into the Identifying and Assessing Phase. Three components are identified at this stage: Exposure, Sensitivity (Vulnerability), and Adaptive Capacity. This study brings together the CBDRI framework with the indicator model by Zarafshani et al. (2016) to create a new framework to assess the social vulnerability to droughts in Nsanje (Fig. 1.4.5). It develops Zarafshani et al.'s (2016) framework and modifies it to reflect the local conditions of Nsanje. The starting point for candidate questions was the CBDRI laid out by Bollin et al. (2003). The CBDRI has been used in Nsanje before by Mwale et al. (2015) for flood risk, where it was used with only minor modifications.

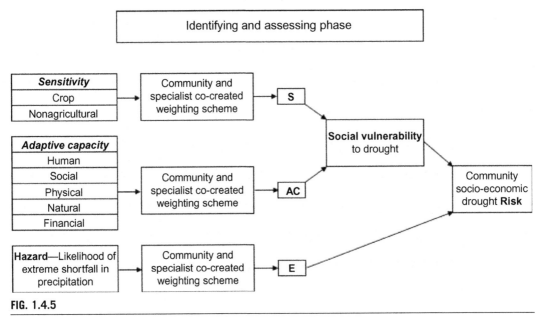

FIG. 1.4.5

Study Risk Framework developed for this study.

Expanded from Zarafshani, K., Sharafi, L., Azadi, H., Van Passel, S., 2016. Vulnerability assessment models to drought: toward a conceptual framework. Sustainability 8(6), 588..

Zarafshani et al. (2016) intended the resolution of their study to be carried out at a household level, which would not have been practical for a district-wide spatial variation study, so the community level assessment was retained. In order to understand social vulnerability to drought at a granular level, this study aimed to gather data at the Group Village (GV) level. At this level, detailed census information was not available. Since any assessment of psychological resilience, as recommended by Zarafshani et al. (2016), is not appropriate at a community level, it was not attempted.

The questionnaire was then adapted from CBDRI, or from other studies of vulnerability (Appendix X). Since all interviewees at the GV level, even members of the VCD and VCPC, were likely to be subsistence farmers and not educated professionals, careful planning and trialing of the method were undertaken before focus groups were attempted. Focus groups were chosen as a means of rapidly gaining a community-level response through group discussion.

In the framework adopted for this study, hazard is defined in strict hydrological and meteorological terms (see earlier definitions). Exposure is confined to the scale of the affected population and agricultural base. This is closer to the definition used in the CBDRI. The resulting methodology is a hybrid of the CBDRI and the Zarafshani et al. (2016) frameworks.

1.4.3.3 **Fieldwork and site selection**

At the time of the study, there were 82 Group Villages in Nsanje, although new Group Village heads were in the process of being appointed. The number of Group Villages is subject to change and is assigned on political "will" (Mwale, 2014). For this study, 34 study communities representing 41%

of the communities were selected to cover a wide and varied cross-section of the district. Logistical and planning support was provided by an expert local facilitator. No formal boundary mapping was available for GV areas, so all results are presented as data points at the approximate center of the GV. The interviews were conducted at the normal meeting point for discussions of the village committees, often a large shady tree, schoolhouse, or the compound of the Group Village Head. For those GVs not accessible to the study team, due to lack of road access, the interview group was met at a suitable accessible location. The location of the respondents' GV was noted on a paper map, in discussion with the expert local facilitator who was familiar with the district.

1.4.3.3.1 Description of data collection

To obtain an effective representation of the village, a focus group discussion (FGD) with representatives of the village was conducted. The purpose of the FGDs was to collect data at a level of granularity below that available in official data. To generate qualitative indicator data, the communities were asked questions such as "What proportion of your fields are in the marsh?" or "What proportion of your community can read?" and after discussion the group would settle on an answer. The groups were generally allowed to reach their own answer, with the facilitator only encouraging quieter members to speak and ensuring the opinions of the women in the groups were heard. If the groups were unsure how to express their responses, they were advised to express the whole in question as "10" and settle on a number from 0 to 10 to express the proportion. Groups were also asked qualitative questions about their livelihoods. The results of the questions were fed into the index, weighting the factors that make the communities vulnerable, split into Crop Vulnerability and Other Livelihood Vulnerability to drought. This was balanced against the Social, Physical, Natural, Human, and Financial Adaptive Capacity scores to give the overall vulnerability scores.

The communities were also asked for a short history of the reliability of their food and daily water systems, and how often they had failed in the last 5 years. The purpose of this data was to provide a calibrating dataset against the main index. Their responses were assigned a value of 0–5, with full supply in all years scoring 5 and total failure scoring 0. Years where food supply was partial were given an appropriate decimal.

A typical interview group is shown in Fig. 1.4.6. The interviews were conducted in Sena, the local dialect of Chichewa. The interviewer worked through the questionnaire, reading the questions aloud in English, and they were then translated by the facilitator into Sena. The questions were asked in a semi-structured format. Discussions would be held between the villagers and facilitator, with occasional clarifications back and forth with the interviewer in English. Some members of the community spoke English to a certain extent, and so some sections of discussions with specific individuals would be conducted directly with the interviewer. Where clarification was required or to expand points ad hoc, follow-up questions would be asked. Several fully qualitative questions were asked, to prompt discussion among respondents or to source background information.

1.4.3.3.2 Group composition

The selection of the individuals in the group followed the guidelines laid out by Bollin et al. (2003) in the CBDRI. Representations were from formal and informal leaders, from community structures, and ordinary representatives of the community, as shown in Fig. 1.4.6. Groups of 10 people were targeted. The ideal composition of the group requested was as follows: the GVH or representative, three members of the Village Development Committee, three members of the Village Civil Protection Committee, and three ordinary members of the community. The membership of these committees was from the

FIG. 1.4.6

Typical group composition.

local communities and not civil servants. A male/female ratio of 1:1 was targeted. This composition was by and large achieved, although the exact format was not always adhered to due to availability and communications. In total 339 people were interviewed, 207 men (61%) and 132 women (39%).

1.4.3.4 Index composition and weighting scheme

Questions to calculate the subcomponents of the index were selected based on review of social vulnerability literature and local expert guidance. Multiple questions were selected for each subcomponent, and questions found to give unsuccessful results were then dropped from the analysis. Indicators were typically dropped where the communities were consistently unsure on how to answer, or where the answers were so universal that they did not help map relative variation (for instance, no group reported the existence of a community drought plan, so the indicator was dropped from the analysis (the index measures only relative social vulnerability, not absolute, so adding a universal zero value is not helpful). The indicators used in the final index are seen in Fig. 1.4.7.

Typically, the questions answered "100%" by the community were given a value for that component of 1 and proportionally scaled for lower responses. Exceptions were questions CS3 and NAS2. Question CS3 is related to the typical crop cycles per "normal" year. The communities usually report obtaining 1–3 cycles of staple crops (rice, maize, millet, etc.) per year, from their land. Three crops per year was the highest recorded value; therefore, a community reporting 3 cycles per normal year was assigned an index value of 1 (the highest score) for that indicator, 2 cycles scored 0.666, and 1 crop 0.333, etc. Alternative livelihoods' dependence on water (question NAS2) was answered by a list of the

Index component	Indicator	Indicator answer to give a component score of 1
Crops sensitivity (CS)	Proportion of cropland in marsh (CS1)	100%
	Proportion of land with access to artificial irrigation (CS2)	100%
	Typical number of crop cycles per year (CS3)	3 cycles = 1
	CS total:	CS = 1–((CS1 + CS2 + CS3)/3)
Non-agricultural livelihood sensitivity (LS)	Proportion of income related to non-agricultural activities (NAS1)	100%
	Proportion of nonagricultural income sources dependent on water (NAS2)	100%
	NAS total:	NAS = 1–(NAS1 * (1–NAS2))
	Sensitivity total (ST):	ST = (CS x 0.7) + (NAS x 0.3)

Index component	Indicator	Indicator answer to give a component score of 1
AC Human (ACH)	Proportion of households who own a radio (ACH1)	100%
	Proportion of households who own a mobile phone (ACH2)	100%
	Proportion of population who are literate (ACH3)	100%
	ACH total:	ACH = (ACH1 + ACH2 + ACH3)/3
AC Social (ACS)	Community participation (voting last election) (ACS1)	100%
	ACS total:	AVS = ACS1
AC Physical (ACP)	Proportion of households owning goats (ACP1)	100%
	ACP total:	ACP = ACP1
AC Natural (ACN)	Proportion of land undegraded (proportion of land subject to recession irrigation) (ACN1)	100%
	ACN total:	CAN = ACN1
AC Financial (ACF)	Proportion of households who own their land (ACF1)	100%
	ACF total:	ACF = ACF1
	AC total (ACT):	ACT = (ACH + ACS + ACP + CAN + ACF)/5

	Final social vulnerability total (SVT):	SVT = ST–ACT

FIG. 1.4.7

Index calculations.

alternative nonagricultural income sources the community had access to, and the score was based on what proportion of the listed activities were reported to rely on available water. The main components of crop sensitivity and noncrop livelihood sensitivity were then calculated, by taking an average of their subcomponent scores.

It should be noted that, while social vulnerability is a negative concept, in the answers to the questions in the social vulnerability section, naturally higher numbers indicate less sensitivity to harm: i.e., use of modern conservation agriculture methods is assumed to reduce sensitivity to harm. Therefore, a reported 90% of the community's land managed through these methods scored as 0.9 is indicative of lower social vulnerability than, say, 10% use of conservation methods, which would result in a score of 0.1. This would result in higher numbers in the matrix sensitivity score representing lower sensitivity, which is not intuitive. To allow for the sensitivity and adaptive capacity scores to work "against" one another, the crop sensitivity and nonagricultural livelihood sensitivity scores are calculated as 1 minus the indicator response value. That is, if the respondents give a value of 0.9 for the livelihood sensitivity score, it is calculated as $1 - 0.9$ in order to give a final value of 0.1, allowing higher values to indicate higher sensitivity.

Weighting is one of the most complex and difficult parts in applying aspects of social vulnerability mapping. Unweighted, the index will give equal importance to all subcomponents, which is unlikely to reflect their relative importance to the communities. Different approaches to this exist; however, in this study qualitative questions were asked to the villagers regarding what the most important types of assistance they can receive in a drought are, and what coping strategies they implement during the drought. This information, along with discussion with local experts, including the local government land improvement, fisheries, water, disaster risk management, and agricultural extension officers, and a representative of the local Red Cross, were used to co-create a weighting scheme for the indicators that reflects which was the most significant. To avoid overparameterization of the model, the weighting strategy was designed to have as light a touch as possible, with only changes that had clear support being made. In practice only two main changes were made to the basic index as part of this process: the 70/30 livelihood split described earlier, and the use of recession irrigation data for natural capital in the adaptive capacity score. Natural capital was intended to represent undegraded land, but no data was available on the relative degradation of the communities' land; local expert advice was that a reasonable assumption was that only land subject to annual flooding in Nsanje is undegraded, as nutrients and soil are introduced by the river from erosion and nutrient run-off from plantations further up the catchment. While this enters the same score into the index twice, the emphasis the communities placed on the importance of the marshland justifies its prominence in the index.

The overall social vulnerability score for a given community was calculated by the total sensitivity score minus the adaptive capacity score. This gives a range of potential values between -1 indicating the lowest social vulnerability and 1 indicating the highest social vulnerability. It would have been possible to present the scores as a numerical range, but some communities would have been given a positive and some a negative score. While a positive score represents greater adaptive capacity than sensitivity in the index, due to the inherent uncertainties of social vulnerability indexing, the presence of the zero gave undue prominence to that score as an inflection point. There is limited value in dividing the communities into two categories surrounding the zero point. It is more appropriate to assess the community scores on a continuum. To this end the range between the highest and the lowest scores was divided into five equal bands titled A to E, representing the least to highest vulnerability.

1.4.4 **Results**

1.4.4.1 **Index results**

The subindex scores were computed for each Group Village as in Table 1.4.1, and plotted in Fig. 1.4.8.

The overall scores are dimensionless numbers, only showing the relative social vulnerability between communities. Crop sensitivity scores are generally higher in the southeast and north of the study area, as these are the communities with access to crop irrigation and report a greater number of crop cycles per year. In general, the plains and highland communities reported lower water dependency for their alternative incomes, perhaps reflecting a lower overall level of water abundance, prompting diversification.

Exploring the results for literacy provides some interesting insights. Female literacy was consistently higher than male across the study region. Notably, literacy was lowest in the villages of Tchapo, Monyo, and Nyachikadza, and the villages located deep in the Ndinde marsh scored the lowest. These communities reported that poor accessibility (i.e., only accessible by water) meant they did not receive access to regular government services like education. However, they did not report particularly low levels of electoral participation, as a polling place had been established in the Marsh, a pattern reflected across the district where reporting of 90% voting participation was not uncommon. Only local religious sects ruling against voting appeared to impact on high turnout.

Natural capital scoring was based on the proportion of land in the marsh that each village had access to. The marsh was consistently referenced as being of importance to livelihoods. The local government land officer stated that the marsh is considered fertile land, as it is replenished by deposition of soil washed away from higher in the Shire subcatchment, and is enriched by fertilized runoff from upstream agricultural activities. Conversely, the upland areas and Nsanje plain are significantly degraded by overfarming and soil loss aggravated by deforestation.

Examining the spatial distribution of social vulnerability, villages in the southeast have a low level of crop sensitivity, but other communities tend to have higher adaptive capacity scores. The communities with access to recession irrigation alongside the banks of the Shire show lower levels of social vulnerability to drought, leading to the intuitive conclusion that the people of Nsanje are highly dependent on water for agriculture. The marsh farmers have access to runoff from the whole Shire/Lake Malawi catchment, whereas the other communities only have access to direct rainfall. All communities reported that their access to the marsh was crucial to food production during drought. Essentially this pattern is borne out in the special presentation of the social vulnerability index results, with the majority of the band A low vulnerability communities being grouped in the marsh areas, and the more inland communities tending to score lower. The pattern is not universal, however, Kamphata is notable in being a band E community close to the marsh; the community there reported significant pressure on land in their area, alongside low land ownership and productivity, perhaps indicating relatively lower availability of resources compared to their neighbors. However, the scores for overall social vulnerability do not correlate well with the reported food production, as plotted in Fig. 1.4.9.

Social vulnerability across the district can be described as severe. During the study period, the district was suffering from a severe drought. However, the marsh communities with the highest levels of spate irrigation in the southeast reported some of the lowest levels of food resilience, with relatively higher levels shown in the hills to the southwest (Ching-Oma, Thunye, etc.). The southeastern higher elevations were generally described as having higher rates of rainfall (i.e., more moist climate without

Table 1.4.1 Index scores by community.

Group Village	Crop sensitivity (CS)	Nonagricultural livelihood sensitivity (NAS)	Total sensitivity (TS)	AC Human (ACH)	AC Social (ACS)	AC Physical (ACP)	AC Natural (CAN)	AC Financial (ACF)	AC Total (ACT)	Social vulnerability total (SVT)	Social vulnerability band
Ngabu	0.41	0.96	0.58	0.53	0.80	0.80	0.80	0.60	0.71	−0.13	A
Mgona	0.34	1.00	0.54	0.63	0.80	0.50	0.50	0.90	0.67	−0.13	A
Monyo	0.45	1.00	0.61	0.43	1.00	0.20	1.00	1.00	0.73	−0.11	A
Mbenje (North)	0.46	1.00	0.62	0.70	0.80	0.60	0.60	0.80	0.70	−0.08	A
Malkaza	0.44	0.90	0.58	0.53	0.70	0.00	1.00	1.00	0.65	−0.07	A
Meza	0.38	0.60	0.45	0.33	0.70	0.40	0.80	0.30	0.51	−0.06	A
Kaudzu	0.40	0.80	0.52	0.50	0.90	0.30	0.60	0.60	0.58	−0.06	A
Nyachikadza	0.45	1.00	0.61	0.28	0.80	0.20	1.00	1.00	0.66	−0.04	A
Chitomeni	0.45	0.96	0.60	0.37	1.00	0.00	1.00	0.70	0.61	−0.01	A
Mbenje (South)	0.54	0.85	0.64	0.63	0.90	0.20	0.71	0.40	0.57	0.07	B
Mbeta	0.47	0.87	0.59	0.28	0.80	0.50	0.30	0.60	0.50	0.09	B
Nyang'a	0.56	0.88	0.65	0.50	0.70	0.40	0.70	0.30	0.52	0.13	C
Tchapo	0.56	1.00	0.69	0.27	0.80	0.20	1.00	0.50	0.55	0.14	C
Nguluwe	0.72	0.85	0.76	0.67	0.80	0.50	0.20	0.80	0.59	0.17	C
Lambwe	0.68	0.76	0.71	0.47	0.90	0.30	0.30	0.70	0.53	0.17	C
Thunye	0.79	0.79	0.79	0.63	0.60	0.90	0.00	0.90	0.61	0.18	C
Mpangira	0.66	0.70	0.67	0.40	0.70	0.50	0.00	0.70	0.46	0.21	C
Mchacha	0.79	0.70	0.76	0.63	0.80	0.40	0.00	0.90	0.55	0.22	C
Tengali	0.77	1.00	0.84	0.70	0.90	0.50	0.25	0.75	0.62	0.22	C
Kawa	0.69	0.78	0.71	0.47	0.90	0.60	0.00	0.40	0.47	0.24	C
Kanyimbi	0.79	0.92	0.83	0.53	0.90	0.50	0.00	1.00	0.59	0.24	C
Kanyama	0.89	0.78	0.86	0.60	0.90	0.40	0.00	1.00	0.58	0.28	D
Ching-Oma	0.79	0.90	0.82	0.53	0.80	0.50	0.00	0.90	0.55	0.28	D
Kamanga	0.82	0.67	0.78	0.27	0.90	0.40	0.00	0.90	0.49	0.28	D
Mkhutche	0.82	0.67	0.78	0.40	0.80	0.30	0.00	0.80	0.46	0.32	D
Kachasu	0.75	0.80	0.76	0.30	0.90	0.40	0.00	0.50	0.42	0.34	D
Bitilinyu	0.69	1.00	0.78	0.27	0.70	0.40	0.30	0.50	0.43	0.35	D

Table 1.4.1 Index scores by community.—cont'd

Group Village	Crop sensitivity (CS)	Nonagricultural livelihood sensitivity (NAS)	Total sensitivity (TS)	AC Human (ACH)	AC Social (ACS)	AC Physical (ACP)	AC Natural (CAN)	AC Financial (ACF)	AC Total (ACT)	Social vulnerability total (SVT)	Social vulnerability band
Mpha Mpha	0.76	0.85	0.78	0.57	0.50	0.40	0.00	0.70	0.43	0.35	D
Davite	0.89	0.75	0.85	0.47	0.90	0.40	0.00	0.70	0.49	0.35	D
Chibuili	0.82	0.78	0.81	0.37	0.80	0.10	0.00	1.00	0.45	0.36	D
Galafa	0.84	1.00	0.89	0.53	1.00	0.40	0.00	0.70	0.53	0.36	D
Kamphata	0.76	1.00	0.83	0.40	0.70	0.30	0.40	0.30	0.42	0.41	E
Mtemba	0.89	0.85	0.88	0.43	0.80	0.30	0.00	0.80	0.47	0.41	E
Ntholo	0.89	0.80	0.86	0.30	0.80	0.40	0.00	0.30	0.36	0.50	E

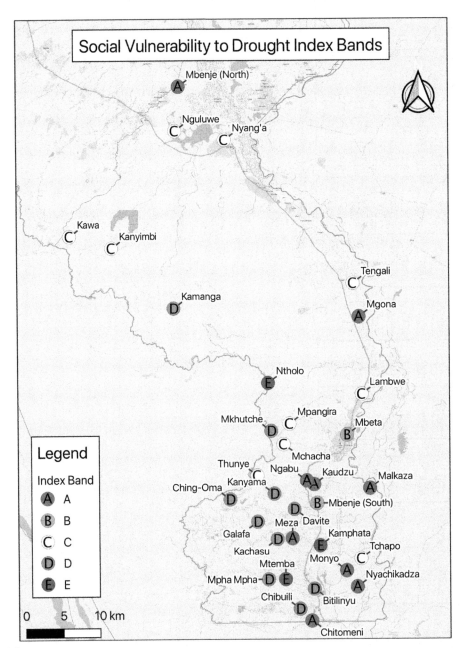

FIG. 1.4.8

Mapped social vulnerability to drought score for Nsanje District.

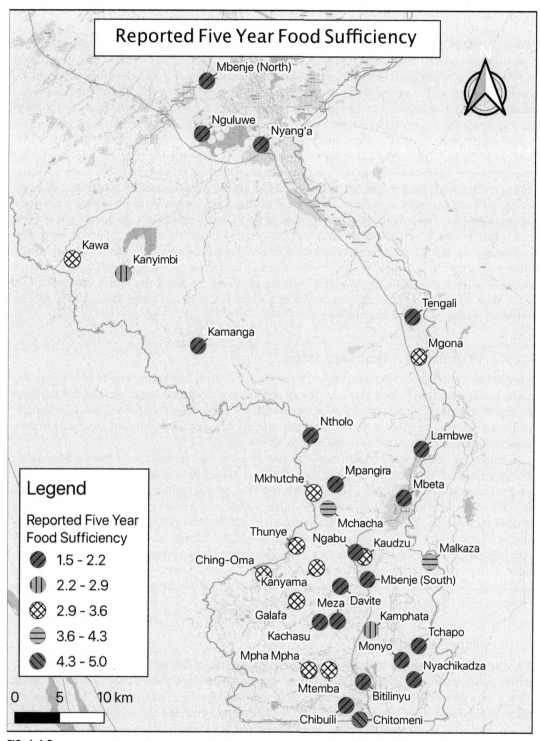

FIG. 1.4.9

Reported sufficiency of food production in the past 5 years.

drought event) locally, which may assist with food crop reliability. Unfortunately, no quantitative precipitation data was available to compare against. Many of these communities reported growing maize, a
relatively moisture-intensive crop. Only one community reported having sufficient food supply in the
previous 5 years, with many reporting that their supplies have been insufficient in all previous years.
Interestingly, Chitomeni, the sole community to report a full score of five on food supply history, was
the only community to report that 100% of its income came from farming, potentially indicating access
to particularly prime land. All communities reported at least a portion of their populations had received
food aid in the past year, and many reported that the aid was not sufficient. The groups often did not
understand how the households receiving aid were selected when crop failures had been near universal
in the previous year. The communities all reported that alternative livelihoods were limited; the most
commonly listed responses were the manufacture of bricks and collection of firewood, with small-scale
fishing in the east. There was limited employment in the district, aside from very limited government
road-building work. Much of the fishing is carried out using illegal mosquito nets, which are so fine
they catch even the youngest fish. Enforcement of net regulations was suspended in the immediate
aftermath of the 2015 floods to support the riverine communities. Many communities reported
engaging in the high-risk "starvation food" activity of diving for water lily tubers in the shallows
of the Shire. These tubers grow in chest deep water, which is also the preferred habitat of the African
crocodile, leading to risk of fatal mauling.

1.4.4.2 Irrigation as a resilience factor

When asked which factors were most critical for resilience to drought, the near-universal response was
access to irrigation. The Ndinde Marsh to the east and the Elephant Marsh to the north were reported as
the important sources of spate irrigation, allowing multiple crop cycles per year and access to agricultural water when the rains failed. Even during the drought, the study team observed apparently healthy
crops and harvest activities in the marshland, as shown in Fig. 1.4.10.

Similarly, providing access to artificial irrigation systems was the most cited response when asked
what government/NGOs could do to assist the population. Aside from a small number of treadle pumps,
mechanical irrigation was absent. Half-constructed or deteriorated infrastructure was observed at
several locations, but little was functional. This infrastructure was primarily located in or near the areas
covered by spate irrigation, aimed at maximizing its effectiveness rather than extending irrigation
significantly into dry areas. A solar groundwater irrigation scheme was visited during the study
(although not in one of the study communities), widely cited as a successful model by other communities, with production continuing during the dry period. Farmers interviewed on site reported theft of
their crops due to failure in the surrounding communities' fields. In response, crops were being
harvested before being fully ripe for secure storage.

1.4.4.3 Adaptive responses

When asked about potential adaptive responses to drought, the communities generally described migration and animal sales. Renting small patches of land in the marshes or seeking work in the marsh
fields was reported as a drought strategy, even among communities located a day's walk from the
fields. While there is little work available in Nsanje, migration to the rest of Malawi, particularly
the closest major city of Blantyre or the sugarcane plantations to the north, was commonly described

FIG. 1.4.10

Harvest activities on the edge of the Ndinde Marsh during drought 2016.

Image author's own.

as an alternative. International migration is also common, with South Africa, and particularly neighboring Mozambique, being frequent destinations. There is significant mining work available for men in Mozambique, with women working in domestic roles or, as several communities described, in prostitution in the mining camps. Animal sales were described as the most significant source of income during drought, with overall ownership of animals having reduced significantly between 2014 and 2016, as herds were sold to raise income, although the glut of available livestock during a drought has been reported as reducing the per-head market price.

Few communities reported any assets beyond livestock. Access to conventional financial services through banks was nonexistent, typically because no assets were held as capital to borrow against. Village Savings and Loans groups (VSL), a form of peer-to-peer lending, were widespread. However, at the time of the study many groups were reported to be nonfunctional. In times of hardship, the need of many members to borrow simultaneously, alongside loan defaults and reduction in saveable income, overwhelms the capital reserves, reducing the effectiveness of this financial model.

1.4.4.4 Drinking water security

Most of the communities are supplied with domestic water from hand-pumped boreholes. These mainly access water through boreholes, with some shallow wells used closer to the river in the east, and natural springs in the hills in the southwest. The difficulty in access for drilling equipment on the rough hill roads was cited as resulting in fewer boreholes being located in these communities. Particularly in the southwest, water conflicts were reported when the yields of springs and boreholes dropped during the dry season. This leads to queuing from early in the morning to access water, with fights breaking out. While no data on the health of the underlying aquifer was available, yield drops during high demand would need to be considered in any plan to expand solar irrigation from groundwater. Most boreholes are managed by a borehole committee, which collects a small fee for access to the borehole, to fund maintenance. However, many committees only collect money for repair on asset failure, rather than maintenance.

1.4.5 Discussion

1.4.5.1 Sources of social vulnerability and adaptive capacity

Social vulnerability in Nsanje stems from a brittle economy highly reliant on rainfed or spate-irrigated agriculture. These findings corroborate the conclusions of Malcomb et al. (2014) in their assessment of drought vulnerability across Malawi and have similarities with drought-related climate change vulnerability in Ghana (Dumenu and Obeng, 2016). Ultimately, population growth is placing increasing pressure on the food supply system of the district, with the population growing 24% in 8 years. As many families report raising six or more children on a small plot of land that is already insufficient, further subdivision between the next generation will only exacerbate the problems. The communities consistently reported that there were very limited alternative sources of income in Nsanje and few villagers have the education or training to find skilled work. This led several groups to recommend the government establish technical schools to provide training. There is little to no irrigation infrastructure to extend irrigation during a precipitation shortfall. Nsanje is caught in a vicious cycle of environmental harm. When rains fail, often the only alternative income available is cutting the forests for firewood either for sale or to be burned to fuel kilns for brick production. The degradation of the forests further destabilizes the water cycle, with increased flash flooding prompting soil loss and further reduction in crop yields. Many ephemeral tributaries of the Shire in the district have become filled with sediment from higher in the catchment due to land degradation, increasing local flood risk. Nearly every community reported that increasing access to irrigation would be an important measure that would give them stability of livelihood during a drought. Similarly, livestock was a resource the communities could dispose of to raise income during agricultural failure. Many groups emphasized the effectiveness of "pass along" programs where an animal (often a goat) was donated, on condition that its first young be passed on to another community member, and so on, and reported them as beneficial.

These findings fit into the larger framework of vulnerability and political ecology (e.g., Blaikie et al., 1994; Wisner et al., 2004). The results indicate that drought vulnerability at local scales is conditioned by underlying political, economic, social, and cultural processes. For instance, environmental degradation in Malawi is directly linked to poverty, since people cut and trade forest resources in order to support their livelihoods and generate income. Similarly, drought has such serious consequences because local people have no access to other types of livelihoods, again connected with wider agendas, such as marginalization, access to markets, and education. Therefore, in order to design effective adaptation options, one must engage with understanding how vulnerability on a local scale is created in the first place and what fundamental changes are needed.

1.4.5.2 Limitations of the method

By its very nature, social vulnerability is unmeasurable (Cutter et al., 2003), and therefore any attempt to measure through proxy indicators will have a degree of inaccuracy. Within this index, there are several factors to be considered. Primarily, the study uses first-hand information from the communities themselves. Nsanje is a donor-reliant environment; while every effort was made to reinforce to the communities that the study was purely academic and their responses would in no way lead to increased aid provision, the possibility cannot be discounted that the communities tailored their responses to appear needier, to be more likely to receive aid. In one community, even after being read the ethics statement, an individual asked the translator what the best response to give to specific questions would be. As another example, one community gave their population as 5042 families, nearly twice the size of the next largest GV. The local facilitator, who had knowledge of government data, questioned the figure, which did not appear to be borne out by the number of dwellings visible. In this instance, the community may have overstated its population to attract greater levels of assistance. It should be noted that even in areas with total crop failure, on average only 30% of families were receiving food aid, which must often be shared with neighbors. When asked what the other families do without aid, one villager replied, "We starve." It is clear there is a significant incentive to seek further aid. Additionally, while many communities were forthcoming about illegal practices, such as felling of protected woodland and fishing with mosquito nets, it may be that fear of legal reprisal sometimes prevented more open answers.

Due to lack of data, the index does not precisely measure the yields and population pressures on the agricultural systems. Furthermore, while the size of the average family plots was discussed with some communities, this data did not form a consistent dataset and so was not used in the analysis. In some communities, respondents at times appeared to use the terms hectare and acre interchangeably, further reducing confidence. Similarly, the nonagricultural income data is weak, in that no attempt was made to rank the relative importance or scale of water dependence of the alternative activities. The demographics of the villages are not recorded in detail. Social vulnerability is linked to age, and it is known that Malawi has a very high dependency ratio in its population (Mwale, 2014). Further improvements would have been made by adding information on the proportion of each community who were below 15 and 65, and not economically active. There were anecdotal accounts of adult children forcing elderly parents to relinquish plots of land, due to increasing population pressure. Similarly, further data on conflicts over water resources would have been beneficial in assessing social strain under drought. An important factor which came up in several of the discussions was the remoteness of some communities from government resources. Many are the best part of a day's walk from the sealed road and distant from many government services, such as police and health posts, with roads unsuitable for bicycles, which are the principal form of land transport available. This is not accounted for by the index. Similarly, even in nondrought times this separates the villagers from the market for their produce, leaving them dependent on visiting traders who may not offer equitable prices. Market access is not accounted for in the index.

1.4.5.3 Comparison with food security

All this data was collected during a period when drought had affected the communities. The comparison between the food production history plotted in Fig. 1.4.9 and the social vulnerability scores in Fig. 1.4.8 is interesting. It is important to note that, to make a comparison between drought vulnerability and food resilience, it is necessary to assume the prevalence of hazard and exposure to give risk. It may be reasonable to assume that drought as a regional hazard will affect the district as a whole. If exposure is also held constant (i.e., the score is not scaled for population) then social vulnerability is approximately equal to risk. Given the recent history of severe drought in Nsanje, we can expect that there should be some correlation between the social vulnerability scores and the reported food production history. However, in general the two measures contradict each other, with regions with low crop sensitivity reporting average-to-low levels of sufficiency of food production. This could reflect potential weaknesses in the index itself. There is no measure of the yield per acre of the farms, so it is not known whether this may be lower for some reason in the marshes. Similarly, it does not consider the plot size of the marshland communities; it may be that the pressure on the land has outstripped supply and the marshes are attempting to feed too many people, leading to supply failure. No community could confidently provide their total acreage and no information was available from government sources.

However, based on qualitative commentary provided by participants, a large portion of the supply failure can be attributed to multihazard influence, both their occurrence and interactions. The communities of the eastern marshes, particularly those located directly within the marsh, had suffered from devastating effects of flooding in 2014 and 2015. This resulted in the destruction of crops, villages, and animals, leading many of the communities to relocate onto the land, as they no longer felt safe in their ancestral villages, due to increased flooding. One village reported that their crops had failed in successive years due to flooding, locust swarms, and a population boom of rats (which they attributed to a low seasonal flood in that year not drowning the rat nests).

1.4.5.4 Future adaptation: The Shire Valley Irrigation Project

Historically, a number of schemes have been proposed to extend irrigation through the plains of Nsanje and Chikwawa to the north, the most recent being the Shire Valley Irrigation Project (SVIP). The project has received recent funding and an environmental impact assessment has been published. Construction activities may be beginning imminently. The scheme proposes to construct a gravity-fed irrigation canal taking water from the Shire at an intake above the Kapichara Dam in the Majete National Park, running through Chikwawa and down to Nsanje, with several feeder canals branching out to deliver water to fields. The project is intended to provide water for a major sugar-producing company, Illovo Sugar, in Chikwawa and small-scale farmers in Nsanje. From the perspective of many Nsanje farmers, the successful delivery of the project will represent the first alternative to rain-fed irrigation they have had access to, with the later phases of the canal system benefiting those in Nsanje's dry northwest. However, as noted by Gwiyani-Nkhoma (2011), there has been a history of irrigation schemes in Malawi having unintended negative consequences and unevenly distributed benefits. The scheme will not increase the volume of water entering Nsanje; in fact, due to increased irrigation in Chikwawa it will reduce it, and any flows taken by the SVIP will be lost to the natural channel of the Shire, and to the wetlands of the Elephant and Ndinde Marshes, both of which are of great importance to their communities. There is a risk of conflict as water resources are diverted from the communities in the east to those in the west.

1.4.6 Conclusions

The district of Nsanje displays near-ubiquitous vulnerability to drought, with a high level of food insecurity, with incomes significantly vulnerable to climatic events and a paucity of robust alternative sources of income. The developed method provides data and subsequently social vulnerability index maps that reflect these findings. Through close feedback with the communities, the co-creation and development of the weighting scheme helps ensure it is relevant to the specific context of Nsanje. The purpose of the social vulnerability index is to serve as a tool for targeting assistance from local decision makers (NGOs/government). The index should be used in an appropriate context, as a rapid assessment tool where household and community census type information are not available, as a first-pass rapid assessment. However, it must be acknowledged that there are inherent uncertainties in modeling social vulnerability through proxy indicators.

The study was able to cover one-third of the communities in Nsanje with one researcher and one translator in less than 3 weeks. Caution is required in a multihazard environment, as the index does not identify those communities that have the greatest shortfall in food supply. While the study represents a step forward in drought social vulnerability mapping, it can be concluded that in the case of Nsanje it would be recommended that the index be expanded to take account of these multihazard aspects, to improve its functionality as a tool for finding and assisting the most vulnerable communities. In the case of hydrohazard in the Southeast African context, where flooding and drought occur in the same location as extremes of the same El Niño/La Niña cycle, more work is required to produce a combined hydrohazard model of social vulnerability covering both drought and flood, particularly in the context of climate change, where greater extremes of the hydrological cycle can be anticipated in the future.

References

Adger, W., 2006. Vulnerability. Glob. Environ. Chang. 16 (3), 268–281.

Antwi-Agyei, P., Fraser, E., Dougill, A., Stringer, L., Simelton, E., 2012. Mapping the vulnerability of crop production to drought in Ghana using rainfall, yield and socioeconomic data. Appl. Geogr. 32 (2), 324–334.

Atkins, 2012. Shire Integrated Flood Risk Management Volume II—Action Plan. World Bank.

Ayantunde, A., Turner, M., Kalilou, A., 2015. Participatory analysis of vulnerability to drought in three agropastoral communities in the West African Sahel. Pastoralism. 5(1).

Birkmann, J., 2007. Risk and vulnerability indicators at different scales: applicability, usefulness and policy implications. Environ. Hazards 7 (1), 20–31.

Blaikie, P., Cannon, T., Davis, I., Wisner, B., 1994. At Risk: Natural Hazards, People's Vulnerability and Disasters, first ed. Routlege, London.

Bollin, C., Cárdenas, C., Hahn, H., Vatsa, K., 2003. Disaster Risk Management by Communities and Local Governments, first ed Inter-American Development Bank, Washington, DC.

Central Intelligence Agency, 2017. World Fact Book, Malawi. https://www.cia.gov/library/publications/the-world-factbook/geos/mi.html. Accessed 21 March 2014.

Chawawa, N., 2018. Why Do Smallholder Farmers Insist on Living in Flood Prone Areas? Understanding Self-Perceived Vulnerability and Dynamics of Local Adaptation in Malawi. University of Edinburgh, Edinburgh, United Kingdom.

Cutter, S., Boruff, B., Shirley, W., 2003. Social vulnerability to environmental hazards. Soc. Sci. Q. 84 (2), 242–261.

Cutter, S., Emrich, C., Webb, J., Morath, D., 2009. Social Vulnerability to Climate Variability Hazards: A Review of the Literature. Department of Geography University of South Carolina, Columbia.

Dumenu, W., Obeng, E., 2016. Climate change and rural communities in Ghana: social vulnerability, impacts, adaptations and policy implications. Environ. Sci. Policy 55, 208–217.

Euroconsult, 2007. Rapid Assessment—Integrated Water Resource Management Strategy for the Zambezi River Basin; 2007. (Print).

Food and Agriculture Organisation, 2018. High risk countries and potential impacts on food security and agriculture. In: FAO: El Niño 2018/19. FAO.

Guimarães Nobre, G., Davenport, F., Bischiniotis, K., Veldkamp, T., Jongman, B., Funk, C., Husak, G., Ward, P., Aerts, J., 2019. Financing agricultural drought risk through ex-ante cash transfers. Sci. Total Environ. 653, 523–535.

Gwiyani-Nkhoma, B., 2011. Irrigation development and its socioeconomic impact on rural communities in Malawi. Dev. South. Afr. 28 (2), 209–223.

Iglesias, A., Cancelliere, A., Cubillo, F., Garrote, L., Wilhite, D., 2009. Coping With Drought Risk in Agriculture and Water Supply Systems. Springer Netherlands, Dordrecht.

Jülich, S., 2014. Development of a composite index with quantitative indicators for drought disaster risk analysis at the micro level. Hum. Ecol. Risk Assess. Int. J. 21 (1), 37–66.

Kim, H., Park, J., Yoo, J., Kim, T., 2015. Assessment of drought hazard, vulnerability, and risk: a case study for administrative districts in South Korea. J. Hydroenviron. Res. 9 (1), 28–35.

Lumumba Mijoni, P.L., Izadkhah, Y.O., 2009. Management of floods in Malawi: case study of the lower shire river valley. Disaster Prev. Manage. 18, 490–503.

Malawian Government, 2016. Malawi Drought 2015-2016 Post-Disaster Needs Assessment (PDNA); 2016.

Malcomb, D., Weaver, E., Krakowka, A., 2014. Vulnerability modelling for sub-Saharan Africa: an operationalized approach in Malawi. Appl. Geogr. 48, 17–30.

Mishra, A., Singh, V., 2010. A review of drought concepts. J. Hydrol. 391 (1–2), 202–216.

MNSO, 2019. Population and Housing Census Preliminary Report. Government of Malawi: National Statistics Office, Zomba.

Mwale, F., 2014. Contemporary Disaster Management Framework Quantification of Flood Risk in Rural Lower Shire Valley, Malawi. (Ph.D.)Heriot Watt.

Mwale, F.D., Adeloye, A.N., Beevers, L., 2015. Quantifying vulnerability of rural communities to flooding in SSA: a contemporary disaster management perspective applied to the Lower Shire Valley, Malawi. Int. J. Disaster Risk Reduct. 12, 172–187. https://doi.org/10.1016/j.ijdrr.2015.01.003.

National Drought Monitoring Centre (NDMC), 2017. Types of Drought. (Online). Available at:https://drought.unl.edu/Education/DroughtIn-depth/TypesofDrought.aspx. Accessed 30 March 2017.

Nillson, A., Shela, O.N., Chavula, G., 2010. Flood Risk Management Strategy: Mitigation, Preparedness, Response and Recovery. Department of Disaster Management Affairs, Lilongwe.

Paulo, A., Pereira, L., 2006. Drought concepts and characterization. Water Int. 31 (1), 37–49.

Šakić Trogrlić, R., Wright, G., Adeloye, A., Duncan, M., Mwale, F., 2017. Taking stock of community-based flood risk management in Malawi: different stakeholders, different perspectives. Environ. Hazards 17 (2), 107–127.

Šakić Trogrlić, R., Wright, G., Duncan, M., van den Homberg, M., Adeloye, A., Mwale, F., Mwafulirwa, J., 2019. Characterising local knowledge across the flood risk management cycle: a case study of Southern Malawi. Sustainability 11 (6), 1681.

Schmidtlein, M., Deutsch, R., Piegorsch, W., Cutter, S., 2008. A sensitivity analysis of the social vulnerability index. Risk Anal. 28 (4), 1099–1114.

The World Bank, 2010. The Zambezi River Basin A Multi-Sector Investment Opportunities Analysis. The International Bank for Reconstruction and Development/The World Bank, Washington, DC.

UNISDR, 2019. Terminology—UNDRR. (Online). Available at:https://www.unisdr.org/we/inform/terminology. Accessed 20 May 2019.

UNICEF, 2019. Massive Flooding in Mozambique, Malawi and Zimbabwe. (Online). Available at:https://www.unicef.org/stories/massive-flooding-malawi-mozambique-and-zimbabwe. Accessed 19 September 2019.

United Nations Development Programme (UNDP), 2012. Reducing Disaster Risk, first ed. United Nations Development Programme, New York.

United Nations International Strategy for Disaster Reduction (UNISDR), 2004. Living with Risk. A Global Review of Disaster Reduction Initiatives. 2004 Version. UN Publications, Geneva.

Wisner, B., Blaikie, P., Cannon, T., Davis, I., 2004. At Risk: Natural Hazards, People's Vulnerability and Disasters, second ed. Routledge, London.

World Bank, 2016. Project Appraisal Document on a Proposed Grant in the Amount of SDR68.1 Million (US$95 Million Equivalent) in IDA Resources and Proposed Grant in the Amount of SDR6.5 Million (US$9 Million Equivalent) in IDA Crisis Response Window Resources to the Republic of Malawi for a Malawi Drought Recovery and Resilience Project. The World Bank.

Yaduvanshi, A., Srivastava, P., Pandey, A., 2015. Integrating TRMM and MODIS satellite with socio-economic vulnerability for monitoring drought risk over a tropical region of India. Phys. Chem. Earth A/B/C 83-84, 14–27.

Zarafshani, K., Sharafi, L., Azadi, H., Van Passel, S., 2016. Vulnerability assessment models to drought: toward a conceptual framework. Sustainability 8 (6), 588.

Urban metabolism and land use optimization: In quest for modus operandi for urban resilience

1.5

Małgorzata Hanzl[a], Andries Geerse[b], Larissa Guschl[b], and Rahul Dewan[b]

Lodz University of Technology, Lodz, Poland[a] WeLoveTheCity, Rotterdam, The Netherlands[b]

CHAPTER OUTLINE

Understanding Disaster Risk. https://doi.org/10.1016/B978-0-12-819047-0.00007-X

1.5.1 Introduction

Disaster risk reduction (DRR) and climate change adaptation (CCA) are recognized as complementary approaches to deal with climate risk (European Environment Agency, 2017; Forino et al., 2015). Sendai Framework for Disaster Risk Reduction 2015–2030 explicitly calls for increased coherence between the two frameworks (UNISDR, 2015[a]; Aitsi-Selmi et al., 2015). There are ever more international policies, which address efficient and sustainable resource management. The United Nations General Assembly announced its Sustainable Development Goals (SDGs) in September 2015 (UN, 2016), and adopted the Paris Agreement (COP21) in December 2015 (UNFCCC, 2016), following on from the IPPC Report (IPCC, 2014). In this document, the UN aimed at improved resilience and climate adaptation (UNEP, 2017). Responding to SDG-11, in October 2016 a New Urban Agenda was proclaimed during Habitat III, which is one of drivers of the change that the document defines in order to "Adopt sustainable, people-centred, age- and gender-responsive and integrated approaches to urban and territorial development" is: "Reinvigorating long-term and integrated urban and territorial planning and design in order to optimise the spatial dimension of the urban form and deliver the positive outcomes of urbanisation" (UN, 2016a, pp. 3–4).

These goals lead to many specific questions, such as that on the desired densities, the ideal proportions of open spaces including green ones, the role and extent of transportation networks, the solutions which answer the requirements of climate resilient development, etc. The debate on resilient planning tends to promote a compact and mixed-use city paradigm as a sort of universal solution (Newman et al., 2017). It reduces journeys, enables the switch to more collective forms of transportation, and this way decreases GHG and energy-source emissions. On the other hand, the necessity to offer ecosystem services and implement nature-based solutions might limit the possibility to densify the urban fabric any further. Therefore, there is a need for a normative framework, which would address the directions of land-use transformations and support good practices in urban design and planning.

The one, which is particularly valid, is the issue of scale. Upgrading the scale of climate-friendly solutions to that of a neighborhood, town, or region may bring added values. The urban metabolism (UM) models, which address flows of resources to and from a settlement, should take into account analyses at various scales. Another perspective is the circular economy (CE) and the potential for the reuse of resources. Although UM models usually address flows of energy or water, land is rarely discussed as a resource, which should be considered in climate change policies. Nevertheless, the theory of a sustainable UM applies to land consumption too. Land consumption should be reduced, multi-sourced, and the land—recycled and recovered—similar to other resources; this "trias ecologica" provides the founding principle for the CE. Despite the slow pace of its transformations, land consumption may be visualized in the form of a Sankey diagram as a particularly viable "urban resource flood."

The conditions of changing climate further strengthen vulnerabilities in urban regions (UN-Habitat, 2011, 2017; Revi et al., 2014; UNFCCC, 2016; UNISDR, 2015; Aitsi-Selmi et al., 2015) and cause planning for increased climate resilience and adaptability to become even more urgent (de Coninck et al., 2018; UNISDR, 2015). The praxis of urban planning and design shows many creative solutions for reducing land consumption and offering resilient environments. Our objective within the current chapter is to review the practices of dealing with the densification of residential and mixed-use development. From the point of view of UM, we are seeking the optimization of future land consumption.

---

[a]UNISDR is now known as UNDRR.

We illustrate our approach with examples coming from urban design practice, this way building a framework for the assessment of urban interventions. The main criteria include density of development, former land use, location, accessibility, etc.

After briefly summarizing the research into resilience thinking, UM and the CE, the current chapter investigates aspects related to land-use transformations. Later, we discuss the examples of land consumption analyses and introduce a methodology of assessment, which uses flow analysis—a Sankey diagram. The method is applied to three case studies coming from the practice of the Dutch firm, We Love the City. The research results are then discussed and observations for the improvements of the assessment methods considered and summarized.

1.5.2 **Resilience thinking**
1.5.2.1 **Theoretical framework**

The numerous definitions of resilience emphasize the need for endurance and continuity (Meerow et al., 2016; Fuller and Quine, 2016; Folke et al., 2010; Walker and Salt, 2006, p. 1; Holling, 1986, p. 296; Tobin, 1999; Adger, 2000; Zhou et al., 2008; Garschagen, 2013). The concept of resilience evolved since its introduction in the 1970s in the field of ecology (Walker et al., 2004; Walker and Salt, 2006; Carter et al., 2015). Davoudi (2012) explains the three main approaches to resilience: engineering, ecological, and socioecological, or evolutionary. From the mechanics perspective, which Holling (1973) adopted while bringing the term to the ecological domain (Alexander, 2013), the emphasis is on the time a system needs to restore to the original state after the disruption. The protection of the local community and the physical structures inferred in this definition (Fünfgeld and McEvoy, 2012) involves "the capacity of a city to rebound from destruction" (Vale and Campanella, 2005). This approach implies that at least some elements of the urban environment should endure. An ecological approach involves more dynamics than the engineering one. In this perspective, the system might be prone to accept some disturbance before reaching a critical threshold (Holling, 1996, p. 33; Wallington et al., 2005); however, the burden of the adaptation in such a case often remains on the side of citizens.

In more recent evolutionary or socioecological approach, the needs of current and future citizens remain the primary driving force of change (Adger, 2000; Tobin, 1999). This comprehensive strategy addresses the transformations of socioecological systems (SES) as a panarchy, in relation to other systems acting at various scales (Walker et al., 2004; Walker and Salt, 2006, p. 38). The resilience thinking extends beyond enduring disasters or adjusting to continuous pressure; contrarily, the socioecological perspective combines adaptation, self-organization, and learning of the local community with their ability to persist and creatively use disturbances to encourage renewal (Folke, 2006).

The stable policies have to be replaced by transformative and adaptive ones (Berkes et al., 2003; Smit and Wandel, 2006), which feature adaptability, robustness (Anderies et al., 2013), and inclusivity. A system might be considered adaptable if its stakeholders demonstrate the capacity to assume a flexible and resilient approach (Walker et al., 2004; Dessai and van der Sluijs, 2007). The consistency, which has to underlie adaptation, should enable the city to keep abreast of changing conditions, even if this leads to transformation into an utterly different framework (Walker et al., 2004). Robustness requires planners to create a framework for citizens to transform the system into the desired direction. In practice, this approach might question static solutions, such as, for instance, zoning; instead, it encourages more flexible conditions, which assume mobilization of citizens investments.

The inclusive approach assumes that all inhabitants both contribute to city development and receive their share of the urban services, which might refer both to policies on general topics and commonly available everyday practices (UN-Habitat, 2010). Thus during the process of transformation, the ongoing changes cannot happen too hastily to make sure citizens can adjust. While endorsing complex adaptive policies, with numerous scenarios and uncertainty embedded, cities become constantly transforming environments (Walker et al., 2010). Davoudi (2012, p. 302) asserts that this approach questions the principles of a system balance, or, in other words, that change remains an integral part of any urban system and happens regardless of the external factors. Therefor resilience does not mean solely a return to the previous state; instead, it defines a broader capacity of a system to transform and adapt to permanently changing conditions (Ibid.).

1.5.2.2 **Resilience in urban design and planning**

Due to climate change, cities and urban regions are becoming vulnerable to various types of dangers; floods and difficulties to manage water, severe storms with heavy rains and violent winds, and hot temperatures followed by drought periods repeat the most often. These conditions further strengthen the already-existing challenges typical for urban settings, such as urban heat island and air pollution (Beaudoin and Gosselin, 2016; Benmarhnia et al., 2016; Lefevre et al., 2015; Mahlkow and Donner, 2017). UM processes and concentration of impervious surfaces alter hydrologic cycles and air flows; these impacts are not limited just to the urban core but spread into the metropolitan zone (Hallegatte et al., 2013; Revi et al., 2014; Wallace, 2017). Climate threats force fast-developing cities to implement adaptation strategies into newly built housing and infrastructure (de Coninck et al., 2018). Whereas the lack of adequate policies combined with changing conditions strengthens the crisis, the growth generates opportunities for creative adaptations if matched with organization capacities (Reckien et al., 2015; Georgeson et al., 2016; Revi et al., 2014; IPCC, 2012). Urban systems maintenance and adaptation to changing climate pertain to building stock and urban services, including ecosystem services, infrastructure and energy systems (Revi et al., 2014, p. 547; Zimmerman and Faris, 2011; Solecki et al., 2018). Newman et al. (2017, p. 6) highlight the immanent features of urban areas, such as various transportation modes, mixed-uses, and variety of renewable resources, which might contribute to successful coping with climate adaptation.

The complexity of urban systems both enhances the challenges of adaptation and at the same time raises the opportunity for synergies at various scales, both locally, due to the cascading character of urban processes, and remotely, when taking into account resources import or human mobility (Lin et al., 2018; Ernstson et al., 2010; Revi et al., 2014; Hunt and Watkiss, 2011; Gasper et al., 2011). The heterogeneous nature of urban systems imposes the need to integrate adaptation strategies into the overall framework of the urban design and planning policy at the local and regional scale. The growing number of adaptation plans (Carter et al., 2015; Dhar and Khirfan, 2017; Mahlkow and Donner, 2017) show that further potentials lay in the synergies stemming from the collaboration of different stakeholders who might include private and public bodies as well as government agencies at various levels of the planning systems (Revi et al., 2014; Araos et al., 2016; Siders, 2017). Joint efforts of authorities and citizens might contribute to the building of local resilience.

Whereas Elmqvist et al. (2019) argue that the concept of urban resilience should not be understood as a normative one, similar to urban sustainability; still it should become more present in the urban planning and design discourse. As explained, in relation to socioecological systems, resilient thinking has acquired normative connotations, being understood as a series of activities, which serve first to increase adaptability through changes of stability landscapes (latitude and resistance) and adjust processes in other scales (panarchy) and second to design plausible transformations and thus raise transformability (Walker et al., 2004). Similarly in practical applications, e.g., Tötzer et al. (2018) or Fuller and Quine (2016), the definition of urban resilience by IPCC (2008) applies, which is: "the ability of a social or ecological system to absorb disturbances while retaining the same basic structure and ways of functioning, the capacity for self-organisation, and the capacity to adapt to stress and change." All the listed possible trajectories enhance the chances of a system—such as, for instance, a local community—to survive and to reduce potential losses, which embeds normative connotations. In practice, in order to evaluate a system, we need to examine its more specific components, and we propose a simplified framework for such an assessment in the current chapter.

1.5.3 Urban metabolism

Wolman (1965, p. 179), in his seminal study, defined the "metabolic requirements of a city" as "all the materials and commodities needed to sustain the city's inhabitants at home, at work and at play." His analysis addressed a hypothetical city of a million inhabitants in the United States. It described the flows of food, water, and fuel into the city and of sewage, air pollutants, and solid waste out of it. In more recent studies, the focus shifted toward the assessment of the anthropogenic impacts on the environment, for instance, Fischer-Kowalski (1998) finds measurement of the flow of energy and materials, which were extracted, produced, used, and disposed of useful for the assessment of the human impact on the environment.

Whereas typically UM research has addressed flows of energy, water, and materials, more contemporary perspective has widened and looks for a broader context of sustainability science (Hoornweg and Freire, 2013). Kennedy et al. (2007, p. 44) define UM as *the sum total of the technical and socioeconomic processes that occur in cities, resulting in growth, production of energy, and elimination of waste.* Currie et al. (2017) highlight the complexity of processes, both sociotechnical and socioecological, that influence the form of a city. These processes generate flows of "materials, energy, people and information," which cater to the needs of the population and influence the environmental hinterland. Dijst et al. (2018) define UM as a network of heterogeneous flows in cities and notice a recent increase in such studies. The activities of mankind are recognized as an integral component of UM and are the principal focus of these studies. Broto et al. (2012) acknowledge the role of patterns of production and consumption, and thus the influence of social processes on circulation and stocks of resources and the ecological environment. A model of material and energy flows, therefore, would create opportunities to comprehend better and optimize the functioning of an urban system (Dijst et al., 2018).

1.5.3.1 Applications of UM studies in urban design

Kennedy et al. (2011) investigate the application of UM studies in the practice of urban design and planning. They list the following ones:

- sustainability indicators,
- urban GHG accounting,

- mathematical models used in policy analyses,
- design tools.

Likewise, Hajer (2016) calls for the adoption of UM frameworks such as Global City Indicators Facility (GCIF 2014) or Large Urban Areas Compendium by the World Bank (Hoornweg and Freire, 2013). He approaches UM studies as a tool for strategic decision making and believes making the imperceptible flows of inputs and outputs discernible, thereby enabling the elimination of negative feedback loops. He calls for the involvement of designers, planners, researchers, and policy makers to examine potential, transformation and transition, and to monitor and evaluate. This view is shared by numerous researchers (for instance, Beloin-Saint-Pierre et al., 2016; Kennedy et al., 2007) who recognize the need to elaborate standards for examining the UM.

As yet, there are not many normative applications of UM studies in urban design and planning. Kennedy et al. (2011) point to the case study in Netzstadt (Oswald et al., 2003) and the analysis of the Paris agglomeration, which refers to the town's core, suburbs, and region (Barles, 2009). Another study by Deilmann (2009) addresses the relationships between UM and the land surface. Kennedy et al. (2011) define the challenge of being able "to design the urban metabolism of sustainable cities."

1.5.3.2 CE—Recent trends

Geissdoerfer et al. (2017, p. 759) define the CE as "a regenerative system in which resource input and waste, emission, and energy leakage are minimised by slowing, closing, and narrowing material and energy loops." They propose the following ways to achieve these goals: design to endure, maintain and repair, reuse and remanufacture, refurbish and recycle. Stahel (2015) divides the CE methods into two categories: either recycling resources or prolonging their serviceable life thanks to improved design or renovation.

Both business and states hope that the implementation of CE helps to deal with the challenges of resource scarcity, a consequence of which is increasing prices (often due to climate change) (Prendeville et al., 2018). The CE approach has been embraced by the European Commission (2015), who, in the document entitled "Closing the loop," has defined an action plan for CE implementation. This plan anticipates increased efficiency if stakeholders limit the use of resources, reuse waste materials, and replace limited resources with renewables or more readily available options. The CE concept, since it is relatively novel, requires further research in order, among others, to locate its assumptions in the broader field of urban sustainability (Prendeville et al., 2018). Regarding land transformations, the "Roadmap to a Resource Efficient Europe" (European Commission, 2011, COM (2011) 571) aims at reaching zero net land take by 2050 (Breure et al., 2018). This requires regeneration of brownfield instead of new greenfield development, a process which Preuß and Ferber (2006) call "circular land use" (Breure et al., 2018).

1.5.3.3 UM analyses—Main topics

The typical approach of UM studies is a material flow analysis (MFA), which measures the flows and stocks of substances and energy (Dijst et al., 2018). The most common themes refer to the management of water and sewage, material objects and waste, energy consumption, and GHG emissions. Most studies, both in the domain of CE and UM, do not directly include land as a resource. For instance, the

European Commission document Closing the loop—An EU action plan for the CE COM/2015/0614 final does not explicitly name land as a resource, which can be subject to recycling.

The processes taking place in the urban environment vary in their pace of transformations (Dijst et al., 2018). Wegener (2004) and Dijst (2013) enriched the initial set of commonly studied topics by adding to it long-term transformations of land use and building stock and the evolution of transportation systems and infrastructure. Other processes they have taken into account concern those human activities, which tend to alter more quickly, such as household composition, employment, and rapid circulation of people and goods. Dijst et al. (2018) also consider the flows of information and money, both in the context of business and personal lives.

Furthermore, taking into account the scale of human impact on the environment, the analyses should also cover natural processes and the extent to which they are affected by human-driven activities (IPCC, 2014). These transformations stem from social and demographic conditions and the extent to which they affect lifestyles. Lyons et al. (2017) name the following features, which, influenced by everyday citizens' behaviors, affect the forms of urban settings: transportation, activities and consumption patterns, residential choices, and ICT use.

The processes, which take place in cities, are interdependent. For instance, urban design influences modal split and transportation modes; these, in turn, affect demand for space, noise levels, and air quality, which further impact the health of citizens, their lifestyles and personal choices. This complexity gets even greater when incorporating issues of urban resilience (Dijst et al., 2018), for example, transparency about natural hazards may indirectly impact housing prices and, in the aftermath, distribution. UM models may become useful when building land-use models incorporating ecosystem services (Haase et al., 2014).

1.5.3.4 Land take as an element of UM studies

The appropriate density of development and population is a commonly recognized feature, which reduces GHG emissions (Dijst et al., 2018). Thus, the planning of dense human settlements becomes an increasingly acknowledged factor to minimize vulnerability to climate change phenomena and enable the planning of regenerative cities (Thomson and Newman, 2018). Notwithstanding the significance of quantitative assessment, researchers also address the role that the forms of the urban fabric and infrastructure play in determining flows of urban resources (Newman and Kenworthy, 2015; Thomson and Newman, 2018). This challenge has been recognized by New Urban Agenda (UN, 2016, p. 3), which calls for the optimization of "the spatial dimension of the urban form." Moreover, as more and more researchers admit, the relationship between lifestyles, urban forms, and their spatial dimension (Thomson and Newman, 2018; Lyons et al., 2017; Davoudi and Sturzaker, 2017) may facilitate a shared understanding of the consequences of everyday choices and, in this way, contribute to the adoption of more sustainable solutions.

While the hitherto land-take, which accompanies population growth has been far too generous (Angel, 2012), the concept of the compact city also has its limitations. Land has to be earmarked to provide infrastructure, including public transportation and ecosystem services—the latter improving the versatility and resilience of the ecosystem structure (Dijst et al., 2018), thereby assuring a better quality of life. Burton (2000) has undertaken an analysis of the pros and cons of the compact city concept from the perspective of social justice. While the advantages do prevail, we need to recognize that there are also certain limitations of increased densities, such as lack of open green spaces and thus possible poorer access to green space in urban zones, reduced dwelling sizes and limited garden space, overcrowding, increased pollution, and worse health conditions (Burton, 2000).

Land belongs to the category of limited resources. Humans use circa three-quarters of the Earth's surface, excluding Antarctica and Greenland, as "*settlements and infrastructures, croplands, grazing and forestry*" (Haberl, 2015). As a finite and nonrenewable resource, land remains subject to competition. Haberl (2015) classifies the phenomena of competition for land into the following categories:

- production versus production, for instance, production of fuel against that of food,
- production versus conservation, such as when food production competes with nature conservation,
- urban or built-up land versus production or conservation.

The environmental stress, which stems from the scarcity of available land (Breure et al., 2018), is a factor, which makes the necessity to address land transformations even more compelling. Firstly, land itself is a finite resource, the transformations of which should be analyzed and tracked in order to clarify the potential in order to reduce demand, constrain its unjustified exploitation, and reuse or regenerate in a more conscious way. Secondly, the patterns of urban land use influence the flow of other resources, such as water, energy, GHG emissions, and the like. Therefore, the analyses of land metabolism should become part of the framework for the assessment of urban redevelopment schemes.

As of now, the number of studies addressing directly land transformations has been limited. The majority of the research concerning land transformations deal with functional changes related to the processes of urban growth on the regional scale: from forest to agricultural and onto anthropogenic ones (Krausmann et al., 2003). Typically, the studies focus on the use and protection of soil for agriculture, in keeping with the UN Sustainable Development Goals (Keesstra et al., 2016). The study by Fischer-Kowalski and Haberl (2007) discusses the concept of a socioeconomic metabolism, which explains the long-term relationships between society and the environment, including their spatial aspects. Furthermore, Krausmann et al. (2003) have highlighted the relationship between the socioeconomic metabolism dynamics and transformations of land use and land cover.

As another example, the assessment framework for UM proposed by Newman (1999) takes into account the flows of water, energy, waste, etc. Among other issues, he has addressed land transformations while referring to themes such as "land, green spaces and biodiversity" and transportation. Dijst et al. (2018) list the following parameters, which refer to land consumption: "(Change in) value per unit of land (m^2, km^2)—compound effect of considered ecosystem services."

1.5.4 **Method of assessment**
1.5.4.1 **Assessment framework**

The relationships between urban resilience (UR) and sustainable development (SD) are subject to numerous theoretical considerations (Elmqvist et al., 2019; Anderies et al., 2013). Whereas researchers differ in their opinions on whether these approaches should be combined or not, they all somehow agree that the assessment should be performed from the point of view of practical applications. Newman et al. (2017, p. 7), looking from a more practical standpoint, highlight the contribution of the resilient approach to the overall sustainability of a city. They emphasize the impact of the reduction of ecological footprint of a city (its "consumption of land, water, materials, and energy, especially the oil so critical to their economies, and the output of waste and emissions") and parallel increase of quality of life—understood as better "environment, health, housing, employment, community." In this perspective, the three concepts discussed: urban resilience, UM and CE, and sustainable development, should be combined together and understood as a common framework, which might help clarify complex urban issues.

In the current study, we propose a simplified framework, which might contribute to the assessment of the specific urban design solutions. The evaluation is based on a list of features specific to urban design projects. This way, looking at more detailed aspects of design applications, such an approach might prove useful in solving the theoretical dilemma whether the concepts should or should not be discussed together or combined. In real life, the scope of each of these theoretical constructs partly overlaps the others; they emphasize different aspects of the same reality, this way making different approaches more relevant and worth attention. The framework includes selected normative components of sustainable urban design, such as mobility and transportation, greenery and natural environment, promotion of social capital through public spaces and social participation, healthy human environment, preservation and reuse of old structure, mixed-use development, compact development, and appropriate densities. These principles reoccur in most documents since Agenda 21, and they offer a commonly recognized assessment framework for urban design. In our proposal we looked for analogous normative axioms in UR and CE frameworks and examined their interrelations (Fig. 1.5.1). The principles have been then combined as a common assessment matrix and applied to the three case studies.

1.5.4.2 Sankey diagrams

Sankey diagrams commonly serve as a way to assess the efficiency of resources utilization. Over a century ago, Irish engineer Matthew Henry Phineas Riall Sankey developed a method to identify inefficiencies and potential savings in steam engines (Schmidt, 2008). His idea became popular in engineering design, especially when dealing with scarce or expensive resources. Ever since, researchers have used this tool to visualize the complexity of flows of materials and energy. The naming of these diagrams often varies, defining either the depicted content, for instance, heat balance, energy, or material flowcharts, or merely remain a Sankey diagram (Schmidt, 2008). Quantitative information, when visualized, becomes far clearer and understandable, which Tufte (2001) emphasized and beautifully depicted. The contemporary works of data visualization specialists, such as David McCandless or Nadia Amoroso, along with numerous others only confirm this thesis. First of all, the visualization makes information accessible for nonexperts, which is essential for participatory urban planning and design. Not only does visualization serve as a useful narrative in evidence-based planning, but it also contributes to the monitoring, understanding, and adjustments of the system. The Sankey diagram, therefore, seems the best choice to optimize consumption budgets of scarce resources, such as land. While limited in number, attempts to address land consumption with the use of the Sankey diagram have been made for regional-scale landscape planning, for instance, by Cuba (2015). His work presented the transformation of land cover in the San Juan area in the period from 1999 to 2003. Such research should be applied through Planning Support Systems and GIS models in order to enable, for instance, assessment of GHG emissions (Blečić et al., 2014).

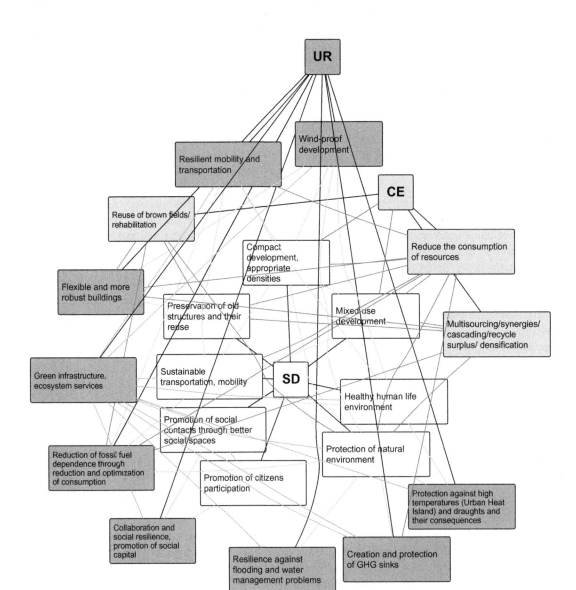

FIG. 1.5.1

Relationships between the three assessment frameworks: SD (sustainable development), UR and CE.

Credit: Own diagram.

1.5.5 Case studies: Objectives, calculations, and discussion

In the current study, we have undertaken the analyses of three case studies of redevelopment done by the Dutch firm, We Love the City (Table 1.5.1). All the concepts focus on satisfying climate change goals, both with regard to climate change mitigation and adaptation strategies, aiming at a more resilient environment. The main features, such as mixed-use development and densification of the urban

Table 1.5.1 Basic features of the three case studies.

	Amstel Station	Harbour Quarter Deventer	Winterswijk
Location	Amsterdam, NL 820,000 inhabitants	Deventer, NL 100,000 inhabitants	Winterswijk, NL 30,000 residents
Client	Amsterdam Metropolitan Region	Coalition of 300 small and midsize investors	National Real Estate Agency and Municipality of Winterswijk
Area of redevelopment	10 ha	21 ha	7 ha (size of exchanged land)

core, as well as concentration on the quality of the public realm, which integrates connected mobility nodes, are commonly recognized as beneficial to regenerative cities' strategies. An approach which encourages public participation serves to enhance active land policies.

1.5.5.1 Amstel Station

Ever since 2003, when the initial concept was elaborated as a result of citizens involvement, We Love the City has been involved in the rehabilitation of the Amstel Station. The Amsterdam City Council approved the master plan in 2009 and the zoning plan in 2012, transforming Amstel Station from simply a transport hub into a strong urban core. The first stage addressed the site's infrastructure, which has been untangled and upgraded. Reducing the space for cars has created more opportunities for cyclists, public transport and mixed-use developments, with shops, workspaces, restaurants, a hotel, and 500 apartments. The increase of residential densities has been possible thanks to the construction of the 100-m Amstel Tower, completed in 2018, which is considered one of the most energy-efficient housing towers in Amsterdam. The housing blocks (B + C) are covered with green rooftop gardens, which will ensure 100% water retention on site. The usage of space has been rethought to achieve a far more efficient and resilient development. The next step will be the restoration of H.G.J. Schelling's 1939 heritage railway station, as part of "The Collection: Special Station Buildings in The Netherlands." The local government's ambition to grow into a more attractive and liveable metropolis has been the driving force behind the development of the site. The profit the Amsterdam Metropolitan Region earns through leasing the land to private developers they reinvest on site to guarantee attractive public spaces and better-connected mobility (Fig. 1.5.2).

1.5.5.2 Harbour Quarter Deventer

The transformation of the former industrial area into Harbour Quarter Deventer introduced various initiatives, which has allowed this site to flourish. An art gallery that the authors proudly describe as the highest in the world, 350 new jobs and the local energy cooperative have been established there. The local cultural heritage served as an inspiration for the redevelopment of the site; this respect for the past led to the reuse of industrial landmarks. The port has been rediscovered as a node of green transport, with mobility seen as service. The densified urban core contains around 250 apartments for first-time buyers on the housing market, entrepreneurs, and students. Two further themes: "Working in the city"

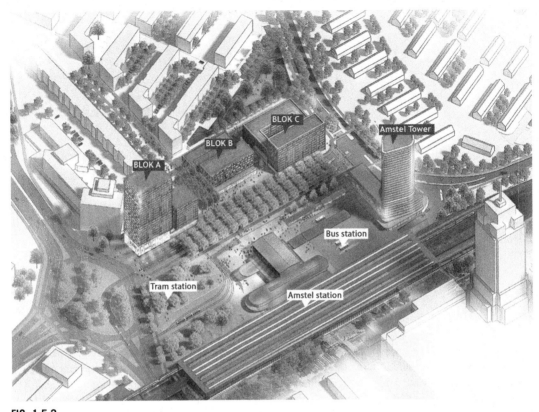

FIG. 1.5.2

Planned development around Amstel Station.

Credit: Project presentation.

and "Place for ideas," stemmed from the involvement of the 300 entrepreneurs who took part in the establishment of an urban business park in the business and residential mixed zone, with 300 small and medium-sized companies. Willing to invest in the area, they set up a local energy cooperative and focused on the promotion of creativity, innovation, and talent (Fig. 1.5.3).

1.5.5.3 Winterswijk

The third case study refers to the medium-sized town of Winterswijk, which, through the process of an urban land swap, created a more compact and efficient solution. Surrounded by a protected national landscape, the urban zone cannot spread beyond its boundaries; thus, the tradition of reallocating and transforming existing properties is strong there. Due to the centralization of public services, many vacant spaces had however emerged. Therefore, the municipality positively responded to the initiative of the National Real Estate Agency to begin a reallotment. Around 70 owners, investors, residents, and other trendsetters participated in this process, which led to a series of land swap deals. Thanks to this,

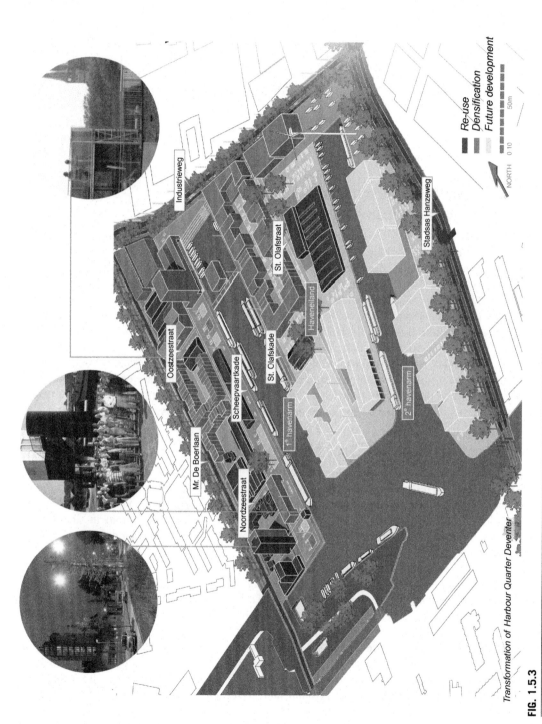

Transformation of Harbour Quarter Deventer

Industrieweg

Oostzeestraat

Mr. De Boerlaan

Noordzeestraat

Scheepvaartkade

St. Olafskade

St. Olafstraat

Haveneiland

1ᵉ havenarm

2ᵉ havenarm

Stadsas Hanzeweg

Re-use
Densification
Future development

NORTH   0 10   50m

FIG. 1.5.3

Harbour Quarter Deventer—redevelopment proposal based on the preservation of the postindustrial heritage.

Credit: Project presentation.

unwanted land in the periphery was withdrawn from the market, and the demand focused on more suitable locations when considered from a spatial and economic perspective. As a result, whereas in 2015 there was 24,000 m^2 of vacant office and public space, by 2018 12,500 m^2 had been demolished and another 8600 m^2 reused. This way the town gained 20,000 m^2 of additional space for greenery and water. Other activities included increasing the energy efficiency of cultural heritage buildings and their transformation to accommodate start-ups, coworking space, and a hotel. The collaboration with landowners, entrepreneurs, and citizens has led to many local initiatives, such as the replacement of a former tax office by new green space and a massive solar canopy serving as the roof for a car park.

1.5.6 Assessment and discussion
1.5.6.1 Normative evaluation—Matrix

The evaluation of various aspects of the three frameworks shows first their interdependence; features considered important for establishing urban resilience are also necessary from the point of view of sustainable development. Similarly, the overlap with the sustainable development framework can be noticed when analyzing the specific aspects of the CE. On the other hand, the direct relationship between the UR and CE frameworks is limited to the three features: flexible and more robust buildings, green infrastructure, ecosystem services, and reduction of fossil fuel dependence through reduction and optimization of consumption. However, the relationships between these two frameworks are less explicit.

The assessment of the three case study shows that all three fulfill many of the qualities commonly recognized as valuable from the three analyzed perspectives. Due to the local specificity, such as the lack of some elements of development, for instance, limited greenery in the case of Harbour Quarter Deventer, some aspects do not apply everywhere. Winterswijk project, which refers to the scale of the whole town, and is limited to land swaps, answers some of the aspects only.

1.5.6.2 Visual representation: Sankey diagrams

Sankey diagrams focus on volumes and directions of flows in a system. The visual approach makes explicit the quantities of resources required and thus enables optimization of how the system functions. For this reason, this method finds application in research and engineering, especially in analyses of GHG emissions, water and energy flows, etc. While the use of a Sankey diagram in studies of land-use transformations remains relatively scarce, it is a topic that has often been discussed due to the role of various activities on GHG emissions (Cuba, 2015).

The case studies presented—Amstel Station and Deventer Harbour Quarter—have been illustrated with Sankey diagrams (Fig. 1.5.4). The diagrams visualize the changes, which have been taking place since 2015 (Amstel Station) and between 2007 and 2016 (Deventer Harbour Quarter), thanks to the implementation of urban design schemes. The bars in the charts represent all land-use categories before and after the implementation of these two projects. The height of each segment corresponds to the size of the respective portion of the land surface in the master plan drawing; categories are arranged thematically.

In the case of the Amstel Station redevelopment, the diagram contains seven categories. Due to the transportation nature of this site, which since its opening in 1939 has been a large-scale multimodal hub visited every day by circa 50,000 train, underground, and bus passengers, the majority of land has been used for transit. This is reflected by the assignment of land surface categories in the diagram, five out of

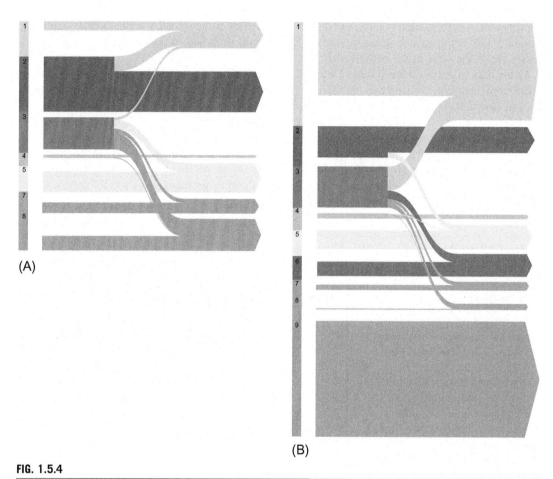

FIG. 1.5.4

Sankey diagrams: Amstel Station redevelopment (A), Harbour Quarter Deventer (B). Categories of land use Amstel Station (A): (1) built-up land, (2) traffic lanes, (3) auxiliary traffic space, (4) parking lots, (5) pavements, (6) cycling paths, (7) greenery. Categories of land-use Harbour Quarter (B): (1) built-up land, (2) traffic lanes, (3) auxiliary traffic space, (4) parking lots, (5) pavements, (6) public space, (7) cycling paths, (8) greenery, (9) water.

Credit: Own diagram.

the seven defining various transportation modes: traffic lanes, auxiliary traffic spaces, parking, pavements, and cycling paths. The diagram (Fig. 1.5.4A) clearly shows the changes, which have taken place; while mobility still remains the main land use, a large share of space has been reassigned to residential/commercial use and greenery. Table 1.5.2 contains the values of land surfaces before and after the redevelopment. Thanks to the reorganization of transportation and the increase of the floor area ratio parameters, it was possible to significantly raise the amount of residential and multifunctional spaces and to add many new green areas. This way, the image of the site has been altered and a vibrant station square equipped with shops, workspaces, restaurants, a hotel, and 500 apartments provided. In addition, the transportation modes have changed, promoting soft modes, such as pedestrian and bike circulation. The optimization of this limited space has satisfied climate change goals, both reducing the amount of GHG emitted and increasing the resilience of the proposed solutions. The Sankey diagram clearly shows the scale of the optimization.

In the case of the Harbour Quarter Deventer (Fig. 1.5.4B), the auxiliary spaces have been transformed into other types of land uses. In this case, the number of categories is higher and numbers nine types of land uses, including built-up areas, transportation of various modes, redesigned public space and water—with the most significant rise being achieved in the category of built-up space. Public space, pavements, cycling paths, and greenery have also increased. This was all possible at the expense of less car traffic, auxiliary vehicle space, and parking spaces. In this case, the increase in the floor area ratio of development led to the growth of available residential and service provision. The distribution of these functions provided the opportunity to achieve greater synergies and thus a more lively and sustainable development. The detailed values for all the categories are listed in Table 1.5.3.

Table 1.5.2 Evaluation of activities within the projects.

Activities/goals	SD	UR	CE	Amstel Station	Harbour Quarter Deventer	Winterswijk
Reuse/rehabilitation	X	–	X	+	+	+
Reduce the consumption of resources	X	–	X	+	+	+
Multisourcing/synergies/cascading/recycle surplus	X	–	X	+	+	+
Resilient mobility and transportation	X	X	–	+	+	+
Wind-proof development	X	X	–	+	–	–
Flexible and more robust buildings	X	X	X	+	+	+
Green infrastructure, ecosystem services	X	X	X	+	+	+
Reduction of fossil fuel dependence through reduction and optimization of consumption	X	X	X	+	+	+
Collaboration and social resilience	X	X	–	+	+	+
Resilience against flooding/water retention solutions	X	X	–	+	+	–
Creation and protection of GHG sinks	X	X	–	+	–	+
Protection against UHI and draughts	X	X	–	+	+	–
Compact development/appropriate densities	X	–	X	+	+	+
Preservation of old structures and their reuse	X	X	X	+	+	+
Sustainable transportation/mobility	X	X	X	+	+	–
Promotion of social contacts through better social spaces	X	X	–	+	+	–
Promotion of citizens' participation	X	X	X	+	+	+
Protection of natural environment	X	X	X	+	+	+
Healthy human life environment	X	X	–	+	+	–
Mixed-use development	X	–	X	+	+	+

Table 1.5.3 Quantitative assessment of land consumption.				
	Amstel Station		Harbour Quarter Deventer	
Categories of land cover	Before (m^2)	After (m^2)	Before (m^2)	After (m^2)
Built-up land	6200	18,180	5400	67,000
Traffic lanes	38,000	28,000	21,500	18,000
Auxiliary traffic space	22,220	0	29,140	0
Parking lots	2280	1800	3660	3000
Pavements	14,000	20,020	12,500	16,500
Public space	–	–	9800	15,500
Cycling paths	7300	10,000	3500	6000
Greenery	10,000	22,000	500	4000
Water	–	–	80,000	80,000

In the case of Winterswijk, due to the scale of the project (which covered the whole area of the city) assessment in the form of a Sankey diagram would require much more specific data on the land use in the whole urban zone, which is available. We have decided however to include this project, which, thanks to the innovative strategic approach of land swap, clearly fits into the current methodology.

1.5.7 Conclusions

In the current chapter, we propose a method of evaluation of urban design interventions based on the normative frameworks of sustainable development, urban resilience and CE. We compared the three frameworks looking for overlaps and mutual relationships. In the next step, we applied the method to the three case studies of urban design projects of different scale of interventions and different specificity, coming from the practice of the Dutch firm, We Love the City. All the chosen urban design projects fulfill a range of criteria within each of the assessment frameworks. These results are consistent with the general perception of climate change risks and adaptation policies in the Netherlands (European Environment Agency, 2017).

In the next step, we analyzed two of the case studies looking for the optimization of the land use. The Sankey diagram is a method of monitoring used to increase the efficiency of applied solutions, fitting for to the studies of the UM and the CE; though, so far, not been broadly used for the analyses of land cover. In the current study, this method has been used to analyze the transformations of land surface in two case studies in the Netherlands. The two projects—Amstel Station and Harbour Quarter—propose urban design solutions at the neighborhood scale, therefore quantifying land use has been relatively easy. The diagrams especially highlight the scale of reuse of auxiliary transportation spaces. Another significant aspect is increase of the pervious surfaces and vegetation, as well as promotion of pedestrian and cycling forms of mobility. On the scale of a whole town, the data are more challenging to acquire. Both projects aim at solutions to ease climate change risks, increase efficiency; with the diagrams proving the success of these goals. The exemplary analyses underline that visual representations of this kind are a convincing tool in the debates on reuse and optimization in urban design and thus may lead to more resilient solutions.

This current study proved that an attempt at the methodology for analyses may become a useful tool for urban design. The current experience gave us the initial thoughts on the conditions, which have to be fulfilled to apply the method to a project. Further steps include application of the similar methodology to other case studies, including larger-scale ones.

References

Adger, W.N., 2000. Social and ecological resilience: are they related? Prog. Hum. Geogr. 24, 347–364.

Aitsi-Selmi, A., et al., 2015. The Sendai Framework for Disaster Risk Reduction: renewing the global commitment to people's resilience, health, and well-being. Int. J. Disast. Risk Sci. 6 (2), 164–176. Beijing Normal University Press.

Alexander, D.E., 2013. Resilience and disaster risk reduction: an etymological journey. Nat. Hazards Earth Syst. Sci. 13 (11), 2707–2716.

Anderies, J.M., et al., 2013. Aligning key concepts for global change policy: robustness, resilience, and sustainability. Ecol. Soc. 18(2).

Angel, S., 2012. Planet of Cities. Lincoln Institute of Land Policy, Cambridge, MA.

Araos, M., et al., 2016. Climate change adaptation planning in large cities: a systematic global assessment. Environ. Sci. Pol. 66, 375–382.

Barles, S., 2009. Urban metabolism of Paris and its region. J. Ind. Ecol. 13 (6), 898–913.

Beaudoin, M., Gosselin, P., 2016. An effective public health program to reduce urban heat islands in Québec, Canada. Rev. Panam. Salud Pública 40 (3), 160–166.

Beloin-Saint-Pierre, D., et al., 2016. A review of urban metabolism studies to identify key methodological choices for future harmonization and implementation. J. Clean. Prod. 163, S223–S240.

Benmarhnia, T., et al., 2016. A difference-in-differences approach to assess the effect of a heat action plan on heat-related mortality, and differences in effectiveness according to sex, age, and socioeconomic status (Montreal, Quebec). Environ. Health Perspect. 124 (11), 1694–1699.

Berkes, F., Colding, J., Folke, C., 2003. Introduction. In: Berkes, F., Colding, J., Folke, C. (Eds.), Navigating Social–Ecological Systems: Building Resilience for Complexity and Change. Cambridge University Press, Cambridge, UK, pp. 1–30.

Blečić, I., et al., 2014. Urban metabolism and climate change: a planning support system. Int. J. Appl. Earth Obs. Geoinf. 26 (1), 447–457.

Breure, A.M., Lijzen, J.P.A., Maring, L., 2018. Soil and land management in a circular economy. Sci. Total Environ. 624, 1025–1030. Elsevier B.V.

Broto, V.C., Allen, A., Rapoport, E., 2012. Interdisciplinary perspectives on urban metabolism. J. Ind. Ecol. 16 (6), 851–861.

Burton, E., 2000. The compact city: just or just compact? A preliminary analysis. Urban Stud. 37 (11), 1969–2006.

Carter, J.G., et al., 2015. Climate change and the city: building capacity for urban adaptation. Prog. Plan. 95, 1–66.

Cuba, N., 2015. Research note: Sankey diagrams for visualizing land cover dynamics. Landsc. Urban Plan. 139, 163–167.

Currie, P.K., Musango, J.K., May, N.D., 2017. Urban metabolism: a review with reference to Cape Town. Cities 70 (6), 91–110.

Davoudi, S., 2012. Resilience: a bridging concept or a dead end? Plan. Theory Pract. 13 (2), 299–307.

Davoudi, S., Sturzaker, J., 2017. Urban form, policy packaging and sustainable urban metabolism. Resour. Conserv. Recycl. 120, 55–64.

de Coninck, H., Revi, A., et al., 2018. Chapter 4. Strengthening and implementing the global response. In: Masson-Delmotte, V., et al. (Eds.), IPCC Special Report on Global Warming of 1.5°C. IPCC.

Deilmann, C., 2009. Urban metabolism and the surface of the city. In: Strubelt, W. (Ed.), Guiding Principles for Spatial Development in Germany. Springer-Verlag, Berlin, Heidelberg, pp. 1–16.

Dessai, S., van der Sluijs, J.P., 2007. Uncertainty and Climate Change Adaptation: A Scoping Study. Copernicus Institute for Sustainable Development and Innovation, Department of Science Technology and Society.

Dhar, T.K., Khirfan, L., 2017. Climate change adaptation in the urban planning and design research: missing links and research agenda. J. Environ. Plan. Manage. 60 (4), 602–627.

Dijst, M., 2013. Space–time integration in a dynamic urbanizing world: current status and future prospects in geography and GIScience. Ann. Assoc. Am. Geogr. 103 (5), 1058–1061.

Dijst, M., et al., 2018. Exploring urban metabolism—towards an interdisciplinary perspective. Resour. Conserv. Recycl. 132, 190–203.

Elmqvist, T., et al., 2019. Sustainability and resilience for transformation in the urban century. Nat. Sustain. 2 (4), 267–273.

Ernstson, H., et al., 2010. Urban transitions: on urban resilience and human-dominated ecosystems. AMBIO 39 (8), 531–545.

European Commission, 2011. Roadmap to a resource efficient Europe; 2011. COM 571.

European Commission, 2015. An EU action plan for the circular economy; 2015. COM 614.

European Environment Agency, 2017. Climate change adaptation and disaster risk reduction in Europe. Enhancing coherence of the knowledge base, policies and practices; 2017. Copenhagen.

Fischer-Kowalski, M., 1998. Society's metabolism: the intellectual history of materials flow analysis, Part I, 1860–1970. J. Ind. Ecol. 2 (1), 61–78.

Fischer-Kowalski, M., Haberl, H., 2007. Socioecological Transitions and Global Change. Trajectoriesof Social Metabolism and Land Use. Edward Elgar Publishing, Cheltenham, UK, Northampton, MA.

Folke, C., 2006. Resilience: the emergence of a perspective for social-ecological systems analyses. Global Environ. Change 16 (3), 253–267.

Folke, C., et al., 2010. Resilience thinking: integrating resilience, adaptability and transformability. Ecol. Soc. 15 (10).

Forino, G., von Meding, J., Brewer, G.J., 2015. A conceptual governance framework for climate change adaptation and disaster risk reduction integration. Int. J. Disast. Risk Sci. 6 (4), 372–384 Beijing Normal University Press.

Fuller, L., Quine, C.P., 2016. Resilience and tree health: a basis for implementation in sustainable forest management. Forestry 89 (1), 7–19.

Fünfgeld, H., McEvoy, D., 2012. Resilience as a useful concept for climate change adaptation? Plan. Theory Pract. 13 (2), 324–328.

Garschagen, M., 2013. Resilience and organisational institutionalism from a cross-cultural perspective: an exploration based on urban climate change adaptation in Vietnam. Nat. Hazards 67, 25–46.

Gasper, R., Blohm, A., Ruth, M., 2011. Social and economic impacts of climate change on the urban environment. Curr. Opin. Environ. Sustain. 3 (3), 150–157.

Geissdoerfer, M., et al., 2017. The circular economy. A new sustainability paradigm? J. Clean. Prod. 143, 757–768.

Georgeson, L., Maslin, M., Poessinouw, M., Howard, S., 2016. Adaptation responses to climate change differ between global megacities. Nat. Clim. Change 6 (6), 584–588.

Haase, D., Frantzeskaki, N., Elmqvist, T., 2014. Ecosystem services in urban landscapes: practical applications and governance implications. Ambio 43 (4), 407–412.

Haberl, H., 2015. Competition for land: a sociometabolic perspective. Ecol. Econ. 119, 424–431.

Hajer, M.A., 2016. On being smart about cities seven considerations for a new urban planning and design. In: Allen, A., Lampis, A., Swilling, M. (Eds.), Untamed Urbanisms. Routledge Taylor & Francis, London and New York, pp. 50–63.

Hallegatte, S., Green, C., Nicholls, R.J., Corfee-Morlot, J., 2013. Future flood losses in major coastal cities. Nat. Clim. Change 3 (9), 802–806.

Holling, C.S., 1973. Resilience and stability of ecological systems. Annu. Rev. Ecol. Syst. 4, 1–23.

Holling, C.S., 1986. The resilience of terrestrial ecosystems; local surprise and global change. In: Clark, W.C., Munn, R.E. (Eds.), Sustainable Development of the Biosphere. Cambridge University Press, Cambridge, UK, pp. 292–317.

Holling, C.S., 1996. Engineering resilience versus ecological resilience. In: Schulze, P. (Ed.), Engineering Within Ecological Constraints. vol. 31. National Academy of Engineering, Washington, pp. 32–43.

Hoornweg, D., Freire, M., 2013. In: Hoornweg, D., Freire, M., Baker-Gallegos, J., Saldivar-Sali, A. (Eds.), Building Sustainability in an Urbanising World. A Partnership Report. World Bank, Washington, DC.

Hunt, A., Watkiss, P., 2011. Climate change impacts and adaptation in cities: a review of the literature. Clim. Change 104 (1), 13–49.

IPCC, 2008. IPCC Fourth Assessment Report: Climate Change 2007.

IPCC, 2012. In: Field, C.B., Barros, V., Stocker, T.F., Qin, D., Dokken, D.J., Ebi, K.L., … Midgley, P.M. (Eds.), Managing the Risks of Extreme Events and Disasters to Advance Climate Change Adaptation. A Special Report of Working Groups I and II of the Intergovernmental Panel on Climate Change. Cambridge University Press, Cambridge, UK/New York, NY, USA.

IPCC, 2014. In: Core Writing Team, Pachauri, R.K., Meyer, L.A. (Eds.), Climate Change 2014: Synthesis Report. Contribution of Working Groups I, II and III to the Fifth Assessment Report of the Intergovernmental Panel on Climate Change. Intergovernmental Panel on Climate Change (IPCC), Geneva, Switzerland.

Keesstra, S.D., et al., 2016. The significance of soils and soil science towards realization of the United Nations sustainable development goals. Soil 2 (2), 111–128.

Kennedy, C.A., Cuddihy, J., Engel Yan, J., 2007. The changing metabolism of cities. J. Ind. Ecol. 11, 43–59.

Kennedy, C., Pincetl, S., Bunje, P., 2011. The study of urban metabolism and its applications to urban planning and design. Environ. Pollut. 159 (8–9), 1965–1973.

Krausmann, F., et al., 2003. Land-use change and socio-economic metabolism in Austria – part I: driving forces of land-use change: 1950–1995. Land Use Policy 20 (1), 1–20.

Lefevre, C.E., et al., 2015. Heat protection behaviors and positive affect about heat during the 2013 heat wave in the United Kingdom. Soc. Sci. Med. 128, 282–289.

Lin, D., et al., 2018. Ecological Footprint accounting for countries: updates and results of the National Footprint Accounts, 2012–2018. Resources 7 (3), 58.

Lyons, G., Mokhtarian, P., Dijst, M., Böcker, L., 2017. The dynamics of urban metabolism in the face of digitalization and changing lifestyles: understanding and influencing our cities. Resour. Conserv. Recycl. 132, 246–257.

Mahlkow, N., Donner, J., 2017. From planning to implementation? The role of climate change adaptation plans to tackle heat stress: a case study of Berlin, Germany. J. Plan. Educ. Res. 37 (4), 385–396.

Meerow, S., Newell, J.P., Stults, M., 2016. Defining urban resilience: a review. Landsc. Urban Plan. 147, 38–49.

Newman, P.W.G., 1999. Sustainability and cities: extending the metabolism model. Landsc. Urban Plan 44 (4), 219–226.

Newman, P., Kenworthy, J., 2015. The End of Automobile Dependence: How Cities Are Moving Beyond Car-Based Planning. Island Press, Washington, DC.

Newman, P., Beatley, T., Boyer, H., 2017. Resilient Cities: Overcoming Fossil Fuel Dependence, second ed. Island Press, Washington, DC.

Oswald, F., Baccini, P., Michaeli, M., 2003. Netzstadt: Designing the Urban. Birkhäuser, Basel, Boston.

Prendeville, S., Cherim, E., Bocken, N., 2018. Circular cities: mapping six cities in transition. Environ. Innov. Soc. Trans. 26, 171–194.

Preuß, T., Ferber, U., 2006. Circular Flow Land Use Management: New Strategic, Planning and Instrumental Approaches for Mobilisation of Brownfields. German Institute of Urban Affairs, Berlin. Occasional Paper.

Reckien, D., Flacke, J., Olazabal, M., Heidrich, O., 2015. The influence of drivers and barriers on urban adaptation and mitigation plans – an empirical analysis of European cities. PLoS One. 10(8), e0135597.

Revi, A., et al., 2014. Urban areas. In: Field, C.B., Barros, V.R., Dokken, D.J., Mach, K.J., Mastrandrea, M.D., Bilir, T.E., … White, L.L. (Eds.), Climate Change 2014: Impacts, Adaptation, and Vulnerability. Part A: Global and Sectoral Aspects. Contribution of Working Group II to the Fifth Assessment Report of the Intergovernmental Panel on Climate Change. Cambridge University Press, Cambridge, United Kingdom and New York, NY, USA, pp. 535–612.

Schmidt, M., 2008. The Sankey diagram in energy and material flow management: Part I: History. J. Indus. Ecol. 12 (1), 82–94.

Siders, A.R., 2017. A role for strategies in urban climate change adaptation planning: lessons from London. Reg. Environ. Change 17 (6), 1801–1810.

Smit, B., Wandel, J., 2006. Adaptation, adaptive capacity and vulnerability. Global Environ. Change 16 (3), 282–292.

Solecki, W., et al., 2018. City transformations in a 1.5°C warmer world. Nat. Clim. Change 8 (3), 177–181.

Stahel, W.R., 2015. Circular economy. Nature 531, 6–9.

Thomson, G., Newman, P., 2018. Urban fabrics and urban metabolism – from sustainable to regenerative cities. Resour. Conserv. Recy. 132, 218–229.

Tobin, G.A., 1999. Sustainability and community resilience: the holy grail of hazards planning? Global Environ. Change B Environ. Hazards 1, 13–25.

Tötzer, T., et al., 2018. Towards climate resilient planning in Vienna. From models to climate services. In: -Hanzl, M., O'Reilly, J. (Eds.), Review 14 Climate Change Planning. International Society of City asnd Regional Planners, Bödo, pp. 190–206.

Tufte, E.R., 2001. The Visual Display of Quantitative Information. Graphics Press, Cheshire, Connecticut.

UN, 2016. Transforming Our World: The 2030 Agenda for Sustainable Development. A/RES/70/1, United Nations (UN), New York, NY.

UN, 2016a. Habitat III New Urban Agenda Draft Outcome Document for Adoption in Quito. October.

UNEP, 2017. The Adaptation Gap Report 2017. United Nations Environment Programme (UNEP), Nairobi, Kenya.

UNFCCC, 2016. Decision 1/CP.21: adoption of the Paris Agreement. In: Report of the Conference of the Parties on Its Twenty-First Session, Held in Paris From 30 November to 13 December 2015. Addendum: Part Two: Action Taken by the Conference of the Parties at Its Twenty-First Session. United Nations Framework Convention on Climate Change (UNFCCC), pp. 1–36. FCCC/CP/2015/10/ Add.1.

UN-Habitat, 2010. State of the World's Cites 2010/2011: United Nations Human Settlements Programme. Earthscan Publications Ltd, London, Sterling, VA.

UN-Habitat, 2011. Cities and Climate Change: Global Report on Human Settlements 2011. Earthscan, London, UK and Washington, DC.

UN-Habitat, 2017. Sustainable Urbanisation in the Paris Agreement. United Nations Human Settlements Programme (UN-Habitat), Nairobi, Kenya.

UNISDR (United Nations International Strategy for Disaster Reduction), 2015. Sendai framework for disaster risk reduction 2015–2030. UNISDR, Geneva, Switzerland.

Vale, L.J., Campanella, T.J., 2005. Introduction: the cities rise again. In: Vale, L.J., Campanella, T.J. (Eds.), The Resilient City. How Modern Cities Recover From Disaster. Oxford University Press, Oxford; New York, pp. 3–26.

Walker, B., Salt, D., 2006. Resilience Thinking: Sustaining Ecosystems and People in a Changing World. Island Press, Washington.

Walker, B., Holling, C.S., Carpenter, S.R., Kinzig, A., 2004. Resilience, adaptability and transformability in social–ecological systems. Ecol. Soc. 9 (2), 5.

Walker, W., Marchau, V.A.W.J., Swanson, D., 2010. Addressing deep uncertainty using adaptive policies: introduction to section 2. Technol. Forecast. Social Change 77 (6), 917–923.

Wallace, B., 2017. A framework for adapting to climate change risk in coastal cities. Environ. Hazards 16 (2), 149–164.

Wallington, T.J., Hobbs, R.J., Moore, S.A., 2005. Implications of current ecological thinking for biodiversity conservation: a review of the salient issues. Ecol. Soc. 10(1).

Wegener, M., 2004. Overview of land use transport models. In: Hensher, D., Button, K., Haynes, K., Stopher, P. (Eds.), Handbook of Transport Geography and Spatial Systems. vol. 5. Emerald Group Publishing Limited, pp. 127–146.

Wolman, A., 1965. The metabolism of cities. Scientific American 213 (3), 179–190.

Zhou, H., Wang, J., Wan, J., Jia, H., 2008. Resilience to natural hazards: a geographic perspective. Nat. Hazards 53, 21–41.

Zimmerman, R., Faris, C., 2011. Climate change mitigation and adaptation in North American cities. Curr. Opin. Environ. Sustain. 3 (3), 181–187.

A comparative analysis of social indicators from the vulnerability cartography between Bragança Paulista and Campos do Jordão (São Paulo—Brazil)

1.6

Franciele Caroline Guerra[a] and Bruno Zucherato[b]

São Paulo State University (UNESP), Institute of Geosciences and Exact Sciences, Rio Claro, São Paulo, Brazil[a] Human and Social Sciences Institute, Federal University of Mato Grosso/Araguaia Campus, Cuiabá, Brazil[b]

CHAPTER OUTLINE

1.6.1 Introduction

The advancement of technology as well as the means of communication observed throughout the 20th century have allowed the creation and systematization of data on the occurrence of disasters that have supported mankind with more accurate means for understanding its occurrence and consequently for its anticipation, prediction, and coping. It is estimated that more than 22,000 catastrophic events have occurred worldwide from 1900 to 2018 (EM-DAT, 2018). Hence, risk and vulnerability studies are vitally

Understanding Disaster Risk. https://doi.org/10.1016/B978-0-12-819047-0.00008-1

important to develop policies and practices to properly identify, understand, analyze, prevent, and mitigate those events.

Vulnerability assessment and mapping, as a scientific area applied to risk studies, seeks a level of spatial differentiation supported by statistical data of a social, cultural, and economic character, which are usually represented from the levels of perception about the ways in which the potentially dangerous processes of the territories manifest themselves, due to both the characteristics of the population and the political and civil protection decisions (Cunha, 2015).

The geographical science, as a science that pursues categorization and spatial differentiation, plays a key role in risk studies, since understanding the social functioning of a community, as well as the geographic context it occupies, will be both the trigger and the field of risk. The correct knowledge of these patterns helps to determine the possibility of disaster manifestations and also to establish policies and practices for their confrontation and the re-establishment of normality. Still, risk studies are also permeated by diverse areas of knowledge, making use of different concepts and categorization.

In this regard, studies on risks and vulnerabilities have been gaining ground in academic, legal, political, and practical discussions around the world. Since it is a recent scientific field of study, whose early systematizations as we know them today dates back to the nineteenth century, there is a great diversity of concepts used and also a limitation of their evaluation and spatial representation. As disasters have different natures, they can have both a more restricted and localized performance and a more extensive and diffuse action, which makes difficult to precisely establish their constraints. For instance, defining the influence area of a drought is a much more complicated task than determining the influence of a hillside collapse or a river flooding, which hampers the task of establishing a geographic scale in risk studies.

Furthermore, the administrative and natural units of data collection and aggregation are often mixed in the available datasets used in risk studies. As they are the result of events occurring in nature and their consequences for human communities, risk encompasses both population data, which are usually made available by censuses collected in political administrative units, and natural data that extrapolate any type of administrative limit established by the human being in an irregular and heterogeneous manifestation.

In Brazil, as of 2011, the federal government established the Law no. 12.608/2012 (Brasil, 2012a) as a basis for federal environmental policies, which includes prevention and mitigation actions. These actions are focused on risk management and response to disasters, promoting the supervision for the good of the urban environmental patrimony, being then improved by the National Plan of Risk Management and Disaster Response (Brasil, 2012b). The governmental plan was conceived to act in four main axes: mapping risk areas; structuring of the monitoring system; alert; and structuring work. It is important to develop and support studies of these thematic axes for the advancement of disaster policies.

Brazil occupies the 123rd position in a world index of the most vulnerable countries to cataclysms (UNU-EHS, 2016), since 85% of the disasters are caused by three types of occurrences: sudden floods, landslides, and prolonged droughts (UNISDR, 2015).[a] Over the last five decades, an estimated 10,225 Brazilians have died in natural disasters, mostly in floods and due to falling hillsides (UNU-EHS, 2016).

---

[a]UNISDR is now known as UNDRR.

This work aims to analyze the spatial patterns over two territorial fragments related to the distribution of population in social vulnerability settled in the Brazilian cities of Bragança Paulista/SP and Campos do Jordão/SP (Fig. 1.6.1), especially the sectors classified as urban by the Brazilian Institute of Geography and Statistics (IBGE), characterizing the population according to some important social variables when exposed to risks. A comparative study of these two areas is important due to the fact that both are located in the southwestern portion of the Mantiqueira Mountains of São Paulo, an area with a complex hydrographic network, presenting shear zones, rugged reliefs, plateau and mountainous areas, granite rocks, shales, and also because they retain some similar economic and social characteristics.

Both study areas shelter large environmental protection and permanent areas, important to the conservation and preservation of local vegetation and fauna. The local climate is characterized by a hot and rainy summer, in opposition to a cooler and drier winter. Based on these aspects, a comparative study of social vulnerability indicators between the two localities was carried out. Eventually, the considerations made here may help to understand the nature and manifestation of the proposed vulnerability,

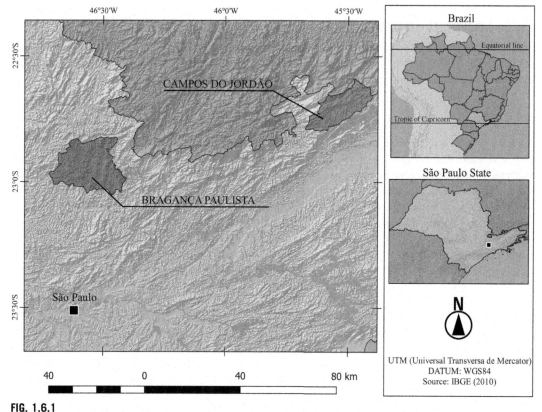

FIG. 1.6.1

Location of study areas.

starting from the perspective of the Brazilian urbanization process and the response capacity, in order to support the elaboration of public policies for disaster risk reduction.

In this context, the results and analysis made here are timely to build the necessary database and assessment to meet national targets related to disaster risk reduction-related global agendas. Expressly, the Sustainable Development Goals—SDGs and the Sendai Framework for Disaster Risk Reduction, which are both advocated by the United Nations. Concerning the Sendai's framework for action, its goal resides in: "Prevent new and reduce existing disaster risk through the implementation of integrated and inclusive economic, structural, legal, social, health, cultural, educational, environmental, techno-logical, political and institutional measures that prevent and reduce hazard exposure and vulnerability to disaster, increase preparedness for response and recovery, and thus strengthen resilience" (UNISDR, 2015). Furthermore, our study focuses on the Sendai's first priority—"understanding disaster risk" –, working on national and local levels, thus supporting stakeholders as described on item (b): "To en-courage the use of and strengthening of baselines and periodically assess disaster risks, vulnerability, capacity, exposure, hazard characteristics and their possible sequential effects at the relevant social and spatial scale on ecosystems, in line with national circumstances" (UNISDR, 2015).

1.6.2 Literature review

To precisely establish the scope of the study and considering the multiplicity of meanings of the terms involved in its field of action, it is important to clarify the concepts of "Risk" and "Vulnerability" for a better understanding of their applications in the proposed study.

1.6.2.1 Risks

The concept of risk has countless meanings, differentiated by the knowledge fields in which it is ap-plied. Aven and Renn (2010) carry out a very extensive discussion about several concepts of risk, for which they established some patterns. In their study, the authors define nine main conceptions of risk that are distinguished by the elements and approaches used (Table 1.6.1; Aven and Renn, 2010).

The first definition group of the term risk is more simplistic and conceives it as the result of the expectations of losses by the occurrence of an unexpected event. Thus, it seeks to establish what can be lost if a disaster occurs, focusing the efforts on the survey of the elements that can be damaged because of the occurrence of an event. The second risk group definition perceives it as the probability of occurrence of an uncertain event. While the first definition group focuses on what can be lost, the sec-ond focuses on the likelihood of something unexpected happening.

The third and fourth groups have fairly similar perspectives, though they are noticeably different. While one seeks the establishment of uncertainty, the other pursues to objectify this uncertainty. The difference is in the means of its knowledge, while one does not concern its measure, only the knowledge or not of its presence, the other tries to carry out its measurement. The fifth group perceives the risk as a function of the possibility of occurrence of something with the potential losses that can be caused. This perspective introduces a more complex analysis in order to define risk, as the result of the interdepen-dence pondering more than one factor.

The sixth, seventh, and eighth groups of risk concepts also consider risk as functions between inter-dependent elements. These groups try to measure the chances of an event occurrence, either weighing

Table 1.6.1 Summary table of risk definition groups.

Risk groups	Definition	Synthesis
Group 1	Risk = Loss expectation	R = E
Group 2	Risk = Probability of an event (undesirable)	R = P
Group 3	Risk = Objectivity of uncertainty	R = IO
Group 4	Risk = Uncertainty	R = I
Group 5	Risk = Possibility/potential of losses	R = Po
Group 6	Risk = Possibility and scenarios/consequences/severity of consequences	R = P&C
Group 7	Risk = Event and consequence	R = C
Group 8	Risk = Consequence/damage/gravity + uncertainty	R = C&I
Group 9	Risk = Effect of uncertainties on objectives	R = ISO

Data from Aven, T., Renn, O., 2010. Risk management and governance. In: I. R. U., Mumpower, J.L. (Eds.), Estimating Impact: A Handbook of Computational Methods and Models for Anticipating Economic, Social, Political and Security Effects in International Interventions. Springer, Berlin, Heidelberg. doi:10.1007/978-3-642-13926-0.

its uncertainty, the prediction of the event or the determination of a scenario, or in the degree of losses as consequence. The construction basis is the forecasting and inventory of goods and values that can be damaged. The ninth and last risk concept groups have a more general and comprehensive constitution. They focus on understanding risks as effects of uncertainties and unpredictability of a given objective.

Aven (2012) sought to group all these conceptual definitions of risk in order to establish trends and understand the chronology of their evolution. His results showed the holistic perspective as the dominant group of trends in the current risk studies, composed of the definition groups 5–9, since they consider broader aspects of disaster understanding.

Furthermore, the conception of risk as a function of spatial, social, and temporal factors, integrating the event uncertainties and the potential consequences, is an adequate manner to provide satisfactory responses to the spatial and geographic approaches of risk.

The International Risk Governance Council defines risk as an uncertain consequence of an event or activity concerning something endowed with human value (IRGC, 2017). In this concept, it is important to highlight the importance of including elements of uncertainty and the human factor. Hence, risk can be synthesized as knowledge of disorders caused by an event that can affect an individual or community. Risk is thus somewhat social.

The United Nations Office for Disaster Risk Reduction, in turn, defines risk as the probability of harmful consequences or expected losses (deaths, injuries, property, livelihoods, economic activity) resulting from interactions between natural or human dangerous processes and vulnerable conditions (UNISDR, 2017). In this definition, we notice the addition of extra dimensions of risk causes.

Following this concept, the risks attributed to human causes are defined: the human being itself, either the possibility of human losses or the number of affected individuals, as their possessions and their survival means (IRGC, 2017). Thus, it is evident a concern to safeguard not only the individual or community itself but also the components of their subsistence.

The consideration of both human and economic losses in risk definitions is not a United Nations particularity. The Center for Research on the Epidemiology of Disasters (CRED) also includes such aspects in its definition of risk, which is defined as expected losses (including lives, injured persons, damaged property, and disrupted economic activity) due to a particular hazardous process for an area and reference period (EM-DAT, 2018). In this case, the exact criteria for damages required to determine the risk are specified.

Another definition of risk that is widely used and based on the holistic perspective was proposed by Rebelo (2003), that seeks to establish a more specific scope of its application in the scientific field. The risk is defined as the result of the sum of risk factors (named Hazard and Vulnerability), being expressed by the equation:

$$Risk = Hazard + Vulnerability$$

In this definition of risk, an attempt is made to quantify the term risk. The division of risk between hazard and vulnerability also implies its concept from the geographical perspective, since spatial variation will be critical for the establishment of both analyzed terms. This same perspective is shared by Wisner et al. (2004), who consider risk as the product of hazard and vulnerability. Also, Willis (2007), who studied the risk of terrorist attacks, defined it as a function of threat, vulnerability, and consequences. Considering the interaction of these three elements, risk suppression is focused on determining the expected loss.

Several authors—Cunha and Dimuccio (2001); Cunha and Leal (2012); Freitas et al. (2013)—have conceived risk in a similar manner to that presented by Rebelo (2003). They consider that risk involves interaction, although not in a simple addition, of two groups of elements: (1) Threat (Álea or Hazard), which due to its spatial or temporal relevance, is formed by Susceptibility and Probability; and (2) Vulnerability, which, in turn, can also be subdivided into Exposed Population, Values of the Goods, and Social Vulnerability.

The comprehension about the concept of risk by studying (sub)concepts allows a more specific approach that will be used in the proposed study through the determination of vulnerability. The concept of hazard can be defined as the possibility of occurrence of a hazardous event, while vulnerability refers to the possible impacts the event can cause when it hits the population and then returns to normal state. Risk, however, is the interaction between hazard and vulnerability, that is, the final results when a disaster occurs, being classified in different intensities and referring to some human value as material goods, natural resources, or even people.

It is well known that the risk field is very complex due to its wide application both in terms of concepts and different case studies. In this sense, certain authors present the risk concept as the result of the product of the threat (in the same sense as hazard) by the vulnerability, considering as interdependent conditions (Dagnino and Carpi Junior, 2007). In this perspective, since it is not a sum but a product, if one of the components reaches the zero level, it cancels the risk. For instance, an area encompassing a community that is very threatened by hazardous processes but does not present any vulnerability is not exposed to risk, as well as a highly vulnerable population that is not threatened by any dangerous processes is also considered out of risk.[b]

---

[b]It is worth mentioning that cases like these are only theoretical, since there will always be something vulnerable or that can be affected by a dangerous process (people, goods, whatever the landscape or environmental aspect or condition). The opposite also does not work so radically, since there are always dangerous processes that can happen, even if there are no slope movements, earthquakes, floods, or droughts, in theory there may always be, for instance, a meteorite impact. That is why it is said that zero risk does not exist. There are always dangerous processes and there are always conditions that can be affected.

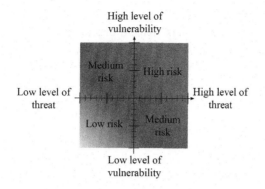

High level of
vulnerability

Medium
risk   High risk

Low level of   High level of
threat   threat

Low risk   Medium
risk

Low level of
vulnerability

FIG. 1.6.2

Relationship between levels of threat and vulnerability in determining the intensity of risks.

Data from Dagnino, R. de S., Carpi Junior, S., 2007. Risco Ambiental: Conceitos e Aplicações. Climatol. e Estud. da Paisag. 2, 50–87.

According to the UNISDR (2017) while the hazard determines the geographical location, intensity, and probability of an event occurring, the concept of vulnerability determines the susceptibility of this occurrence. Hazard is the harm cause or situation that causes loss and threats people; consequently, hazard will be synonymous of existing conditions that give rise to a risk situation (Mendes et al., 2010).

Fig. 1.6.2 shows the notion of risk intensity (high, medium, low) depending on the combination of threat level (horizontal axis) and vulnerability level (vertical axis). It can be noticed that risk is present in all sections, even though levels of vulnerability and threats are low, increasing as variables grow.

According to Dagnino and Carpi Junior (2007), the risk (R) complex formula comprehends the relation $R = F(T,V)$, presenting risk as a function of threat (T) and vulnerability (V). This function depends on the analyzed problem and can be related to:

- Land use and occupation over the region;
- Water exchange;
- Basin or river morphometry;
- Impermeability Index;
- Engineering works to retain or reduce risks.

By involving several factors that may decrease or increase the risks human beings are exposed to, this risk conception implies the existence of nonrisk situations. The same is true for vulnerability since it is an indissoluble condition for the risk occurrence.

In summary, we have seen there are several paths to approach risk assessment. In this research, risk is considered as function or product (times) and, consequently, the absence of one factor will imply the lack of risk.

1.6.2.2 Vulnerability

The concept of vulnerability has a well-defined meaning in the scientific community, although it has been used and appropriated by different fields of knowledge, generally related to risk studies. The common sense usually defines vulnerability as the weak side of a subject or issue, also as the spot where someone or something can be hurt and/or attacked, frequently used in the same sense of weakness.

Regarding vulnerability, it is a complex concept that imposes the necessity of its division into typologies, which is the result of several interactions. It implies the development of both physical processes and the human dimension (social, economic, and environmental), and can be a destabilizing phenomenon, with natural or anthropogenic origin, that will increase the susceptibility to the hazard impacts (Mendes et al., 2010).

The UN definition of vulnerability considers it as the conditions determined by physical, social, economic, and environmental factors or processes, which increase the susceptibility of an individual, a community, assets, or systems to the impacts of hazards. (UNISDR, 2017). The conception focused on compensation is also suggested by Wisner et al. (2004), considering the definition of vulnerability as a model of pressure and relief. In this perspective, vulnerability is the result of the set of situations that intensify and attenuate it; thus, the risk situation is present when the pressures exerted are greater than the conditions that relieve it.

Another important definition of vulnerability is that proposed by Cutter (1996, 2003, 2011) who considers vulnerability as the search for a rational and experimental explanation for defining, describing, explaining, and predicting damages—caused by events that may cause harm to an individual, community, structure, or object—and to identify their fragile points. The author considers that there are basically three broad groups of conceptions regarding vulnerability: as a pre-existing condition; as an attenuated reaction; and as the dangerousness of places (Cutter, 2003). The latter agrees to what was previously defined as "threat" or hazard. In this sense, if we merge the vulnerability of places and the social vulnerability, we would have the Risk.

Vulnerability as a pre-existing condition is based on the assumption that its identification and measurement are based on conditions that put people and places in a vulnerable situation. This vulnerability measurement modeling focuses on examining the sources of hazard (or potential exposure to risks) for both biophysical disasters and technological disasters. The major focus of the related studies is on the identification of the hazards and the observation of the human occupation of these places, as well as on the degree or potential of losses associated with the identified risks. For the researchers that follow this vulnerability approach, the key concepts for their understanding are the identification of the magnitude, duration, impacts, frequency, and speed that characterize the risk exposure of a community or individual (Cutter, 1996).

Another vulnerability conception line is the understanding of vulnerability as an attenuated reaction. This line of research regarding the risk and vulnerability area focuses on responsiveness that includes social resistance and resilience to risk (Cutter, 1996). The nature of a hazardous event, or the conditions that commonly trigger, must be considered within this understanding from a point of view that results from social construction and not from biophysical conditions. Some disaster types (such as drought and starvation) are not only linked to natural conditions but have historical, social, cultural, and economic roots that limit the individual or community to adequate response when exposed to disasters. This conception of vulnerability also recognizes the importance of social differences in the recovery capacity and response to disasters, but it has a more pronounced sociological character and does not give the necessary importance to the spatial occupation or processes distribution. Therefore, it does not consider the socioeconomic differences that define the location of different social strata in space, justifying their occupation closer or distant from potential risk sources.

The third vulnerability design model presented by Cutter (1996), known as the vulnerability of places, consists in the integration of two predecessor models in a more integrated and geographic conception of vulnerability, focused on the bond between society and the environment. Hence, vulnerability is conceived as the set of processes of environmental nature (such as biophysical risks) and human nature (such as social and economic structures) capable to help communities and individuals to respond to disasters they are exposed. In this case, vulnerability of places is equal to risk. The result can be both the geographic space where vulnerable people and places are located, and the social space where the most vulnerable places are located (Cutter, 1996).

Cutter's studies pointed out the importance of geographic understanding in risk studies, understanding that the connections between society and nature are indissoluble and thus the contribution of these two components is important for the systemic understanding of the risk issue. Other perspectives of vulnerability eventually ignore the social factor and address policies or infrastructures, mainly explored in management and engineering studies. In the other hand, there are also perspectives that consider only the social factor and neglect the spatial distribution of society and how this can affect risk studies.

Cutter (2003) proposes the statistical technique of Exploratory Factor Analysis (AFE) as alternative to identify variables of interest for the determination of vulnerability. These statistical procedures allow to organize and identify factors that can explain the proposed vulnerability dimensions. These techniques are widely used and have been replicated by several studies in varied locations, including United States (Schmidtlein et al., 2008), Brazil (Freitas et al., 2013; Bortoletto et al., 2014; Freitas and Zucherato, 2015), and Portugal (Cunha et al., 2011; Mendes et al., 2010; Cunha and Leal, 2012).

1.6.2.3 Methodological approaches and indicators of risk, vulnerability, and disaster resilience

Searching for an overview regarding the concepts and methods addressed here, we expand the existing set of indicators of risk, resilience, and vulnerability to disasters. During the past several years, we have noticed an increasing attention to issues in the matter of vulnerability, capacity, and resilience in disaster management. Since then, a growing corpus of literature has sought to expand this theoretical understanding and how to measure it empirically. Beccari (2016) has identified a variety of practices for the construction of composite indicators of risk, vulnerability, and resilience, using the most common hierarchical or deductive indexes.

Since Briguglio published an index examining the economic vulnerabilities to disasters of small island states in 1995 (Briguglio, 1995) and Cutter and colleagues developed the Social Vulnerability Index (SoVI) in order to better understand disaster social drivers (Cutter et al., 2003), there has been a dramatic increase in the number of methodologies in order to measure disaster risk, vulnerability, or resilience aspects (Beccari, 2016). However, during the last 5 years, there has been a noteworthy increase in the publication rate reaching about two-thirds the indicator methodologies developed since 2010. It can be noticed in Fig. 1.6.3 the escalation of the number of methodologies regarding new disaster risk, vulnerability, and resilience indexes published each year.

Notwithstanding the challenges and limitations identified by the authors to determine the current state of practice in developing indicators for the study of disaster risk, vulnerability and resilience, a broader understanding of how the indicators are being constructed and how the variables are being used is valuable to those who are building these indices. It is also relevant to identify the current common practices and gaps, as well as the extent of practice and contribution to policymakers, tailoring their needs.

According to Beccari (2016), several authors seek to compare diverse indicators related to disaster studies. The statistical variables used by experts came from existing statistical datasets and were combined by simple addition, applying the same weight. The number of variables used in the analyzed

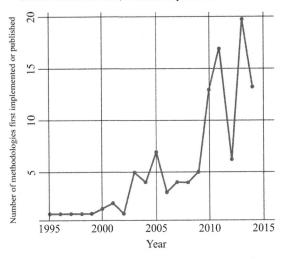

FIG. 1.6.3

Historical increment in the number of methodologies worldwide after the index published by Briguglio in 1995. In each year, the number of new composite indicators of disaster risk, vulnerability, and resilience implemented.

Data from Beccari, B.A., 2016. Comparative analysis of disaster risk, vulnerability and resilience composite indicators. PLOS Currents Disasters. Edition 1. doi:10.1371/currents.dis.453df025e34b682e9737f95070f9b970.

Table 1.6.2 Variables most used in all methodologies and their respective number of data combinations.	
Variables	**Number of methodologies**
Population density	33
Unemployment rate	31
Population, 65 years and older	19
GDP per capita	19
Percent female population	18
Doctors per population	16
Literacy rate	15
Total population	14
Beds in hospitals per population	14
Percent individuals below poverty line	12
GINI index	12

Data from Beccari, B.A., 2016. Comparative analysis of disaster risk, vulnerability and resilience composite indicators. PLOS Currents Disasters. Edition 1. doi:10.1371/currents.dis.453df025e34b682e9737f95070f9b970.

studies varies between 2 and 235 variables, although two-thirds of the methodologies adopted less than 40 variables. According to 106 studied methodologies, the availability of information and the general and specific study objectives, 2298 single variables were used. The most frequent variables are common statistics, such as population density and unemployment rate (Table 1.6.2).

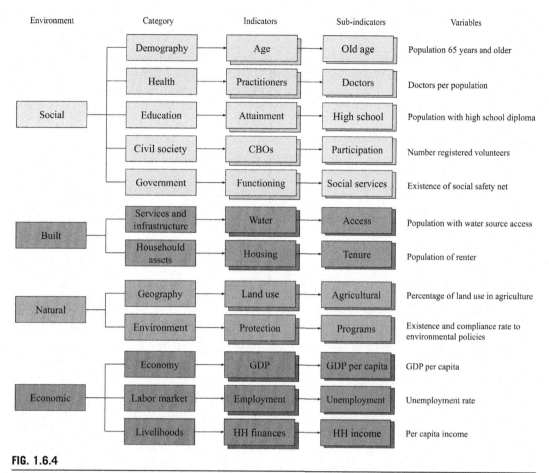

FIG. 1.6.4

Scheme employed in order to classify variables in composite indicator methodologies.

Modified from Beccari, B.A., 2016. Comparative analysis of disaster risk, vulnerability and resilience composite indicators. PLOS Currents Disasters. Edition 1. doi:10.1371/currents.dis.453df025e34b682e9737f95070f9b970.

Fig. 1.6.4 shows the hierarchy of classification and frequency of use regarding different concepts in different methodologies, as well as the composition of each methodology, which was analyzed and grouped into subindicators, indicators, categories, and environments, based on the phenomena measured by each variable.

It was verified by the survey that, on average, 34% of the variables used in each methodology are related to the social environment, 25% to the disaster environment, 20% to the economic environment, 13% to the built environment, 6% to the natural environment, and 3% to other vulnerability-related dimensions. However, variables that specifically measure actions related to disaster mitigation or preparedness represent only 12%, on average, of the total of variables in each index. Merely 19% of the methodologies make any sensitivity or uncertainty analysis and only one case was comprehensive.

Table 1.6.3 Frequency of methodologies distributed in all 15 categories.

Category	Number of methodologies
Demography	87
Education	67
Health	64
Services and infrastructure	61
Economy	59
Disaster hazards and impacts	59
Labor market	47
Livelihoods	47
Housing and household assets	47
Disaster resilience	41
Civil society	39
Geography	37
Environment	28
Government	24
Indices	21

Data from Beccari, B.A., 2016. Comparative analysis of disaster risk, vulnerability and resilience composite indicators. PLOS Currents Disasters. Edition 1. doi:10.1371/currents.dis.453df025e34b682e9737f95070f9b970.

According to the classification hierarchy (Beccari, 2016) shown in Fig. 1.6.4, the variables were grouped into 334 subindicators, an average of 6.9 variables per subindicator. The most common subindicators are strongly influenced by the most frequent variables. The number of methodologies that included variables from each of the categories is shown in Table 1.6.3. It shows that most methodologies included some measure of demography, education, and health, with existing indices and measurement of government aspects and environment.

This overview shows the diversity of variables used in the composition of vulnerability indexes. On the one hand, it is necessary an organization of these variables and their availability at the scale of the proposed study. On the other hand, it is necessary that variables can be useful and interpretable, and that they reflect at some level the dimensions involved in the measurement of vulnerability. Principal Component Analysis is the most popular statistical dimension reduction method, being implemented in 17 methods, and typically implemented using the procedure developed by Cutter et al. (2003) in the Social Vulnerability Index (SoVI).

Regarding this review, a wide range of practices are noticeable in the development of indicators for disaster risk measurement, vulnerability, and resilience at global and local levels, in many different countries. There is a significant diversity in the literature about variable selection approaches, data collection methods, standardization methods, weighting methods, and aggregation approaches. However, current practices have two main limitations that may restrict their use or potentially lead to wrong decisions—low usage of direct measures of disaster resilience and lack of sensitivity and uncertainty analysis (Beccari, 2016).

The literature review verified the extent of the vulnerability analysis, the continuing desire to quantify vulnerability, and the many methodologies employed. However, the review revealed several widespread methodological limitations in measuring and mapping vulnerability, such as:

(1) the focus on population concentrations within known hazard zones and not entire populations, potentially omitting vulnerable populations;

(2) it has been limited by scale—either too large to uncover local-level nuances or too insular to allow for useful comparisons at a country or larger level, often utilizing site-specific data or indicators;

(3) the focus on income as the key variable for measuring vulnerability, deprivation or resilience; and

(4) the reliance on proprietary data and/or methodologies. (Garbutt et al., 2015, p. 163).

The review found a plethora of large-scale attempts (country to near-global) in order to assess vulnerability with a focus on relative vulnerability measures based predominantly on economic aspects of life and idiosyncratic views on the weighting of indicators. This issue can be a result of data availability, which usually shows a superior availability at more general and less-detailed levels, and also the potential of social impact of their studies. Small-scale studies cover a vast population exposed to vulnerability and therefore have a considerable appeal. On the other hand, they reflect less the essential peculiarities that can be found in the contexts to which they are concerned.

1.6.3 **Material and methods**

The research method proposes a systemic investigation and adopts the concept of vulnerability (social and environmental) by applying the Exploratory Factor Analysis (EFA), which statistically organize the dataset in order to properly contrast the variables that most contributed to the study objectives. Yet, as a statistical procedure, it requires caution during the data treatment in order to effectively express reliable information about the vulnerability.

Hence, the methodological basis for the proposed study –"social vulnerability"–, in the systemic view, adopted the considerations of Cutter (2003) and Mendes et al. (2010) included the proposals recently published in Brazil by Freitas et al. (2013), Freitas and Zucherato (2015), Bortoletto and Freitas (2016), Bortoletto (2017), and Zucherato (2018).

As a means of adapting the technique proposed by Cutter (2003), it is proposed the formulation of a simplified index of social vulnerability based on the use of statistical variables that revealed as relevant for understanding the vulnerability in other studies. Thus, for the multicriteria analysis, the methodological procedures presented by Zucherato (2018) were implemented, which resulted in a simplified methodological model, as shown in Fig. 1.6.5.

It is also worth mentioning the originality of the debate, even if incipient, the adaptation of Cutter's approach to the Brazilian context. Since the publication of the Cutter and colleagues' Social Vulnerability Index (SoVI), it was observed the development of numerous indexes using Principal Component Analysis (PCA). These procedures are focused at the subnational level for the reason that PCA usually requires a large number of study units to produce reliable results. As a data reduction technique, it is also suitable for data-rich sites such as those found in developed countries, where statistical agencies routinely collect comparable data covering small regions as villages and watersheds.

The PCA can be implemented in a number of ways, with a key choice being the rotation method used in the construction of the principal components. Where the rotation method was listed, all of them

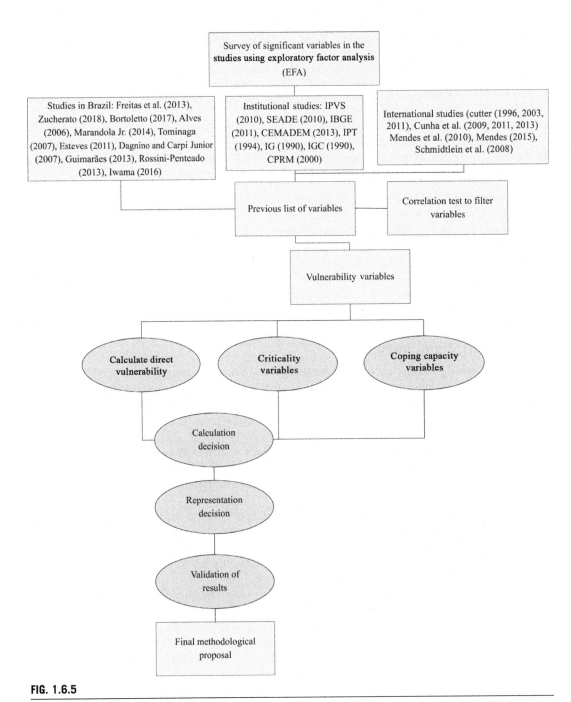

FIG. 1.6.5

Methodological chart.

used the varimax rotation in order to reduce the number of factors due to the desire to try to explain the conceptual significance of each factor. However, it is also likely that the use of a varimax rotation in Cutter's SoVI had some influence.

PCA has only been applied at the subnational level, but it is likely that a nation's comparison will not offer significant advantages over other methods, for the reason that to be statistically valid only a small number of variables could be included. According to Beccari (2016), stakeholder-based methods were less popular than the deductive or PCA methods; however, they seemed to be further oriented with more variables directly related to disaster resilience.

Based on the dataset provided by IBGE statistical base, it was decided to include variables that reflect both the support capability and the criticality in order to create only one analysis model for all those variables. This decision relied on the study scope, which aimed to determine the final social vulnerability rather than the detailed dimensions of its formation. In order that the resulting data could be compared, the created model included both the census tracts of Bragança Paulista and those of Campos do Jordão, thus building a common base of both municipalities. The city of Bragança Paulista is divided into 298 census tracts, from which 268 presented effective data for the social vulnerability calculation. Also, Campos do Jordão city is divided into 92 census tracts, from which 83 were used to calculate the social vulnerability, comprising a total of 360 analysis units inserted in the current study. It was only considered in our analysis the census tracts classified as urban by the IBGE, as we understand that the available variables better represent the urban aspects of the social vulnerability.

The previously conducted literature review led to the identification of three categories or dimensions of social vulnerability: (1) Economic wealth, material and well-being; (2) Health and social support and (3) Accessibility. These categories are further divided into seven subcategories that reflect the necessary conditions for the identification of areas with social and environmental vulnerability (Table 1.6.4). The process of identification and categorization of variables followed the conceptualization of social vulnerability of Mendes et al. (2010) and Tavares et al. (2018), which considers the components of criticality and support capability in the factorial analysis of principal components (PCA).

After obtaining a thematic list of variables, we proceeded to the data collection stage. For this purpose, a query was made in the statistical database available at IBGE (2011). The selection criterion was the variables that had some relation with the thematic axes used. For the analysis, the census sectors classified by IBGE (2011) as rural were disregarded. Initially, 13 variables were selected and submitted to the EFA. In order to validate the proposed model, 3 variables were suppressed, and 10 variables remained valid for the determination of the final social vulnerability (cf. Table 1.6.4).

The chosen variables (cf. Table 1.6.4) were submitted to collinearity tests and the ones with very high correlation values were excluded (values greater than 0.8 for direct correlations and less than −0.8 for inverse correlations). This procedure aims to eliminate redundant variables and introduce in the index calculation the variables that presented a more diverse nature. After the tests, as aforementioned, three variables were deleted from the proposed simplified social vulnerability calculation.

Subsequent to the redundant variables' exclusion, the remaining variables were evaluated in terms of their contribution to increase or reduce vulnerability. For the final index calculation, all the variables that contribute to decrease the vulnerability were multiplied by (−1) in order to present a correct cardinality. This procedure basically placed all the variables in the same direction of contribution for the final calculation of the social vulnerability values. The variables were standardized by the z-score method (Abdi, 2007) to present an average value of 0 and a value of standard deviation

Table 1.6.4 Evaluation of social vulnerability indicators based on data from the Brazilian Institute of Geography and Statistics (IBGE).

Dimension of social vulnerability	Category	Subcategory (Thematic Ax)	IBGE 2010 variables	Influence nature[a]
Criticality	Economic wealth, material, and well-being	Household composition	(V01) Number of residents per household	Increase
		Income	(V02) Average monthly income	Decrease
			(V03) Variance of monthly income (Suppressed)	Increase
	Health and support	Age	(V10) % of people under 18	Increase
			(V11) % of people over 70	
		Genre	(V07) % of female household heads	
		Ethnicity	(V09) % of nonwhite residents (Suppressed)	
		Education	(V08) % of literate residents aged 5 years and over	Decrease
	Accessibility	Living condition	(V04) % of residents in rental houses (Suppressed)	Increase
			(V05) % of dwellers in occupied homes	
Supportability		Basic services	(V12) % of dwellers with garbage collection	Decrease
		Infrastructure	(V06) % of residents in houses with 3 or more bathrooms	
			(V13) % of households with semiadequate housing	Increase

[a]*Increases: The higher the value of the variable, the greater the vulnerability; Decreases: The higher the value of the variable, the lower the vulnerability.*

of 1 (−1 for values below one standard deviation). This procedure allows a data comparison of amplitudes and different natures through normalization. After this process, the data were inserted into the SPSS program to build the explanatory model. The KMO values were used as validation criteria, as well as the variance to determine their explanatory factors (Table 1.6.5).

The resulted KMO values showed a reasonably robust model (Hair et al., 2006; Martinez and Ferreira, 2010), appropriate for model validation. Using the validated model, it was possible to verify the identified factors by scree-plot analysis (Fig. 1.6.6). Four factors bearing eigenvalue higher than 1.0 were considered for the final social vulnerability calculation. These four extracted factors allowed us to comprehend that 70% of the constructed model (Table 1.6.6) is suitable for studies focused in human nature data (Pestana and Gageiro, 2014).

Table 1.6.5 Results of the Kaiser-Meyer-Olkin and the Bartlett's sphericity tests.

KMO and Bartlett's test		
Kaiser-Meyer-Olkin measure of sampling adequacy	0.639	
Bartlett's sphericity score	Approx. chi-square	1431.017
	df	55
	Sig.	0

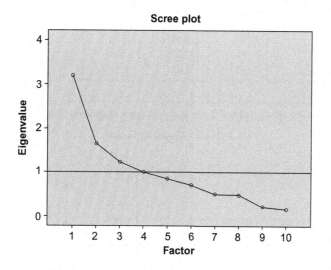

FIG. 1.6.6

Scree plot showing the factors considered in the statistical analysis.

Afterwards, the rotated component matrix (Table 1.6.7) was verified and determined the contribution of each variable related to the obtained factors, thus verifying the social vulnerability structure calculated for the study area.

The final index of the simplified social vulnerability was calculated for the census tracts of the municipalities of Campos do Jordão/SP and Bragança Paulista/SP. All factors were considered with equal contribution (same weight) to the social vulnerability score. The cartographic representations of both municipalities were prepared using GIS to organize the obtained results. It was adopted the cartographic patterns presented by Bertin (1999) and Bertin et al. (1987) with respect to the chromatic scale: red color shades for high and very high vulnerability, yellow for the average vulnerability, and green shades for the low and very low vulnerability. The classification method for the cartographic representation is the method of the standard deviation referenced to the mean. The obtained mean values are identified, and the five classes are determined adding and subtracting half standard deviation for the

Table 1.6.6 Distribution of variance results from the constructed model.

Total variance explained			
	Initial eigenvalues		
Factor	Total	% of Variance	Cumulative %
1	3.202	32.015	32.015
2	1.641	16.405	48.421
3	1.224	12.239	60.66
4	1.002	10.023	70.683
5	0.852	8.522	
6	0.706	7.062	
7	0.506	5.06	
8	0.485	4.847	
9	0.221	2.205	
10	0.162	1.621	

Table 1.6.7 Contribution of variables from the rotated component matrix.

Rotated component matrix				
	Component			
Major variable	1	2	3	4
(V11) % of people over 70	0.838			
(V08) % of literate residents aged 5 years and over	0.606			
(V01) Number of residents per household	−0.771			
(V10) % of people under 18	−0.874			
(V06) % of residents in houses with 3 or more bathrooms		0.915		
(V02) Average monthly income		0.799		
(V12) % of dwellers with garbage collection			0.794	
(V07) % of female household heads			0.502	
(V05) % of dwellers in occupied homes				0.848
(V13) % of households with semiadequate housing				0.672

Extraction Method: Principal Component Analysis
Rotation Method: Varimax with Kaiser Normalization
Rotation converged in 6 iterations

average social vulnerability class, adding and subtracting one standard deviation for the high and low classes and twice for the very high and very low classes, respectively. This classification was performed separately for both municipalities. After mapping, the areas under each vulnerability class were calculated in both municipalities. It is important to notice that these maps are particularly useful for each area to describe the social vulnerability patterns regarding the municipalities separately, nondependents. This study does not compare the differences regarding the mapped classes between the municipalities; however we assessed the social vulnerability spatially and discussed the classes' idiosyncrasies for each area.

1.6.4 **Results and discussion**

The results obtained from the constructed EFA model reveal its feasibility in explaining social vulnerability for the area in a simplified manner. The KMO, as well as the explained variance and eigenvalues, expose the practical application in determining the social vulnerability. The results regarding social vulnerability of Bragança Paulista and Campos do Jordão are comparable due to their geographical similarities. However, the comparison must be done with caution due to the difficulties in establishing a valid model for heterogeneous areas, which usually require more tests to verify its statistical behavior.

We verified the existence of four explicative factors and the variable related to individuals over 70 years old (expressed in percentage form) presented more strength to explain the social vulnerability. This result may reveal the importance of the productive forces to determine the social vulnerability over the region. Besides, the elderly is a stratum that naturally presents a financial dependence, such as pensions (public or private) and family support. The walking difficulties and lack of accessible risk information also need attention when planning a disaster management for this stratum. The other variables that are important to explain the extracted factors are related to housing conditions, such as number of bathrooms, waste collection and people living in irregularly occupied houses. This product reveals that housing conditions are much more than a reflection of society's purchasing power, but also determinant to place the social vulnerability values.

The results exposed both municipalities presenting similar values in percentage of the municipal area classified as rural sectors by IBGE: 45.13% in Campos do Jordão and 44.82% in Bragança Paulista. Afterwards they were excluded from the analysis process. The southeastern Brazil is characterized by presenting concentrated large urban areas, in general encompassing one municipality, with a clear distinction between rural and urban areas. The last census (IBGE, 2011) reported a rural population representing 0.62% of the total population in Campos do Jordão and 3.06% in Bragança Paulista. Despite the extensive rural areas, these data expose the proportionally small population living there.

Although the analyzed municipalities are located in relatively near areas, and therefore have a certain similarity in their geographic context, the social and economic dynamics where their populations are inserted present quite diverse results. The simplified social vulnerability values showed Bragança Paulista bearing higher mean value of SV (0.72) than Campos do Jordão (0.43).

A closer look at the vulnerability areas in Campos do Jordão (Fig. 1.6.7), for example, shows that neighborhoods located in areas of very high social vulnerability are basically divided into two categories. The first presents neighborhoods with historical problems in their occupation, such as Morro do Britador (SV 0.86), Vila Eliza (SV 0.73), and Jardim California (SV 0.69). These neighborhoods present a very poor socioeconomic condition and are constantly affected by disasters such as landslides.

The Morro do Britador area, for example, comprises an occupation area started in the 1920s where a quarry used to operate. The anthropic intervention in a structural way in this zone allied to irregular occupation and population with socioeconomic deficiency resulted in a set of factors that facilitated

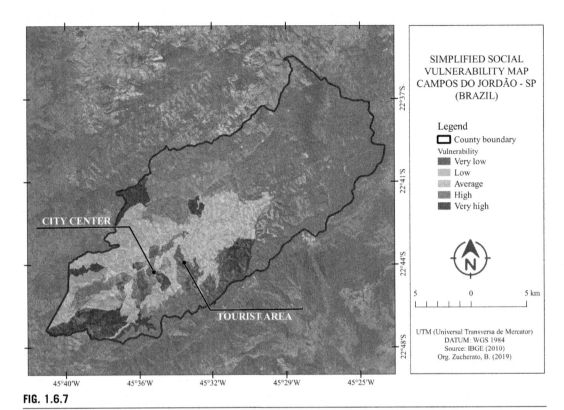

FIG. 1.6.7

Simplified social vulnerability map of Campos do Jordão/SP.

and made possible the systematic occurrence of landslides and debris flow. A well-known landslide occurred in the area was related to a massive rain in January of the year 2000, which is estimated to have destroyed about 300 family houses.

In Campos do Jordão county, the area where Vila Eliza is located also comprehends a neighborhood that is quite fragile socioeconomically and with very alarming violence rates for a municipality of the proportion of Campos do Jordão. A large portion is composed of irregular occupation and the area also constantly suffers with the occurrence of disasters, where landslides also gain prominence. One of the most destructive events that occurred in the area was the landslide in August 1972, when 60 houses and 17 people were buried.

Other city areas that were considered as areas of very high vulnerability refer to neighborhoods composed of farms in consolidation processes that often do not have basic services available, such as those located in the southern portions and also in the extreme north near Campos do Jordão state park (cf. Fig. 1.6.7).

Bragança Paulista (Fig. 1.6.8) is characterized by an area of very high vulnerability as a belt that extends to the north and west of the urban area, as it also has a higher concentration of inhabitants. The northern region is constituted by popular housing occupations, financed and subsidized by the

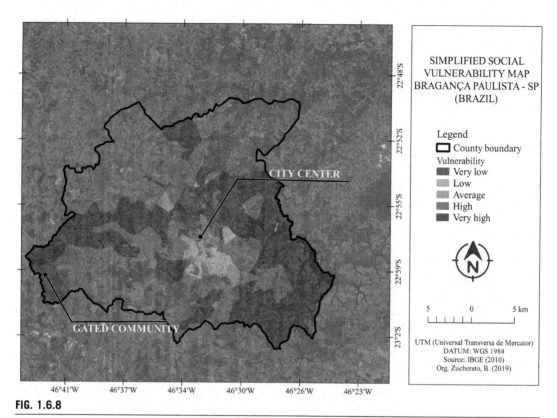

FIG. 1.6.8

Simplified social vulnerability map of Bragança Paulista/SP.

government. The social housing with implementation of housing complexes during the 1970s to the 1990s was built to meet the migrants' flow from rural exodus. It is important to take into account their precarious and outdated infrastructure that form large outlying areas.

Some subdivisions have also arisen in the southern, eastern, and western neighborhoods such as Jardim Santa Helena and Jardim América (located in tracts with SV scores of 0.32 and 0.41), closer to the center, intended to the middle class, are essentially gated communities. The old neighborhoods near the center undergo a recovery. Southwards prevails houses and high standard condominiums, reserved to the population of high purchasing power. The central region has old houses and also concentrates the commerce of the city. The areas closest to the rural area are still in the expansion process.

The most recent expansion of the city toward the peripheral area clearly is associated with the highest vulnerability classes, as a result of socioeconomic segregation policies. In opposition, it is worth highlighting the low vulnerability presented in the tracts located on the extreme west of the municipality, which are characterized by a luxurious gated community. The vulnerability spatialization showed that the central areas of the city present low vulnerability and the urban peripheral areas high vulnerability, which appears to be a pattern among other cities where there is dispersion and/or exclusion center-periphery.

From the reading of the social vulnerability map and the identification of the risk areas in the municipality, it was possible to note the consequences of the urban expansion in the Bragança Paulista periphery (north, east, and west regions), accommodating a massive population flux due to the process of rapid urbanization, and therefore interfering with the natural environment, particularly by land subdivisions and housing programs. Other areas of the city that were considered to be exposed to high and very high vulnerability, based on variable indicators, refer to high slope areas and natural floodplains adjacent to rivers, being defined as areas of risk to slope instability, erosion, and flooding. According to a survey by the Geological Institute from 2002 to 2016, 74 residents were affected by disaster events registered in Bragança Paulista.

Moreover, in an IBGE (2018) study regarding the population exposed in disaster risk areas in Brazil, some variables were categorized at the national level for the analysis: age groups of people most vulnerable to disasters; access to the water supply network; access to sanitary sewage; and access to garbage collection. Based on this assessment, the municipality of Bragança Paulista presented 1200 households at risk and 3934 inhabitants at risk, for a population of 146,744 in the 2010 demographic census. According to the same study, Campos do Jordão owned 2925 households in risk areas and 10,298 inhabitants at risk, from a total population of 47,789 (IBGE, 2011) in the same census. In order to understand a more detailed behavior of the social vulnerability isolated in each analyzed area, would be interesting to build explicative models for each municipality. Hence it could be possible to properly verify the differences regarding social vulnerability deep inside the cities.

It is noteworthy that the database (2010 Brazilian census) adopted in this study to elaborate a simplified social vulnerability index is the only data source. Since the census is available countrywide, it is possible to replicate the procedures to other areas and municipalities over the nation. Additionally, to achieve a more refined analysis would be required complement some other variables that can reflect aspects regarding the vulnerability, e.g., natural features such as slope values and river network, or even including other data sources and fieldwork collection.

1.6.5 Final remarks

The development and analysis of social vulnerability within the studied municipalities clearly showed the most vulnerable areas and those with the lowest vulnerability, which allowed us to establish the urbanization patterns that impelled and stopped the processes. The spatial distribution of social vulnerability revealed a tendency identifying less vulnerable areas and peripheral areas with high vulnerability following the pattern of center-periphery opposition.

The presented methodological approach provides a simplified mechanism adopting key indicators of vulnerability in order to develop a comparative study of two geographically similar areas. These indicators were used to create a vulnerability index that provides local and regional information from different geographic areas with a proper resolution to identify compartments of vulnerable communities.

The method used is scalable and adaptable, which allows to work with easily accessible public data and enables public agencies to share information. It potentially improves local knowledge and supports structural and non-structural measures to reduce local vulnerability. The current study results may support public authorities with a source of reliable information to adopt and establish public policies to prevent, alert, mitigate, and recover vulnerable areas, thus converging into the Sendai's first priority.

The proposed research has a range of possibilities to be extended, such as a refinement of the adapted methodology by including more variables and possibly indicators, as well as the technique replication in other areas and other scales. There is also the possibility to examine and cross social data, then identifying their changes in space and time.

Moreover, an agreement was found between the variables used in this study and those discussed in the literature, in understanding the communities' vulnerability processes. This finding has the potential to feed a broader international discussion on measuring progress in the Sendai Framework for Disaster Risk Reduction and the Sustainable Development Goals, both featuring a substantial focus on data acquisition and measurements.

It is worth to highlight that erroneous use of variables not directly related to disaster risk reduction and resilience can undesirably impact the sensitivity and uncertainty of the analysis. This situation may lead policymakers to believe that the results of the indexes are more accurate than they really are. The incorrect use of a comparative index by local decision makers to allocate resources for disaster risk reduction, without regard to its reliability, may lead to wastage of government resources or possibly to generating even greater risks.

The lack of a sensitivity analysis implies that the exclusion of disaster-related variables may not be questioned by policymakers or researchers which use this index, thus increasing the risk of misuse. Therefore, it is necessary to properly evaluate its quality and reliability. Moreover, we emphasize that greater efforts are required to develop these indices in order to ensure they fit appropriately with the decision maker's necessities, owning then high quality and increasing the understanding of vulnerability and resilience.

According to the consulted literature, it is still not clear which variables are most important for disaster resilience studies. Current approaches are broadly adapted to specific contexts and mostly incompatible with one another. This scenario may represent a significant barrier to the accomplishment of the Sendai Framework for Disaster Risk Reduction, the disaster-related goals of the Sustainable Development Goals and other elements of the post-2015 development agenda, as the parties seek agreement on indicators to properly measure performance toward these compromises. Further research is desirable to enhance the knowledge related to the variables that are most indicative of disaster risk, vulnerability, and resilience, as well as the different circumstances in which they apply.

Acknowledgments

FAPESP and CAPES for the master's and doctoral scholarships awarded (process n° 2017/005642, 2018/11369-9, and BEX 9537/13-9). The authors would like to thank Prof. Dr. Lúcio Cunha for his support in the statistical methodology and Roger Dias Gonçalves for the careful review. We would like to thank the editor Pedro Pinto Santos for providing highly constructive comments that helped tremendously in improving the quality of the chapter.

References

Abdi, H., 2007. Z-scores. Sage Publications, Thousand Oaks (Technical note).
Aven, T., 2012. The risk concept—historical and recent development trends. Reliab. Eng. Syst. Safe. 99, 33–44. https://doi.org/10.1016/j.ress.2011.11.006.

Aven, T., Renn, O., 2010. Risk management and governance. In: I. R. U., , Mumpower, J.L. (Eds.), Estimating Impact: A Handbook of Computational Methods and Models for Anticipating Economic, Social, Political and Security Effects in International Interventions. Springer, Berlin, Heidelberg. https://doi.org/10.1007/978-3-642-13926-0.

Beccari, B.A., 2016. Comparative analysis of disaster risk, vulnerability and resilience composite indicators. In: PLOS Currents Disasters.https://doi.org/10.1371/currents.dis.453df025e34b682e9737f95070f9b970 Edition 1.

Bertin, J., 1999. Sémiologie graphique : les diagrammes, les réseaux, les cartes, 3. ed L'École des hautes études en sciences sociales, Paris (in French).

Bertin, J., Carrión, A., Bonin, S., 1987. La gráfica y el tratamiento gráfico de la información. [s. l.].

Bortoletto, K.C., 2017. Estudo das Vulnerabilidades Social e Ambiental em áreas de riscos de desastres naturais no município de Caraguatatuba SP. PhD thesisUniversidade Estadual Paulista, Instituto de Geociências e Ciências Exatas, Rio Claro. 217pp (in Portuguese).

Bortoletto, K.C., Freitas, M.I.C., 2016. Vulnerabilidade Socioambiental e Histórico de Desastres Naturais na Área Urbana de Caraguatatuba SP, Período de 2000 a 2015. In: International Plant Production Symposium. Anais… Encontro de Pós-Graduandos da Unesp, p. 1 (in Portuguese).

Bortoletto, K.C., Freitas, M.I.C., Rossetti, L.A.F., Oliveira, R.B.N., Leite, A., 2014. Indicadores socioeconômicos e ambientais para análise da vulnerabilidade socioambiental do município de Santos - SP. In: Anais do XXVI Congresso Brasileiro de Cartografia V Congresso Brasileiro de Geoprocessamento XXV Exposicarta. Anais… Gramado: UFRGS/FAURGS (in Portuguese).

Brasil, 2012a. Lei nº 12608, de 10 de abril de 2012. Lei Nº 12.608, de 10 de abril de 2012. Brasília, DF, 10 abr. http://www.planalto.gov.br/ccivil_03/_ato20112014/2012/lei/l12608.htm accessed 26.07.18 (in Portuguese).

Brasil, 2012b. Plano Nacional de Gestão de Riscos e Resposta a Desastres Naturais 2012–2014. Ministério do Planejamento. http://www.planejamento.gov.br/apresentacoes/apresentacoes-2012/120808_plano_nac_risco_2.pdf/ (accessed 05.09.19).

Briguglio, L., 1995. Small island developing states and their economic vulnerabilities. World Dev. 23 (9), 1615–1632.

Cunha, L., 2015. Vulnerabilidade e riscos naturais: exemplos em Portugal. In: Freitas, M.I.C., Lombardo, M.A., Zacharias, A.A. (Eds.), Vulnerabilidades e riscos: reflexões e aplicações na análise do território. UNESP-IGCE-CEAPLA, Rio Claro (SP) (in Portuguese).

Cunha, L., Dimuccio, L., 2001. Considerações sobre riscos naturais num espaço de transição. Territorium - Coimbra 9, 37–51 (in Portuguese).

Cunha, L., Leal, C., 2012, Natureza e sociedade no estudo dos riscos naturais. Exemplos de aplicação ao ordenamento do território no município de Torres Novas (Portugal). As novas geografias dos países de língua portuguesa, paisagens, territórios e políticas no Brasil e em Portugal, Geografia em Movimento; 2012, pp. 47–63 (in Portuguese).

Cunha, L., Mendes, J.M., Tavares, A., Freiria, S., 2011. Construção de modelos de avaliação de vulnerabilidade social a riscos naturais e tecnológicos. O desafio das escalas. In: Santos, N., Cunha, L. (Eds.), Trunfos de uma Geografia Activa. IUC, Coimbra, pp. 627–637 (in Portuguese).

Cutter, S.L., 1996. Vulnerability to environmental hazards. Prog. Hum. Geogr. 20 (1), 529–539.

Cutter, S.L., 2003. The vulnerability of science and the science of vulnerability. Ann. Assoc. Am. Geogr. 93 (1), 1–12.

Cutter, S.L., 2011. Revista Crítica de Ciências Sociais 9. Rev. Crít. Ciênc. Soc. 93, 59–69 (in Portuguese).

Cutter, S.L., Boruff, B.J., Shirley, W.L., 2003. Social vulnerability to environmental hazards. Soc. Sci. Q. 84 (2), 242–261.

Dagnino, R. de S., Carpi Junior, S., 2007. Risco Ambiental: Conceitos e Aplicações. Climatol. e Estud. da Paisag. 2, 50–87.

EM-DAT, 2018. The International Disaster Database: Center of Research in Epidemiology of Disaster; 2018. http://www.emdat.be/. [Accessed 9 May 2019].

Freitas, M.I.C., Cunha, L., Ramos, A., 2013. Vulnerabilidade socioambiental de concelhos da Região Centro de Portugal por meio de sistema de informação geográfica. Cadernos de Geografia 32, 313–322 (in Portuguese).

Freitas, M.I.C., Zucherato, B.A., 2015. Técnica de Dasimetria aplicada ao Mapeamento da Vulnerabilidade Socio-ambiental para a Área Insular de Santos SP. In: Freitas, M.I.C., Lombardo, M.A., Zacharias, A.A. (Eds.), Vul-nerabilidades e Riscos: Reflexões e Aplicações na Análise do Território. IGCE/CEAPLA, Rio Claro, pp. 133–155 (in Portuguese).

Garbutt, K., Ellul, C., Fujiyama, T., 2015. Mapping social vulnerability to flood hazard in Norfolk, England. Environ. Hazard. 14 (2), 156–186. https://doi.org/10.1080/17477891.2015.1028018.

Hair, J.F., Black, W.C., Babin, B.J., Anderson, R.E., 2006. Multivariate Data Analysis: A Global Perspective, 6th. ed Pearson Prentice Hall, Upper Saddle River, NJ.

IBGE, 2011. Resultados do censo 2010. http://censo2010.ibge.gov.br/resultados.html (accessed 07.02.2017). (in Portuguese).

IBGE, 2018. População em Áreas de Risco no Brasil/IBGE. Coordenação de Geografia, Rio de Janeiro, ISBN: 978-85-240-4468-7 (in Portuguese).

IRGC, 2017. Introduction to the IRGC Risk Governance Framework. revised version, EPFL International Risk Governance Center, Lausanne. https://doi.org/10.5075/epfl-irgc-233739.

Martinez, L., Ferreira, A., 2010. Análise de dados com SPSS: primeiros passos. Escolar editora, Lisboa (in Portuguese).

Mendes, J.M., 2015. Sociologia do Risco: Uma breve introdução e algumas lições. Imprensa da Universidade de Coimbra, Coimbra (in Portuguese).

Mendes, et al., 2010. Social vulnerability to natural and technological hazards: the relevance of scale. In: Briš ,Soares, G., Martorell, (Eds.), Reliability, Risk and Safety: Theory and Applications. Taylor and Francis, London, pp. 445–451.

Pestana, M.H., Gageiro, J.N., 2014. Análise de dados para ciências sociais: A Complementaridade do SPSS, 6ª ed Edições Sílabo Ltda, Lisboa (in Portuguese).

Rebelo, F., 2003. Riscos naturais e acção antrópica. Estudos e reflexões. Imprensa da Universidade de Coimbra, Coimbra (in Portuguese).

Schmidtlein, M.C., et al., 2008. A sensitivity analysis of the social vulnerability index. Risk Anal. 28 (4), 1099–1114.

Tavares, A.O., Barros, J.L., Mendes, J.M., Santos, P.P., Pereira, S., 2018. Decennial comparison of changes in social vulnerability: a municipal analysis in support of risk management. Int. J. Disaster Risk Reduct. 31, 679–690. https://doi.org/10.1016/j.ijdrr.2018.07.009 (accessed 01.09.2018).

UNISDR, 2015. Making Development Sustainable: The Future of Disaster Risk Management. Global Assessment Report on Disaster Risk Reduction. United Nations ISDR, Geneve.https://archive-ouverte.unige.ch/unige:78299.

UNISDR, 2017. Terminology. https://www.unisdr.org/we/inform/terminology (accessed 01.09.2019).

UNU-EHS, 2016. World Risk Report, 2016. United Nations University-Institute for Environment and Human Security. United Nations University – EHS. Platz der Vereinten Nationen.. ISBN: 978-3-946785-02-6 https://collections.unu.edu/eserv/UNU:5763/WorldRiskReport2016_small_meta.pdf (accessed 01.09.2019).

Willis, H.H., 2007. Guiding resource allocations based on terrorism risk. Risk Anal. 27 (3), 597–606.

Wisner, B., et al., 2004. At Risk: Natural Hazards, People's Vulnerability and Disasters, second ed. Routledge, New York.

Zucherato, B., 2018. Cartografia da vulnerabilidade socioambiental no Brasil e Portugal: estudo comparativo entre Campos do Jordão e Guarda. Universidade de Coimbra, PhD thesis, 00500: 369 pp (in Portuguese).

PREVENTION

Developing guidelines for increasing the resilience of informal settlements exposed to wildfire risk using a risk-based planning approach

2.1

Constanza Gonzalez-Mathiesen[a,b,c] **and Alan March**[a,b]

University of Melbourne, Melbourne, VIC, Australia[a] *Bushfire Natural Hazards Cooperative Research Centre, Melbourne, VIC, Australia*[b] *Universidad del Desarrollo, Concepción, Chile*[c]

CHAPTER OUTLINE

2.1.1 Introduction

Internationally, disasters have increased in number and impact, especially since the 1950s (CRED, 2016). Between 1995 and 2015, the vast majority (90%) were caused by weather-related events—which include hydrological, meteorological, and climatological hazards—producing great social, environmental, and economic impacts that include more than 4 billion people affected, 606,000 lives claimed, and US$ 1891 billion costs adjusted at 2014US$ value (CRED, 2015). Furthermore, these trends can be expected to continue to rise due to climate change (CRED, 2015; Davis and Alexander, 2016; Intergovernmental Panel on Climate Change, 2012), suggesting that there is an increasing need for disaster-related research, policy, and action.

Understandings of disaster risk need to take into consideration that disasters result from the combination of social factors with hazards. The Sendai Framework for Action's priority one is "understanding disaster risk" (UNISDR, 2015, p. 15[a]). In disaster contexts, risk is often described as the relation between the likelihood of a natural hazard taking place; and the likely consequences on vulnerable systems and people. This implies that natural and human processes are bound together in almost all disasters; therefore, disasters cannot be simplistically understood as *natural* (Blaikie et al., 2004) as they are primarily social (Benton-Short and Short, 2013). In fact, disasters can be described as the consequence of the combination of three social factors: (1) exposure to a hazard; (2) vulnerability; and (3) insufficient coping capacity (UNISDR, 2009). First, hazards are potentially dangerous phenomena, activities, substances, or conditions that may cause social, economic, or environmental losses, damages, or disruptions (UNISDR, 2009). Disasters are associated with systems' exposure to a hazard; this refers to people, properties, or other elements present in areas (UNISDR, 2009) where hazard events have the potential to cause harm and losses (Coppola, 2015). Therefore, despite their potential harm, hazards alone do not cause a disaster. Second, vulnerability refers to the characteristics and circumstances of an individual, community, country, system, or asset that make it susceptible to the damaging consequences of a hazard (Coppola, 2015; UNISDR, 2009). This perspective relates to the conceptualization of social vulnerability, distinct from biophysical vulnerability, which is used to describe systems' susceptibility to climate change and ability to adapt or not to those impacts (Birkmann, 2006; WBGU, 2004). Usually, vulnerability is correlated to socioeconomic status (Benton-Short and Short, 2013; Blaikie et al., 2004) and losses are often disproportionally concentrated mainly according to demographics, poverty, and lack of political voice (Blaikie et al., 2004; Hewitt, 1998). Among these, communities living in informal settlements are often more vulnerable to disasters' impacts (Sharifi and Yamagata, 2018). Third, coping capacity refers to people, organizations, and systems' capacity to deal with—or respond to—the impact of a natural hazard (UNISDR, 2009) to overcome their vulnerabilities (Bogardi, 2006). Vulnerability and coping capacity are dependent on each other and can be considered opposites (Billing and Madengruber, 2005); thus, the lack of capacity to cope can be regarded as a characteristic of vulnerability (Blaikie et al., 2004). It can be related to both the resources as well as the abilities to face adverse consequences and be addressed at the individual or institutional level (Billing and Madengruber, 2005).

Disaster resilience refers to systems', communities', or societies' capacity to resist the impact of a hazard and to adapt to them and transform into new states to innovatively adapt, change, and evolve

---

[a]UNISDR is now known as UNDRR.

into more desirable development pathways for long-term risk reduction and prevention. This requires learning from the experience gained from previous disasters and transferring disaster risk management knowledge within different places in a contextualized manner. The extension of the concept of resilience to the disaster discourse changed the focus to the development of resilience, offering new ways of approaching the management of disaster risks. Manyena (2006, p. 434) supports that its incorporation into disaster discourse can be seen as "the birth of a new culture of disaster response," contributing to improving the understanding of risk and vulnerability, and changing the focus from emergency management to development of resilience, which allows communities to make appropriate choices within their context.

2.1.1.1 Wildfires interacting with settlements

Wildfires are a particular type of hazard. They refer to grass, scrub, or forest vegetation burning out of control over a large area of land. Fires can be ignited by human activities or natural processes such as lightning depending on the ecosystems' fire regimes (Moritz et al., 2014). Fires in vegetation are transmitted or propagated to other flammable materials causing them to progress ignition via heat, direct contact, or by embers that carry the fire directly to new fuels. The overall speed, intensity, direction, and means of ongoing transmission of fires in vegetated areas are influenced by topography, vegetation or available fuel, and weather conditions.

Wildfires bring about greater disaster risks at the urban–rural interface of wildfire-prone areas, where lives and properties are more exposed (Gill and Stephens, 2009). The main ways that fires interact with housing and by which houses are destroyed in wildfires are: ember attack, heat radiation, direct flame, and fire-driven wind (CFA, 2012; Ramsay and Rudolph, 2003). Usually, the risks and consequences of wildfires are even greater in contexts of informality, where settlements have been built with little consideration of risks. As previously mentioned, communities living in informal settlements are often more vulnerable to disasters' impacts (Sharifi and Yamagata, 2018), and wildfires are no exception.

2.1.1.2 Spatial planning dealing with wildfires

Spatial planning deals with challenges to achieve improved settlements. It is concerned with the spatial impacts of different problems and with the spatial coordination of different policies (Hall and Tewdwr-Jones, 2010), bridging the gap between spatial and a-spatial policies (Bracken, 2014) by considering the spatial component of broader issues, such as social or economic policies. The concept "spatial" is thus used its wider sense (Hall and Tewdwr-Jones, 2010), extending to notions such as economy or psychology, rather than just being limited to three-dimensional geometrical spaces.

Spatial planning capacities for effectively mainstreaming and applying DRM knowledge in a contextualized manner are widely accepted by both academic and international organizations. The academic discourse strongly argues for the integration of DRM considerations into planning (such as Burby, 1998; King et al., 2016; March and Kornakova, 2017; Yamagata and Sharifi, 2018). Furthermore, the need to improve interdisciplinary practices to deal with disaster risk, including spatial planning, has been consistently highlighted by the Hyogo Framework for Action (UNISDR, 2005) and its successor, the Sendai Framework for Disaster Risk Reduction (UNISDR, 2015), as well as by the Sustainable Development Goals (UNDP, 2015), among others.

Spatial planning can be approached as an evidence-based procedure. The most enduring planning process (March et al., 2017b) is known as Rational Comprehensive Planning (RCP). RCP approaches planning as a continuous process to gather and analyzes information to devise sensitive systems of guidance and control, and whose effects can be monitored and modified if necessary (Hall, 2014). This implies an evidence-based approach to planning. It tries to examine alternative actions by tracing their likely consequences for helping decision makers and communities to address their problems in a clear and logical manner while acknowledging underlying matters such as growth or equity, without claiming unique expertise or ability to solve complex problems instantly (Hall and Tewdwr-Jones, 2010). Despite that, RCP—and in its turn, evidence-based planning—is periodically criticized for overlooking the complexities involved in the process and the internal and external forces that influence it (for instance Davoudi, 2006; Hillier, 2007; Weiss, 2001), all current planning methodologies are usually variants of RCP and involve gathering, analyzing, and applying knowledge, and require monitoring and updating. Furthermore, the research-generated evidence is valuable for decision making and should have a role in it (Krizek et al., 2009). Thus, any pragmatic inquiry of planning-related issues must consider that this is the methodology commonly used, acknowledging its assumptions and limitations.

Evidence-based planning and risk-based planning have procedural similarities that allow them to work together (March et al., 2017b), as summarized in Table 2.1.1. ISO 31000 (2018) is the first Risk Management International Standard intended for wide application. The process established by ISO 31000 (2018) considers three overall stages: (1) establishing the context; (2) risk assessment; and (3) risk treatment. The context is established by considering the external and internal context. The overall process of risk assessment identifies, analyzes, and evaluates risk. The stage of risk treatment selects the alternatives for modifying risk and their implementation. It involves a cyclical process of assessing treatments, deciding whether residual risk is acceptable, if not tolerable, creating new risk treatments, and assessing the effectiveness of those treatments (ISO 31000, 2018).

Table 2.1.1 Evidence-based and risk-based planning procedures parallels.

Evidence-based planning procedure	Risk-based planning procedure
Analysis of the circumstances and problem/opportunity identification	Establishing the context
Identification of goals and objectives	Risk assessment—identification
Design of alternatives	Risk assessment—analysis
Evaluation and selection of goals	Risk assessment—evaluation
Implementation	Risk treatment
Monitoring effects and adjusting.	Communication and consultation & Monitoring and review (to take place during all previous stages)

Authors' own development based on ISO 31000, 2009; March, A., Kornakova, M., Handmer, J., 2017. Urban planning and recovery governance. In: Urban Planning for Disaster Recovery. Butterworth-Heinemann, Boston (Chapter 2).

2.1.1.3 **Acting upon wildfire risk in informal settlements**

Implementing wildfire risk reduction policies presents considerable challenges for spatial planning systems. A crucial issue is developing ways of reducing the vulnerability of existing settlements. Physical mechanisms to increase resilience to wildfire are best established before settlements are in place. Existing settlements often have limited capacity to provide adequate separation from fire sources; thus, retrofitting measures might be used to modify fuel levels or increase buildings' resistance (Gonzalez-Mathiesen and March, 2018).

Acting upon urban risks such as wildfire in informal settings adds a layer of complexity to this already-challenging issue of retrofitting. A key tension for spatial planning is the need to allow, facilitate, require, and even actively bring about certain outcomes. In contrast, individual rights to property provide land occupiers and users certain rights to autonomy of use and self-determination. In the context of informality, additional challenges and possibilities exist. The context of lesser economic resources, mistrust of authority, tenure insecurity, location in often risky areas is frequently contrasted with considerable self-organization abilities, innovation, adaptability, strong networks, and a strong sense of personal responsibility. Thus, there is a need to explore practical ways that can contribute to developing resilience to wildfires for communities living in informal settlements exposed to wildfire risk.

This research reports on the production of guidelines to develop resilience to wildfires for communities living in informal settlements exposed to wildfire risk. It argues that a way for spatial planning to deal with informal settlements wildfire risks is taking a risk-based procedural approach. This is expected to contribute to the exploration of practical ways that promote the development of resilience to wildfires for communities living in informal settlements exposed to wildfire risk.

2.1.2 **Examining and improving the risks in one informal settlement: Agüita de la Perdiz, Concepción, Chile**

This section describes the approach used to analyze the case study of the informal settlement of Agüita de la Perdiz, Concepción, Chile, examining the risks associated with wildfires, proposing key actions to improve the risks faced in that settlement. The approach taken is aligned with the Sendai Framework for Action's priority one "understanding disaster risk" (UNISDR, 2015, p. 15), which supports the need to use, develop, and disseminate local knowledge and practices for disaster risk assessment and conceptualization of risk reduction measures. This section first describes the research methodology used. The investigation uses an action research strategy within the context of a seminar and three-day workshop. The case is analyzed by participants applying a risk-based approach. Second, the principles of fire behavior and house destruction discussed during the workshop are presented. Third, a summary of the analysis of the case's wider context, likely fire behavior, interactions with settlement and houses, risk and consequences, is set out. Lastly, the key actions for house and site improvements identified are synthesized and listed.

2.1.2.1 Producing design guidelines using a risk-based approach

The investigation is approached through participatory action research. This is applied research and people within the community actively participated in the research process (Whyte et al., 1991). It is the result of a collaboration during the seminar "Prevention of Forest Fire Risks in Urban Settlements and Buildings: A Planning and Design Approach." The Universidad del Biobío and the Nodo de Arquitectura Sustentable organized the seminar, which was facilitated by academics with relevant expertise from the University of Melbourne and Bushfire and Natural Hazards Cooperative Research Centre (BNHCRC). The aim of the activities was to spread knowledge about general design and planning strategies to mitigate wildfire risk as well as to develop local capacities. The 25 participants were architects from firms in partnership with the university, architecture students, public servants, community representatives, and the general community.

The study case is Agüita de la Perdiz in Concepcion Chile. Chile is characterized by its extensive forests, and wildfire hazards represent an important and risk (CONAF, 2017). In 1958, Agüita de la Perdiz commenced as an illegal settlement (Vega, 2015) located in an enclosed gully close to the inner city of Concepcion. Over time, residents have gradually gained land rights by receiving individual property titles (Municipalidad de Concepcion, 2016). The settlement and its physical context pose several challenges to the management of wildfire risks. Nonetheless, the community has demonstrated self-organizing capacity, which implies that there is potential to build upon this capacity to engage in wildfire mitigation activities.

The seminar included a three-day workshop and a visit to the study site in Agüita de la Perdiz. The workshop took place in August 2017 in the facilities of the Universidad del Biobío. The facilitators encouraged the discussion and engagement of all the participants and the inclusion of their different perspectives. Activities were developed using a risk-based approach, informed by evidence-based planning and risk-based planning (ISO 31000, 2018) procedures, as summarized in Table 2.1.1. To do so, a protocol of three stages was used: (1) context, establishing circumstances and problems; (2) assessment, identifying the risks, objectives, and design alternatives; and (3) specification, selecting the actions to implement for treatment of risk.

The first day of the seminar focused on providing participants with core concepts and understandings about wildfires and the urban challenges they imply, including general concepts of built environment resilience, characteristics of wildfires, the landscape and weather conditions influences, their interaction with settlements, initiation and progression of fire in structures, among others.

The second day consisted of a visit to the study site and a workshop. The visit was guided by a community representative and supported by local officials. The representative guided a walk in the neighborhood and its surrounding landscape, described the fire history of the area, reported on recent fires, and referred to the community organizations and to their response to past fires. The workshop focused on exploring approaches to site responsive design to wildfire. Based on the general principles of fire behavior and house destruction previously discussed, the group conducted a strengths, weaknesses, opportunities, and threats (SWOT) analysis of the site.

On the third day, the workshop focused on identifying design and community actions to reduce wildfire risks in Agüita de la Perdiz. Spatial mapping was used for integrating the wide range of evidence relevant to understanding and acting upon wildfire risks in the site. Based on the characterization of the site, the principles of fire behavior and house destruction were applied to identify key actions for house and site improvement, settlement improvement, and community development. These actions were curated, detailed, and grouped collaboratively as guidelines.

The product of the seminar was condensed by the facilitators in a report titled "*Manual de diseño ante incendios forestales*"[b] (March et al., 2017a). The report explores the settlement's design and mitigation to reduce wildfire risks in Agüita de la Perdiz. It is structured in two sections: (1) general principles that influence the behavior of fire and house destruction; and (2) applying the principles by analyzing wildfire risk and developing a design response.

2.1.2.2 Principles of fire behavior and house destruction

The key elements that influence fire behavior in a typical forest and how this translates to principles of house destruction were presented and discussed during the workshop as the first step of the workshop. This provided general understandings and knowledge that were then applied.

Fires in forests or other forms of vegetation follow general principles that provide a strong basis for developing likely fire behavior scenarios of fire progression and intensity. Fires need a combination of oxygen, fuel, and ignition to start and to continue burning. Fires can be ignited by intentional or accidental human activities human activities or natural processes such as lightning. Once fires are ignited, they may in some cases self-extinguish. However, under certain conditions, fires may continue and progress through the landscape, sometimes with significant impacts. It is difficult to know in advance where and when fires will be initiated and exactly how they will progress in a given year or fire season. Nevertheless, it is possible to examine the range of likely scenarios that might eventuate in the long term and to plan to mitigate and to avoid the worst effects of these. Developing a range of likely fire scenarios in advance has several important uses for the reduction of risks in and around human settlements.

2.1.2.2.1 General principles that influence fire behavior in a typical forest

The overall speed, intensity, direction, and means of ongoing transmission of fires in vegetated areas are influenced by topography, vegetation or available fuel, and weather conditions. As set out below, these elements provide the basis for developing understandings of likely risk scenarios levels in various landscapes. It is important not to just prepare for "normal" fires, but to consider the possibility of extreme conditions that might lead to large scale disasters in which capabilities are overwhelmed.

The topography has significant impacts upon fires and the way they progress. As slope increases, fires move more quickly, doubling in speed for every 10° increase in slope. On a slope, a fire transfers heat, predrying, and by direct flame contact to adjacent fuels more readily, and consequently advances more quickly. As fires move more quickly, they consume the available vegetation more rapidly. Accordingly, the intensity of the fire will increase. The same amount of fuel is being consumed more quickly, and the heat and flame outputs will increase in turn. This can be described in flame length, and kilowatts per m^2 of heat outputs.

The vegetation or available flammable material provide considerable fuel for fires. The fuels can be understood as latent energy and the greater the amount of suitable fuel in a given location, the greater the potential for increased fire intensity, often described as tonnes per hectare of fuels. In addition, the arrangement of these fuels is a significant factor in fire ignition and progress. In general terms, vegetation that is fine, dry, and loose—such as dry leaves, grasses, litter, and many scrubs—ignites easily

---

[b]Design manual facing wildfires.

and burns quickly, and often is the first to ignite and propagate fire to other elements. In contrast, large, dense, and heavy logs and solid vegetation may, while being flammable, be difficult to ignite, but once it commences burning it may continue to do so with intensity and for a considerable time (note that this varies considerably between species). The vertical arrangement of vegetation must also be considered. Fuels may be located on the ground in the case of leaf and stick litter, in the understory and midstory in the case of shrubs, small trees, and the lower foliage of trees, in the canopy and leaves of the trees themselves, and flammable bark of some species may go from ground to canopy level. Generally, it is most common that fires are ignited and continue to burn at the ground plane level in grasses or leaf litter or through the understory, but more intense fires may sometimes "ladder" upwards. The possibility for human-made structures such as housing, sheds, vehicles, gas bottles, or even wood piles to be "fuel" for fires also needs to be considered. As fires transition from forests to urban areas, they may encounter these other fuels and continue to progress accordingly.

Weather conditions—temperature, humidity, and wind—is another fundamental factor to consider. Hotter weather means that materials will ignite more readily, due to the curing and preheating effects upon flammable materials. This interacts with humidity as the percentage of water in the air, as a function of temperature and the level of water content, whereby higher temperatures are directly related to lower humidity. In direct association with temperature and humidity are winds. Hot and low humidity winds increase the propensity for materials to dry out. Winds increase fire intensity by providing greater amounts of oxygen to fires, causing them to burn more intensely and to propagate flames to nearby fuels. Accordingly, winds generally cause fires to move in the direction of the wind, and may carry embers to new fuels ahead of the existing fire. When fires become extreme, the heat-induced updrafts from the fire tend to create new winds that in extreme fires are extremely strong. A good understanding of prevailing and extreme weather conditions in a given area may assist in the development of likely fire behaviors.

2.1.2.2.2 General principles of house destruction

Understanding of the ways fires interact with housing provides a key starting point for improvements to the resilience of communities. The main ways that fires interact with housing and by which houses are destroyed are ember attack, heat radiation, direct flame contact, and fire-driven wind.

Embers are small burning fuels such as twigs, leaves, and bark that are borne by the wind. When embers are blown into a flammable receiving environment, they are often the key point of ignition and propagation of a wildfire. Accordingly, embers may precede the main fire front by a considerable distance, complicating the process of defense, catching occupants and fire-fighters by surprise. Accordingly, embers are now understood as a key threat to housing and other structures. Embers can enter structures through very small openings, such as roof and wall vents, poorly made or maintained building elements, underfloor areas and around windows, doors or other structural components. If these embers meet flammable materials in these spaces, they may burn undetected for some time, often to the point that the structure is unable to be saved.

The heat radiation generated by a wildfire is considerable, causing flammable materials to ignite, and discomfort and injury or death to humans, livestock and fauna. A typical wildfire can reach 12–19 kWm^2, through to $40 kWm^2$ in extreme cases. Many materials are susceptible to ignition or failure at these levels of heat. Plastic materials may melt, fail or ignite, and many wood products will ignite and burn when exposed to considerable heat. Accordingly, the choice of materials in housing that may be exposed to wildfire is important. The heat generated by fires decreases with distance from flames, so a key factor of the intensity a building will be subjected to is its distance from flammable vegetation.

Direct flame contact significantly increases the chances of ignition and propagation of fire to structures. It is difficult and expensive to design structures to a standard that can withstand intense direct flame contact in a wildfire. Accordingly, the use of other design approaches above such as separation and vegetation management are typically key elements in achieving an appropriate standard, in association with choice of appropriate materials.

Fires are typically more intense and dangerous in windy conditions due to increased intensity, speed, likelihood of ember attack and their unpredictability. Larger fires may develop localized updrafts and winds associated with the intense heat produced, particularly as they interact with slopes. As a result, there may be particularly strong winds associated with intense fires. These winds can cause damage to structures or trees and other elements such as unsecured materials to fall or to become airborne, or to damage electricity lines and connections. This damage and flying objects can damage structures and impede firefighting efforts. Roofs can lift, or windows can break, allowing access for embers to find flammable materials and for houses to ignite.

2.1.2.3 Analysis of wider context, likely fire behavior, interactions with settlement and houses, risk and consequences

The principles of fire behavior and house destruction previously described were applied to the case to analyze the site at two main spatial scales. The wider context of the study case is analyzed considering the likely fire behavior and the likely risk and consequence outcomes. In addition, an understanding of the possible ways that fires might progress and impact upon settlements is fundamental to develop of risk reduction design approaches.

2.1.2.3.1 Applying the principles to the wider site context

Spatial mapping was used to integrate the diverse evidence and wider contextual aspects relevant to understanding and acting upon wildfire risks, developing understandings of relevant risk factors (see Fig. 2.1.1). Elements considered were:

- Fire history (including general footprint of previous events, nature of impacts, and fuel reduction)
- Typical fire weather and worst-case scenarios
- Response services (including water plugs, distance to response agencies, road network)
- General vegetation and fuel load

By considering these elements and applying the general principles that influence fire behavior in a typical forest presented in the previous section, the group identified that the site's wider context poses a number of challenges to the management of wildfire risks. The site is effectively separate from the main settlement, and the majority of its boundaries are adjacent to forests. While the site does not have a dedicated fire station, there are some located to the north, relatively close by. However, on a high-fire danger day, these stations are likely to be dealing with threats to the extensive urban - forest perimeters in the wider area. The settlement has extensive forested areas to the east and south, meaning that long fire runs could occur, allowing a fire to develop significant speed and heat prior to arrival at the settlement. The strong possibility for wind shifts during the progress of fires may mean the settlement faces multiple fire fronts and ember attack directions. The only road access to the site is from the north

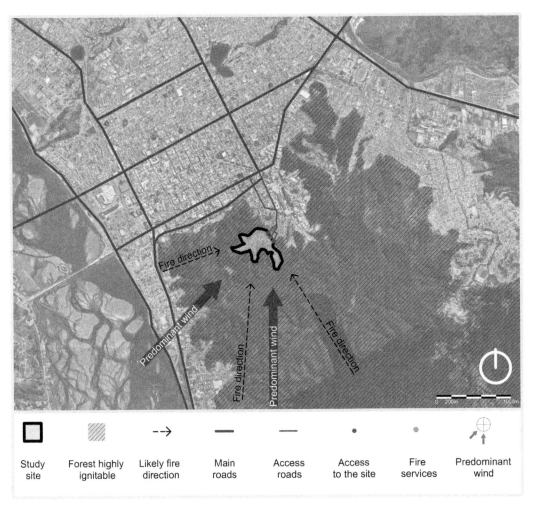

FIG. 2.1.1

Spatial mapping applying the principles to the wider site context.

Authors' own development based on March, A., Gonzalez-Mathiesen, C., Poblete, C., García, R., Wegertseder, P., 2017a. Manual de diseño ante incendios forestales. Nodo de Integración y Proyección de Empresas Especializadas en Arquitectura y Construcción Sustentable. U. Bio-Bío, COPEVAL, CORFO. ed. Concepcion, Chile.

and includes a section whereby it narrows to a single road. This means that traffic, accidents, or fallen power lines could slow or prevent emergency services from attending, and slow or prevent evacuation.

2.1.2.3.2 Identifying likely fire behavior

Spatial mapping was also developed to understand the possible ways that fires might progress and have impacts upon settlements (see Fig. 2.1.2). Elements considered are:

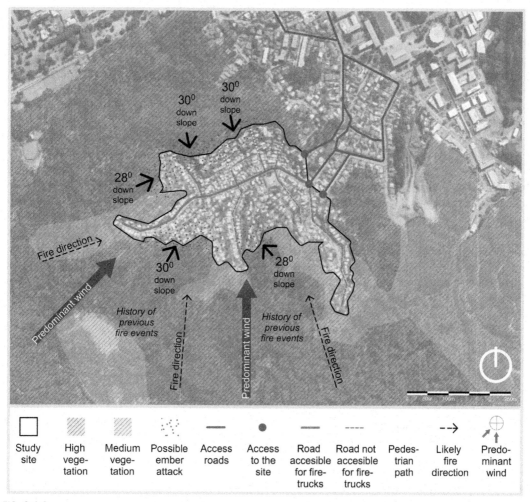

FIG. 2.1.2

Spatial mapping identifying likely fire behavior.

Authors' own development based on March, A., Gonzalez-Mathiesen, C., Poblete, C., García, R., Wegertseder, P., 2017a. Manual de diseño ante incendios forestales. Nodo de Integración y Proyección de Empresas Especializadas en Arquitectura y Construcción Sustentable. U. Bio-Bío, COPEVAL, CORFO. ed. Concepcion, Chile.

- Predominant wind directions and changes, such as sea and land breezes
- Potential long fire runs
- Possible ignition from embers from peripheral areas
- Challenging response scenarios in remote or "separate" areas, such as those with limited road access and redundancy

By considering these elements and the general principles of fire behavior and house destruction presented in the previous section, participants identified that the likely fire behavior affecting the site and the challenges associated with it. With the exception of a small percentage around the entrance roads at the north east, most of the site's perimeter is directly adjacent to and surrounded by vegetation. Surrounding land is held under various other ownerships that challenge fuel reduction. Further, any removal of larger trees could lead to land-slip risks. If house fires occurred at the north-east boundary, it could be isolated by road. Most of the land adjacent to the boundary is steep and difficult to access. Only the northern portion of the boundary is accessible by light vehicles, with the rest accessible only by a narrow pedestrian paths directly adjacent to vegetation. The site lies in the "bowl" of a small valley, meaning that, despite a possible speedy advance to the settlement, the rate of approach of the fire will probably slow once it reaches the peak of the surrounding hills. However, the potential for ember attack is very high, evidenced by previous fire history and the type of vegetation in the surrounding area. The site has only one main road accessible by fire trucks, the remaining roads are narrow, and these often narrow to pedestrian only access widths and steepness. Further, many of these paths are adjacent to significant vegetation. This further limits the active defense of these areas. Further, there are limited turning points for fire trucks. There are a limited number of fire hydrants in the settlement. Many of the structures and surrounds have significant vulnerabilities to direct flame and ember attack, and there is significant risk of house-to-house transmission once houses start burning. There are limited water storage and other facilities suitable for firefighting. However, there is potential to build upon demonstrated existing community self-organizing capacity. There are few open spaces available for refuge and assembly points if needed, and current capacity for official warnings is uncertain.

2.1.2.4 Key actions for house and site improvements

Based on the wider context and likely fire behavior analyses undertaken, the principles of fire behavior and house destruction were applied to defining key actions. Several key actions that can improve upon house and community survival are set out below, grouped around house and site improvement, settlement improvement, and community development. It is worthy of note that while the emphasis set out here is upon physical design, the integration of government, community, business, and other relevant parties in an integrated way is key to improved risk profiles in any community. Furthermore, resilience to wildfire requires that all vulnerabilities in the system are dealt with—in extreme fire events weak points in systems are often exposed and lead to catastrophic outcomes. The initial actions taken in the reported project were based upon providing community members with insights into the fundamental drivers of the threats they face. These insights were formed partly based on consultation with the community, and the report made several recommendations to improve the community's wildfire risk profiles.

The following key actions for improvement at the site and house scale were identified and discussed by the participants:

- *Use of nonflammable external materials (and firebreak materials between structures).* These include noncombustible wall materials such as brick, concrete, concrete fiber sheeting and iron; roofs made from corrugated iron or tiles; good quality of glass.

- *Protect from ember attack (into the dwelling, under the structure; into the roof and vents).* Embers must not get within or under the primary building skin. Vents, windows, gaps between slats or weatherboards and other openings must be sealed or protected with appropriately dimensioned gratings to reduce the risks of ember attack.
- *Design housing forms that reduce ember accumulation.* Simpler building forms are often better in wildfire-prone areas where ember collection points—such as under eaves, corners, inside gutters, under houses or the corners of windows and doors—are avoided.
- *Store water for response.* Storage of water in tanks—located on private or public land—can be an important component of stopping fire progression in the early stages of a fire.
- *Ensure access to basic services.* The provision and correct connection of high-quality utility services to properties—of electricity, gas, and water—can reduce the likelihood of fire ignition and may assist in response activities. Good quality, legal connections reduce the chances of accidental ignition and are better prepared for response actions in case of and emergency.
- *Manage vegetation near to buildings.* Vegetation management appropriate to the situation reduces the chances of direct flame, heat, and ember attack.
- *Maintenance.* Rubbish and other flammables or ember trapping materials are cleared from around buildings; vegetation is managed; avoid wood piles and other flammables close to buildings; clean leaves from gutters and roof systems; and repair damage.
- *In preparation for a fire's possible arrival, move vehicles and other flammables away from houses or other sensitive locations.* Vehicles, garden furniture, gas bottles for barbecues, floor mats, and other common household items may be flammable and should be removed and stored where they do not pose a threat or impeded the movement of people or personnel.

The following key actions for improvement at the settlement scale were identified and discussed by the participants:

- *Ensure access for emergency services.* Develop a plan to undertake an assessment of improvements—such as changes to the road network, surfacing of roads, vehicle turning circles and passing bays or assessment of the maximum grades—that would allow improved access for emergency service vehicles, in conjunction with consideration of the range of water points available.
- *Improve pedestrian access.* Improved access for pedestrians will mean that civilian and professional response will be improved, including preparation and active firefighting, and it can be a key aspect of escape and redundancy in catastrophic conditions.
- *Control surrounding vegetation.* The management of vegetation within and surrounding the settlement, on public and privately held land is a key aspect of the management of fire risks.
- *Develop low fuel perimeter areas (roads; community parks).* A key mechanism for protection of the community would be the development and enhancement of an improved perimeter around the settlement.
- *Provide water for response purposes in strategic locations (community water tanks using storm water; water plugs).* Additional water sources—from roof stormwater captures and from natural stream fed gravitationally—strategically located to provide the best access points in risky areas will considerably improve the ability to actively defend properties across a greater area of the settlement.

- *Provide signals.* Signage and a system of clear signals operated by trained individuals should be developed to assist the community with warning, response, and evacuation processes.
- *Provide adequate distance between structures.* Consider the separation of structures if the site size and arrangement allow, in combination with other elements.
- *Establish safe areas for refuge (adequately equipped).* A process can be undertaken, in conjunction with relevant stakeholders, to identify and develop suitable refuge and safe places.

In addition to the actions set out here, several longer-term community actions can continue to develop community resilience. While each of these activities will require further investigation, the participants identified and discussed that the following actions be taken:

- *Establish a steering committee in collaboration with key stakeholders.* It can provide a connection into community needs and understandings and can contribute to develop greater legitimacy and strength of local resilience.
- *Acquire equipment and community training for improvements and response.* On the basis of developing community members' capabilities, the provision of basic firefighting equipment, in parallel with appropriate training can significantly improve risk profiles.
- *Develop family emergency plans and kits.* If families are provided with a "starter kit" from which they can subsequently develop their own specific responses they will have considered the range of challenges they will face, and the realities of how they can deal with these, significantly improving their ability to understand and act upon the risks faced in a range of scenarios.
- *Establish school education programs and pamphlets.* School education programs provide opportunities to build fundamental skills and understandings in children. It is also typical that many adults learn from the activities and information that children bring home, allowing for a range of groups to be targeted.
- *Establish community and home improvement projects in consultation with key government agencies.* Develop a multiagency and community risk-based assessment program for improvement of homes and community facilities, including the development of a program for the allocation of funding and information to bring about improvements.
- *Fuel Management Community Plan.* This provides a range of opportunities for development of community spirited action and learning about fire risk management, while providing significant expertise in monitoring of vegetation and the identification of potential problem areas.
- *Alert System.* The development of an appropriate alert system, in conjunction with allocation of responsibilities and community learning can significantly reduce fire risks, parallel with signals an (mentioned above).
- *Evacuation Plan.* In parallel with development of community safe places and wider understanding of emergency signals, an evacuation plan can significantly reduce the risks faced when fires do occur.

2.1.3 Transferability to other settlements facing wildfire risks

This section explores the transferability of practical ways to promote resilience to wildfires for communities living in informal settlements exposed to wildfire risk presented. First, it elaborates on the procedural attributes of the risk-based planning practically promote resilience in other settlements facing wildfire risk. Second, it outlines some of the challenges in the context of informality.

2.1.3.1 **Risk-based spatial planning—Iterative processes to identify, analyze, and treat risk**

The guidelines set out offer a set of key actions to increase resilience to wildfires for informal contexts as well as for settlements that have been built with limited risk consideration or are constrained in the provision of adequate separation from fire transmission sources. They can contribute to improve the resilience to wildfires ex-post, acknowledging that physical mechanisms to increase resilience to wildfire are best established before settlements are in place. It is important to point out that:

- Aspects of risk and action should be approached in a comprehensive manner.
- Hazards and their treatments are contextually and site specific and need to deal with nonplanning complexities that impact on other matters as well as risks.
- Multiple time frames, return period, and temporal scales associated with risk treatment need to be dealt with, including short-term, medium-, and longer-range strategic actions.
- Links into financial, community, business, government, and other actions and programs are needed to bring about integrated and effective approaches.
- Shared benefits should be found where possible, such as energy-rating benefits associated with risk reduction in buildings for bushfire based on reducing ember attack.
- Actions consciously balanced, integrated, traded off, understood, and integrated with wider planning and other processes and practicalities.
- There are no "complete" failsafe mechanisms, and (in association),
- Untreated weak points in systems can lead to catastrophic outcomes.

The guidelines are expected to contribute to developing general understandings about design and planning strategies to mitigate wildfire risk.

The process of producing these guidelines using a risk-based approach contributes to the dissemination of wildfire risk management knowledge, and importantly the building of local capacities. Resilience requires learning from previous disaster's experiences and transferring disaster knowledge to different places. A risk-based planning approach allows for continuous improvement, learning from previous experiences and transferring knowledge, approaching wildfire risk management as a repetitive process to identify, analyze, and treat risk; learning and incorporating feedback. In this case, the process followed provided general trajectories for action based on the current knowledge that should be reviewed and adjusted by reiterating the process. Furthermore, it facilitated and legitimized the discussion of key concepts within the community, which is expected to be better prepared to act on wildfire risk in the future and to have the more technical knowledge to address risk in a comprehensive way.

The process undertaken could be replicated in other places to address context-specific issues. This case demonstrates that there are certain core scientific, methodological, and professional principles that can be transferred to other settings. Understandings of the locally particular aspects of hazards and resultant risks need to be developed locally, in parallel with solutions themselves that are relevant to local communities. Understanding the way fires interacts with housing provides a key starting point for improvements to the resilience of communities. Wildfires, in common with all potential natural hazards, occur in spatially specific ways. Spatial mapping provides a fundamental mechanism for integrating the wide range of evidence that is relevant to understanding and acting upon wildfire risks. Improvements to housing can protect assets, and allow families, communities and response agencies to prepare and defend communities more effectively. However, direct "solutions" should not be copied from one case to another, transferring them without modification. A risk-based procedural approach can contribute to developing trajectories for action that are sensitive to the culture, needs, and context in a case-specific manner.

A key principle of risk management expressed in the Sendai Framework (UNISDR, 2015) is the need to integrate land-use and development activities that have impacts upon risk profiles. This relates to the joining of governance processes and matters of collective concern with mechanisms of actual settlement location, management, growth, and change. Accordingly, it is inescapable that some forms of guiding and directing the location, design, materials, and characteristics of individual buildings and wider settlements' morphology is required. Further, there is a need to build upon and apply a range of knowledge sets that extend across a range of expertise and action areas.

2.1.3.2 Challenges to reducing wildfire risks in informal communities

The case demonstrates the range of challenges that need to be addressed to improve the wildfire risks faced by an informal community. A key tension for spatial planning is the need to allow, facilitate, require, and even actively bring about certain outcomes. Within contexts of informality, this implies that challenges generally extend along a spectrum between the general autonomy and self-organizing abilities of residents, and the need to draw on wider systems external to the settlement that usually operate "within" the formal systems of risk management and settlement management. Accordingly, the ideas of collective concerns and understandings focused around wildfire risk inherently challenge assumptions of inside and outside formality.

Developing ways to reduce the vulnerability of existing settlements built with little or no initial consideration of wildfire (or other) risk is a major challenge. As previously pointed out, physical mechanisms to increase resilience to wildfire are best established before settlements are in place. If the risks are taken into the account before settlements are established, development in areas where risks are considered unreasonable could be limited (Burby, 1998). However, historically, settlements have been established with limited consideration to wildfire hazard exposure. This is exacerbated in informal settlement contexts. It is common that informal settlements, such as the case of Agüita de la Perdiz, prioritize "good" locations in term of proximity to the city and the opportunities it offers, yet vacant land available—where people can settle informally—is usually exposed to greater risks. Furthermore, existing settlements often have limited capacity to provide the adequate separation from the fire source. This is evidenced in this case study; most of the site's perimeter is directly adjacent to and surrounded by vegetated land under various other ownerships, challenging fuel reduction and limiting the capacity to provide separation. Moreover, structures themselves are often built with limited consideration of fire risk. Even more so, housing structures in informal contexts are often built with limited resources using whatever materials available, with little planning, and often illegal connection (if any) to services. To illustrate the implications these might have, the implication of illegal connections to services is briefly described. Low-quality electricity connections to housing and to power lines, combined with extreme winds and dust, often cause sparking or fires. In combination, fallen power lines (particularly illegal connections without safety cut offs) during wildfire events may impede firefighting response activities. Similarly, poor-quality gas line connections or the location, storage, and connections of gas bottles may cause or worsen fires as they impact upon a settlement. Reticulated water services are often important to fire suppression, even if they may lose pressure during a fire event. Good quality, legal connections stand the greatest chance of continuity in fires, even if this may mean prioritizing water to the response agency's hydrants.

Understandings of the dynamism and incremental self-organized improvement capacities of communities allow exploring retrofitting measures intended to modify fuel levels or increase buildings' resistance, which can provide a practical way to promote resilience to wildfires. More than four decades ago, Turner (1972) coined the concept of "Housing as a Verb," arguing that housing is a process rather than a product, especially for low-income households. The context of lesser economic resources, mistrust of authority, tenure insecurity, location in often risky areas is frequently contrasted with considerable self-organization abilities, innovation, adaptability, strong networks, and a strong sense of personal responsibility. This is important to consider as it relates to the autonomy of use and self-determination that can drive the incremental development of the settlements, incremental improvement of the structures, and overall incremental process of becoming resilient by retrofitting the neighborhood, its surroundings, and the structures within it, reaching some level of formalization. In fact, this has been the case so far in Agüita de la Perdiz. It started as an illegal settlement in 1958 (Vega, 2015) and has undergone a gradual neighborhood formalization by receiving individual property titles (Municipalidad de Concepcion, 2016), street improvements, and provision of services. The community demonstrated self-organizing capacity coupled with security of land tenure and overall improvements to the area, indicating that the community is capable of gradually implementing retrofitting measures for developing resilience to wildfires. Importantly, the success of most of the subsequent actions will depend upon the community taking several steps that are at least partly self-driven in concert with support from government agencies and other external parties.

In addition, there is a need to integrate a range of actions across diverse groups that include individuals, multiple agencies, and land holders internal and external to the site itself. The integration of government, community, business, and other relevant parties in a cohesive way is key to improved risk profiles, and that resilience to wildfires requires that all vulnerabilities in the system are dealt with (Cutter et al., 2003). In fact, in extreme fire events, weak points in systems—physical or social—are often exposed to potential failures and lead to catastrophic outcomes. In this case, the presence of existing community leaders with experience in fighting fires, along with an apparently sympathetic local government and other agencies, provided a strong base for initial action. Connections with a range of other stakeholders and interest groups provide the possibility for wider links with a range of other urban actors and institutions. Social learning, testing of truths, collaboration, and other "joined up" activities can occur through collaborative approaches. The production of a clear engagement and progress plan for activities, funding, setting of responsibilities, and timelines would improve the likelihood of implementation and collaborations.

2.1.4 **Conclusion**

This chapter has investigated a case of spatial planning and design approaches that are directed to reducing wildfire risks. The case of Agüita de la Perdiz was chosen as a way of examining and improving on the risks in an informal settlement, using participatory action research. The case highlights the challenges faced in informal settlements where many aspects of the location, morphology, design and material decisions have been made incrementally over time without considering risks associated with wildfire. The production of guidelines to improve resilience to wildfires in consultation with the community and other key stakeholders demonstrates a number of key principles that can, with care, be generalized to other settlements facing the risks of wildfires.

This risk-based planning process can be replicated in other contexts, contributing to further developing local capacities and to disseminating wildfire risk management knowledge. As previously mentioned, evidence-based planning and risk-based planning have procedural similarities that allow them to work together. The three-stage protocol used in this research—(1) context, establishing circumstances and problem; (2) assessment, identifying the risk, objectives, and design alternatives; and (3) specification, selecting the actions to implement for treating risk—could be replicated in other places to address context-specific issues. This allows a risk-based approach for developing trajectories for action that are sensitive to the culture, needs, and context in a case-specific manner, contributing to achieve the priority one of the Sendai Framework for Action (UNISDR, 2015).

The core elements of integrated action and application of wider data-driven spatial understandings of risks are typically ignored in the processes of informal settlement establishment and incremental growth, based on self-organization and autonomy. Accordingly, the introduction of new approaches drawing on external governmental or corporate parties may take time and the acceptance of local leaders who can facilitate change from within. In many ways, the introduction of risk reduction actions is a blend of self-organization and acceptance of external inputs that may eventually lead to a level of formalization. Further, it introduces a level of acceptance that the development and application of action research based on spatial analysis of fire behavior and risk is advantageous to both individual and collective risks—suggesting that some level of autonomy may be beneficially lost if willingly given up. In parallel, the development of guidelines provides a way to disseminate the value of taking a wider view of overarching and individuals' action to manage risks subsequently applied to place specific risks.

References

Benton-Short, L., Short, J.R., 2013. Cities and Nature. [electronic resource]. Taylor and Francis, Hoboken.

Billing, P., Madengruber, U., 2005. Coping Capacity: towards overcoming the black hole. In: (ECHO), E. C. D.-G. F. H. A, (Ed.), World Conference on Disaster Reduction, Kobe, Japan.

Birkmann, J., 2006. Measuring vulnerability to promote disaster-resilient societies: conceptual frameworks and definitions. In: Birkmann, J. (Ed.), Measuring Vulnerability to Natural Hazards. [electronic resource]: Towards Disaster Resilient Societies. United Nations University, New York.

Blaikie, P., Cannon, T., Davis, I., Wisner, B., 2004. At Risk: Natural Hazards, People's Vulnerability and Disasters. Taylor and Francis, Hoboken.

Bogardi, J.J., 2006. Introduction. In: Birkmann, J. (Ed.), Measuring Vulnerability to Natural Hazards. [electronic resource]: Towards Disaster Resilient Societies. United Nations University, New York.

Bracken, I., 2014. Urban Planning Methods. [electronic resource]: Research and Policy Analysis. Taylor and Francis, Hoboken.

Burby, R.J., 1998. Natural hazards and land use: an introduction. In: Burby, R.J. (Ed.), Cooperating with Nature. Joseph Henry Press, Washington, DC.

CFA, 2012. Planning for Bushfire Victoria, Guidelines for Meeting Victoria's Planning Requirements. Country Fire Authority, Victoria, Australia.

CONAF, 2017. Incendios Forestales en Chile [Online]. Ministerio de Agricultura. Available: http://www.conaf.cl/ incendios-forestales/incendios-forestales-en-chile/. (Accessed 18 December 2017).

Coppola, D.P., 2015. Introduction to International Disaster Management. [electronic resource]. Elsevier Science, Burlington.

CRED, 2015. The human cost of weather related disasters; 2015. Centre for Research on Epidemiology of Disasters.

CRED, 2016. The International Disaster Database; 2016.

Cutter, S.L., Boruff, B.J., Shirley, W.L., 2003. Social vulnerability to environmental hazards. Soc. Sci. Q. 84 (2), 242.

Davis, I., Alexander, D.E., 2016. Recovery From Disaster. Abingdon, Oxon; New York, NY, Routledge, Taylor & Francis Group.

Davoudi, S., 2006. Evidence-based planning. disP Plan. Rev. 42, 14–24.

Gill, M., Stephens, S., 2009. Scientific and social challenges for the management of fire-prone wildland–urban interfaces. Environ. Res. Lett. 4 (3), 1–10.

Gonzalez-Mathiesen, C., March, A., 2018. Establishing design principles for wildfire resilient urban planning. Plan. Pract. Res. 33 (2), 97–119.

Hall, P., 2014. Cities of Tomorrow: An Intellectual History of Urban Planning and Design Since 1880, fourth ed. Wiley-Blackwell, West Sussex, England.

Hall, P., Tewdwr-Jones, M., 2010. Urban and Regional Planning. United Kingdom, Routledge, London.

Hewitt, K., 1998. Excluded perspectives in the social construction of disaster. In: Quarantelli, E.L. (Ed.), What Is a Disaster?: Perspectives on the Question. Routledge, New York, p. 1998.

Hillier, J., 2007. Stretching Beyond the Horizon: A Multiplanar Theory of Spatial Planning and Governance. Ashgate, Aldershot, Hants.

Intergovernmental Panel on Climate Change, 2012. Managing the risks of extreme events and disasters to advance climate change adaptation. In: Field, C.B., Barros, V., Stocker, T.F., Dahe, Q. (Eds.), Special report of the Intergovernmental Panel on Climate Change. Cambridge, New York, Melbourne, Madrid, Cape Town, Singapore, São Paulo, Delhi, Tokyo, Mexico City.

ISO 31000, 2018. Risk Management—Principles and Guidelines; 2018. Switzerland.

King, D., Gurtner, Y., Firdaus, A., Harwood, S., Cottrell, A., 2016. Land use planning for disaster risk reduction and climate change adaptation: operationalizing policy and legislation at local levels. Int. J. Disast. Resil. Built Environ. 7, 158–172.

Krizek, K., Forsyth, A., Slotterback, C.S., 2009. Is there a role for evidence-based practice in urban planning and policy? Plan. Theory Pract. 10, 459–478.

Manyena, S.B., 2006. The concept of resilience revisited. Disasters 30, 434–450.

March, A., Kornakova, M., 2017. Urban Planning for Disaster Recovery. Elsevier Ltd., Butterworth-Heinemann, Oxford.

March, A., Gonzalez-Mathiesen, C., Poblete, C., García, R., Wegertseder, P., 2017a. Manual de diseño ante incendios forestales. Nodo de Integración y Proyección de Empresas Especializadas en Arquitectura y Construcción Sustentable. U. Bio-Bío, Copeval, Corfo. ed. Concepcion, Chile.

March, A., Kornakova, M., Handmer, J., 2017b. Urban planning and recovery governance. In: Urban Planning for Disaster Recovery. Butterworth-Heinemann, Boston (Chapter 2).

Moritz, M.A., Syphard, A.D., Batllori, E., Bradstock, R.A., Gill, A.M., Handmer, J., Hessburg, P.F., McCaffrey, S., Leonard, J., Odion, D.C., Schoennagel, T., 2014. Learning to coexist with wildfire. Nature 515, 58–66.

Municipalidad de Concepcion, 2016. Vecinos de Agüita de la Perdiz recibieron títulos de dominio [Online]. Concepcion. Available: https://www.concepcion.cl/noticia/vecinos-de-aguita-de-la-perdiz-reciben-titulos-de-dominio/. (Accessed 21 June 2018).

Ramsay, C., Rudolph, L., 2003. Landscape and Building Design for Bushfire Areas. [electronic resource]. CSIRO Publishing, Melbourne.

Sharifi, A., Yamagata, Y., 2018. Resilience-oriented urban planning. In: Yamagata, Y., Sharifi, A. (Eds.), Resilience-Oriented Urban Planning: Theoretical and Empirical Insights. Springer, Cham, Switzerland.

Turner, J.F., 1972. Housing as a verb. In: Turner, J.F., Fichter, R. (Eds.), Freedom to Build, Dweller Control of the Housing Process. Collier Macmillan, New York, pp. 148–175.

UNDP, 2015. Transforming Our World: The 2030 Agenda for Sustainable Development; 2015.

UNISDR, 2005. Hyogo Framework for Action. The United Nations Office for Disaster Risk Reduction, Geneva, Switzerland.

UNISDR, 2009. 2009 UNISDR Terminology on Disaster Risk Reduction. United Nations, Geneva, Switzerland.

UNISDR, 2015. Sendai Framework for Disaster Risk Reduction. The United Nations Office for Disaster Risk Reduction, Geneva, Switzerland.

Vega, M.E., 2015. Los 53 años de la emblemática y combativa Agüita de la Perdiz en las décimas de El Canela. Tribunal del Biobio; 2015.

WBGU, 2004. World in Transition. Fighting Poverty through Environmental Policy, London.

Weiss, C.H., 2001. What kind of evidence in evidence-based policy. In: Keynote Paper Presented at the Third International, Interdisciplinatry Evidence-Based Policies and Indicator Systems Conference. University of Durham, United Kingdom. Verfügbar unter: http://www.cemcentre.org/Documents/CEM%20Extra/EBE/EBE2. 284–291.

Whyte, W.F., Greenwood, D.J., Lazes, P., 1991. Participatory action research. In: Whyte, W.F. (Ed.), Participatory Action Research. Sage Publications, Newbury Park, CA.

Yamagata, Y., Sharifi, A., 2018. Resilience-Oriented Urban Planning: Theoretical and Empirical Insights. Springer, Cham, Switzerland.

Analyzing city-scale resilience using a novel systems approach

2.2

Kerri McClymont, Melissa Bedinger, Lindsay C. Beevers, Guy Walker, and David Morrison

School of Energy, Geoscience, Infrastructure and Society, Heriot-Watt University, Edinburgh, United Kingdom

CHAPTER OUTLINE

Understanding Disaster Risk. https://doi.org/10.1016/B978-0-12-819047-0.00011-1

179

2.2.1 Introduction

2.2.1.1 Understanding disaster risk and resilience

The Sendai Framework for Disaster Risk Reduction 2015–2030 is a global agreement to reduce disaster risk, and strengthen the social and economic resilience to ease the negative effect of climate change and man-made hazards (European Commission, 2019). Under Priority Action 1, "disaster risk management should be based on an understanding of disaster risk in all its dimensions of vulnerability, capacity, exposure of persons and assets, hazard characteristics and the environment. Such knowledge can be used for risk assessment, prevention, mitigation, preparedness and response to disasters" (UNDRR, 2019). This chapter contributes to Priority Action 1 by taking a deeper look at conceptualizations of resilience within the framework of hazard, exposure and risk. The overall aim of this study is to utilize a systems tool (called an abstraction hierarchy) to capture exposure of a city system to flooding by identifying interdependencies between technical and social systems, and how the impacts of such an event propagate through the more abstract aspects of that system. It will illustrate how the abstraction hierarchy can be used to explore all dimensions of exposure—tangible exposure of assets, intangible exposure of cities, and hazard characteristics—in order to capture risk information to develop and implement disaster risk reduction policies. The abstraction hierarchy can be used for multiple hazards, but this chapter will explore its application in a flood hazard context.

Disasters often follow natural hazards, and their severity depends on how much impact a hazard has on society and environment (UNDRR, 2019). The Sendai Framework connects resilience with disaster risk reduction by framing resilience as a global policy goal to reduce vulnerability and minimize risk (Handayani et al., 2019). Risk is a function of hazard, exposure, and vulnerability, all of which are interrelated and contribute to resilience. Exposure can be separated into two categories: tangible and intangible. Tangible exposure is concerned with the physical exposure of the built environment to a particular hazard. Intangible exposure is concerned with the impact of a hazard on the wider system beyond the spatial distribution of exposure as a result of complex interactions in the built environment. The Sendai Framework urges for more integration of infrastructure effects in assessments which identify areas of highest risk and most vulnerable assets or subjects (Fekete et al., 2017). In practice, this requires plans which focus on ways to reduce the impact of a hazard by accounting for the complex interactions between social and physical infrastructure and hazard environments, which has been identified as a significant research gap (Martin, 2015). An improved understanding of system interdependencies can therefore reduce tangible and intangible exposure to hazards. Exposure is an important aspect of vulnerability. Cutter et al. (2009, p. 2) clearly articulate the distinction between vulnerability and social vulnerability definitions and how they relate to exposure: "Vulnerability is the susceptibility of a given population, system, or place to harm from exposure to the hazard and directly affects the ability to prepare for, respond to, and recover from hazards and disasters. Social vulnerability explicitly focuses on those demographic and socioeconomic factors that increase or attenuate the impacts of hazard events on local populations." Social vulnerability therefore heightens the exposure to harm from hazards. Resilience aims to reduce the risk of disasters by focusing on adaptation and creative transformation as opposed to resisting change. While vulnerability tends to take an actor-oriented approach, resilience tends to prefer a systemic approach, which has advanced our understanding of system dynamics and interconnections, socioecological relations, and feedback loops (Miller et al., 2010). Based on the literature, there is an urgent need to operationalize resilience to support more practical operations (Handayani et al., 2019).

Resilience is therefore an important aspect of understanding disaster risk. Resilience as a concept accepts that shocks may occur, which are not necessarily preventable, and therefore focuses on the ability to enhance city performance during a shock event and transform into a better future state (ARUP, 2015). Central to resilience thinking is an ability to confront complexity by considering cross-scale interactions and interdependencies (Folke, 2006). In doing so, processes and structures can be understood to manage the interplay between gradual and rapid change, not simply resisting to change (Folke, 2006). In other words, resilience requires policies, which mitigate the initial impact of a fast shock event coupled with longer, transformative, systemic change. This transformative aspect of resilience is key and should be reflected in disaster risk reduction policies, as attempting to maintain the status quo risks reinforcing existing inequalities.

Resilience is gathering momentum across a range of different disciplines and government reports (see, e.g., ARUP, 2015). This is also evident in Flood Risk Management (FRM), where the Chief Executive of the Environment Agency in the United Kingdom has called for a new approach, which is "less concrete, more resilience" (UKPOL, 2018). Despite decades of technical approaches to environmental management, problems still persist in Flood Management at least partially because we continue to ignore properties of complexity in our approaches. This is exemplified by recent work in FRM for the United Kingdom carried out by Sayers et al. (2017). Their report illustrates that areas in the United Kingdom identified to be at "flood disadvantage"—when flood exposure and social vulnerability coincide—are likely to experience intensifying conditions of flood disadvantage in the future (i.e., the cycle of flood disadvantage is reinforced and magnified as climate change evolves). The future projections account for climate change and assume current adaptation strategies are continued, clearly signifying that a new approach to FRM is required. This chapter coins the term "stubborn disadvantage" as an extension of the term "flood disadvantage", which builds on the idea of intersecting social disadvantages (see O'Hare and White, 2018) and the concept of structural violence (see Whittle et al., 2015), to mean *when the socioeconomic and institutional structure systematically fosters vulnerability so that flood disadvantage is reinforced over time*. Developing more transformative disaster risk reduction strategies that can overcome this stubborn disadvantage requires acknowledging how communities and city systems interact in the context of flood risk.

However, the polysemic nature of resilience has sparked debate in the literature leading to ambiguity in terms of conceptualizing, measuring, and applying the concept across different disciplines. In order to help inform the 100 Resilient Cities framework, Martin-Breen and Anderies (2011) reviewed over 50 years' worth of resilience research to produce a resilience spectrum of increasing complexity, based on three interdisciplinary frameworks: *Engineering Resilience, Systems Resilience, and Complex Adaptive Systems*. It was determined that Engineering Resilience is considered as an ability to *prepare* for any disturbances pre-event and to *resist* the impacts. Systems Resilience is considered as the ability to *cope* with the event effects and *maintain* functionality throughout the disturbance. Complex Adaptive Systems are characterized by the ability to then *learn* and *transform* postdisturbance, to increase future resilience. A systematic review of the FRM literature was undertaken by McClymont et al. (2019) in order to understand where flood resilience is currently perceived along this spectrum.

2.2.1.2 **Resilience in FRM: Current understanding**

By considering flood resilience along this spectrum, analysis of current research shows that our application of resilience in FRM is not as integrated as we might expect (McClymont et al., 2019). Results showed a meager 15% of reviewed papers consider resilience aspects that account for all three frameworks. The majority of articles fell into either the Engineering or Systems frameworks, indicating that they do not consider resilience to be an iterative, adaptive process as required under the Complex Adaptive Systems framework. Recent work (e.g., Restemeyer et al., 2015; Nelson et al., 2019) argues that these adaptive processes are critical to the study of resilience. Thus resilience should be considered as a tripartite concept, requiring theory and approaches from all three frameworks (McClymont et al., 2019). Defining resilience is a source of much contention in the literature (Fisher, 2015). However, its polysemy is in fact useful in conceptualizing the different aspects of resilience. An overemphasis on a universal definition of resilience fails to acknowledge the fluidity of the concept and its associated complexity (McClymont et al., 2019). The authors therefore recommend acknowledging the different aspects of resilience that align with each framework, which will vary in importance depending on context. Acknowledging resilience as a tripartite concept compels us to consider the interactions between social and technical systems across multiple spatial and temporal scales. The review found a lack of methodologies in FRM, which acknowledge these interactions.

2.2.1.3 Resilience in cities: Complex adaptive systems

In order to build more resilient cities, they need to be treated as having characteristics of complex adaptive systems, which include emergent behavior; self-organization; nonlinearity; adaptive behavior; and hierarchical ordering (Patorniti et al., 2018). Consequently, outputs of a complex system such as a city are not easily obtained, based on an understanding of system components in isolation (Beevers et al., 2016). Despite this, cities have usually been defined from a top-down perspective, where sufficient detail of individual system components and their bottom-up interactions (e.g., across natural and human systems, connecting global-scale dynamics to local realities, capturing cascading effects though the system, modeling not just impacts but also feeding back and testing interventions) are not fully detailed (Patorniti et al., 2018; Bedinger et al., 2019). Moreover, such interactions are a major complexity gap identified in a review of methods to study hydrohazards (Bedinger et al., 2019).

2.2.1.4 The need for complexity-smart approaches to flood resilience

This chapter will address the gaps identified by applying the abstraction hierarchy to a city system to understand the interdependencies between infrastructure, flooding, and human vulnerability, which determine a city's resilience, and address stubborn disadvantage. This will directly achieve the stated aim of the chapter and demonstrate the applicability of the method to advance the Sendai Framework goals. The first step of operationalizing the resilience concept will be presented by focusing on tangible and *intangible* exposure to flooding. The purpose of this study is to analyze how flood exposure propagates through wider city functioning by identifying interdependencies between technical and social systems at varying spatial scales. Two case studies with different population densities and urban layouts will be used to explore the role of redundancy on city exposure. In doing so, the following research questions will be addressed:

1. Which nodes are most interdependent to city functioning during normal conditions, compared to when a city is subject to flooding?
2. How is a city with a high population density impacted during a flood compared to a city with low population density?

2.2.2 **Method**

2.2.2.1 **The abstraction hierarchy**

The systems approach used in this chapter utilizes the "abstraction hierarchy" (AH). Originally developed by Rasmussen (1986), an AH acknowledges adaptive processes by modeling how a system *could* function, through five layers of abstraction. Abstraction in this context means the qualities of objects, which are considered separate from the objects themselves. The theoretical reasoning behind the AH was that, in order to support adaptive problem solving, designs must be specified in terms of the constraints that shape behaviors that are possible in any situation (Naikar, 2017). Rasmussen (1986) sought to understand why major industrial accidents occurred despite high levels of hardware reliability. Analysis revealed that human error was the result of around 75% of these accidents, and these errors arose when workers were faced within unfamiliar events, which could have been prevented if the actual state of the system had been known to them (Naikar, 2017). This work signified the importance of providing workers with the information they need about the system (including purposive properties as well as physical properties) in order to adapt their behavior to the demands of a wide range of situations (Naikar, 2017). The underlying theory behind the AH resonates with the resilience concept developed in the systematic review. If resilience is solely perceived as being hazard-specific and deterministic, where designs are implemented to reduce a specific risk and restore conditions to a specific conditions, then problems will continue to occur. Tripartite resilience acknowledges that shocks cannot be fully anticipated; therefore, cities should be treated as complex sociotechnical systems where the constraints of a city at multiple levels of abstraction can be identified for a more complete understanding of city functioning.

Bedinger et al. (2020) presented one of the first applications of an AH to an entire city. This study applies the generic version to specific locations. Individual buildings, households, and other infrastructure objects form the bottom layer ("Physical Objects"). The next level ("Object-Related Processes") explains what each of the objects can physically do. The third level ("Generalized Functions") shows what tasks can be accomplished by using these physical processes. The fourth level ("Values & Priority Measures") represents criteria by which we can determine if the city is fulfilling its "Functional Purposes"—the top level of the hierarchy that identifies why the city exists. In this work, Values & Priority Measures are based on the 12 goals of the 100 Resilient Cities project (Arup, 2015).

Nodes at each level are connected through means-ends links in order to capture functionality—with the functions being increasingly abstracted with each layer. These links are as important as the entities themselves, as they represent "the 'means' that a system can use in order to achieve defined 'ends' " (Beevers et al., 2016, pp. 203). Fig. 2.2.1 illustrates a simplified excerpt of the full AH used in this paper, focusing only on the Object-Related Process of *enabling transport*. This example shows that Physical Objects (e.g., *rail stations*) enable the Object-Related Processes (e.g., *enable transport*). The Object-Related Processes afford a number of tasks or Generalized Functions (e.g., *support distribution of goods*), which contribute to Values & Priority Measures (e.g., *create & maintain a sustainable economy*) and fulfill Functional Purposes (e.g., *economic opportunity*). The AH is bidirectional: working from the bottom of the hierarchy upwards answers the question of why something exists; traveling from the top downwards answers the question of how something can be achieved. For example, enabling transport exists to provide emergency services, and it is achieved by the use of rail stations and roads. The AH is essentially a systems map, where system parts are interconnected between levels

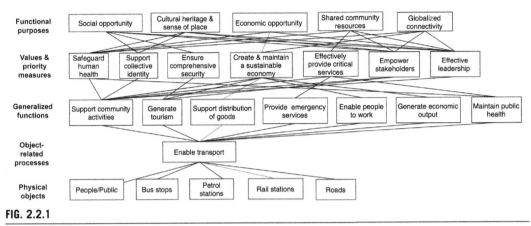

FIG. 2.2.1

Simplified excerpt of a city abstraction hierarchy.

(not within), which are interrelated through their functionality, rather than their physical connectivity. A validation study of the AH is ongoing, which utilizes the knowledge of subject-matter experts in order to agree consensus on the inclusion of nodes at each level of the AH for city functioning. The AH used in this study are based on the pilot results of the validation study. Appendix includes a list of all nodes included in the AH for this study.

2.2.2.2 Case study selection

What is unique about the AH is that it not only captures physical exposure to flooding, it also models how the impact of a flooded physical object propagates through the wider system and therefore captures *intangible* exposure. Since this is the first-time application of a novel methodology, smaller case studies were selected with the aim of scaling up the methodology to city-scale application in the future. Two real-world case study locations were selected: the Hulme council ward of Manchester, United Kingdom, and Natchez City in Mississippi, United States. These locations were selected because they have a similar population of around 16,000, yet have very different population densities of 7671 people/km^2 and 356 people/km^2, respectively. Consequently, they also have very different urban layouts (Fig. 2.2.2). It is expected that a higher population density will increase system exposure to flooding, with a greater number of system components impacted as a result. Therefore this work explores this hypothesis using the AH, and whether it is sensitive to differences in urban layouts.

In order to test this, a baseline AH was modeled for each case study based on the pilot results of the validation study (Appendix), which accounted for all physical objects within a city boundary. These objects were then categorized, depending on their functions to form the bottom layer of the AH. The pilot AH was adapted at the Physical Objects layer as not all nodes were within each of the case studies boundary, and were therefore removed to reflect local context. All other layers of the AH remained the same for each case study. A "flooded" AH was also modeled to determine how a city functions during a flood event. Environment Agency (England) and FEMA (United States) flood maps for a 1:100 year flood event were used to identify physical objects at risk of flooding in both locations. The number of buildings "knocked out" by the flood was accounted for by normalizing the baseline and reducing the

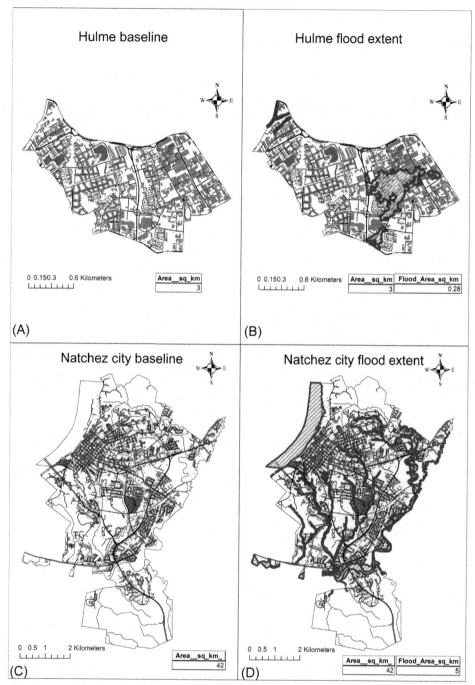

FIG. 2.2.2

Comparison of baseline and flood AH boundary for Hulme and Natchez City. Part (A) illustrates the system components within Hulme boundary to form the bottom layer of the AH. Part (B) illustrates the system components within Hulme identified in the flood risk zone which have reduced functionality to form the bottom layer of the flooded AH. Part (C) illustrates the system components identified within the Natchez City boundary to form the bottom layer of the AH. Part (D) illustrates the system components within Natchez City identified in the flood risk zone which have a reduced functionality to form the bottom layer of the flooded AH. Hulme map contains data from Ordnance Survey (2018) and Environment Agency (2018). Natchez City map contains data from Census Bureau (2019) and Microsoft (2019).

weighting of links connected to that particular physical object. For example, if three churches were flooded out of a possible nine, the weighting changed from 1 in the baseline to 0.66 in the flooded model. Again, only the Physical Object layer was altered to introduce the flood hazard.

2.2.2.3 Network analysis

As AHs represent a type of network, to explore network resilience this chapter uses the network metric-weighted betweenness centrality (BC). This determines which nodes were most central to city functioning during baseline and flooded conditions. The following equation was used to measure the weighted BC:

$$C_B^{wa} = \frac{g_{jk}^{wa}(i)}{g_{jk}^{wa}}$$

where g_{jk} is the number of shortest paths between two nodes, $g_{jk}(i)$ is the number of those paths, which go through node i and wa is the tuning parameter, which equals 0 for an unweighted network (baseline conditions) and 0.5 for a weighted network where tie number and tie weight are given equal importance (flood conditions) (Opsahl et al., 2010). The higher a node's BC, the higher its interdependence, meaning that if that node were to be removed, there would be a greater risk of eliminating the indirect connections among other nodes and disconnecting sections of the network. Moreover, a node with a higher BC than the average would play an important role in that network in a disturbance situation, if alternatives were not provided (Kermanshan and Derrible, 2017).

It is not advisable to directly compare raw BC values from two different networks as BC is sensitive to network size; however, relative change can be compared between two networks for both pre- and postflooding conditions. Due to the structure of the AH, each node has been imposed at a certain level of the hierarchy, so they can only connect to nodes in the tier above or below, and they are not connected to other nodes in their level. Therefore results from each level have been treated independently. Results are presented from four networks, each modeling a different scenario: Hulme (Baseline), Hulme (Flooded), Natchez City (Baseline), and Natchez City (Flooded). From herein, these scenarios will be referred to as: HB, HF, NCB, and NCF, respectively.

2.2.3 Results

2.2.3.1 Changes in node rank between the four scenarios

This section discusses BC results in terms of each node's rank within its own level of the AH. If a node increases in BC from baseline to flood conditions, the node becomes more interdependent (and critical to overall city functionality) during flood conditions. If a node decreases in BC from baseline to flood conditions, the node becomes less interdependent during flood conditions. If a node with a high interdependence is no longer able to function, the rest of the system would be more significantly affected (compared to the removal of a node with low interdependence). Essentially, BC values tell us how system functioning has changed after a hazard.

It is interesting to note that the five most interdependent nodes for Natchez City remain constant between NCB and NCF. However, some variation can be identified between: NCB and HB; NCF and

HF; and HB and HF. For the Functional Purpose, Values and Priority Measures, and Generalized Functions levels, the five most interdependent nodes remain the same during all four scenarios. However, at the Values and Priority Measures and Generalized Functions level we start to see some variation in the rankings between Natchez City and Hulme. For example, "Diverse livelihoods and employment" is the third most interdependent node for NCB and NCF, while it is ranked as fourth for HB and HF. This suggests that the AH is sensitive to differences in urban layout under baseline conditions.

Variation in node ranking becomes more profound at the Object-Related Processes layer between Natchez City and Hulme. In particular, between: NCB and HB; NCF and HF; and HB and HF. For example, "Provide help/support for vulnerable people" has the fifth highest BC for object-related processes for both NCB and NCF. Instead, "Transmit and receive communications" appears fifth for HB, which increases in interdependence for HF, switching to fourth most interdependent. This suggests that when physical objects become flooded in Hulme, other functions that are connected to "Transmit and receive communications" become more central to compensate for a loss in functionality elsewhere, and act as an intermediary between nodes in the network in order to achieve a city's functional purpose. While node rankings between NCB and NCF, and between NCB and HB, remain the same at the Physical Objects layer, there is variation between HB and HF. In HF, "Further Education," "Schools," and "Community Centers" drop to 31st, 28th, and 43rd in terms of interdependence ranking, respectively, and have been replaced by "Police departments," "Hospitals," and "Parks."

Although the most interdependent nodes for each layer have been discussed, the distribution in BC values for each tier of the AH is not even. For example, "Shared community resources" has a much higher BC value compared with the other Functional Purposes, as does "Minimal human vulnerability" compared with the other Values and Priority Measures. These nodes remain dominant for both baseline and flood conditions in each city and are therefore important to be aware of—as they pose a greater risk of disconnecting sections of the network if they were to be removed/reduced by a hazard event. However, when we look closer at the changes in interdependence during flood conditions, we can see there is variation among the middle-range values, especially for lower levels of the AH.

2.2.3.2 Changes in node centrality values across the four scenarios

Although some node rankings may remain consistent across scenarios, BC values change. Fig. 2.2.3 illustrates the variability in change in BC between baseline and flood conditions for both cities, which becomes much less profound at higher levels of the hierarchy. While BC changed between NCB and NCF, and HB to HF, there was much greater variation between the latter (Figs. 2.2.3 and 2.2.4. This is reflected in the change in node rankings for HF compared with HB for the Physical Objects and Object-Related Processes layer.

Moreover, Fig. 2.2.4 illustrates the variability in BC from baseline to flood in more detail. While Fig. 2.2.3 illustrates the percentage change in BC from baseline to flood on the same scale for all levels of the hierarchy, Fig. 2.2.4 has variable scales for each layer to illustrate the percentage change more clearly at each layer along with any outliers. For all layers, Natchez City has a much smaller interquartile range compared to Hulme, but has more outliers indicating that the impact of a flood is more concentrated on a fewer number of nodes. The most profound difference between the two cities is in the Physical Object level, where the range of percentage change from HB to HF illustrates that Hulme is more impacted after a flood event. This could be attributed to Hulme having a higher population density, as there were more physical objects, which were at risk of flooding compared to Natchez City. As a

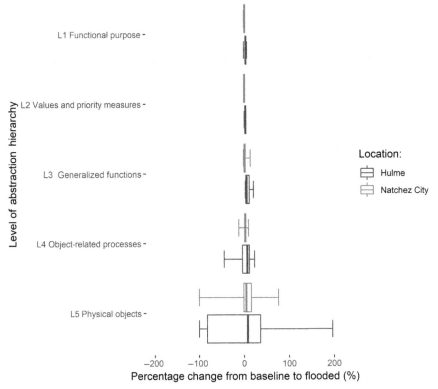

FIG. 2.2.3

Quantile box plot of the percentage in BC from baseline to flood for each case study. The whiskers represent the maximum and minimum percentage change values.

result of Natchez City covering a much more extensive area than Hulme, not only were there less physical objects in the flood extent, there were more redundancies within the city system. For example, in Hulme there were 28 restaurants, compared to 66 for Natchez City. Consequently, the physical objects identified within the flood zone were under 10% of the physical objects identified in the baseline for Natchez City, thus explaining the lower variation compared to Hulme. Despite these redundancies, the impact of the flooded objects still propagated through the wider system, as illustrated in Fig. 2.2.4.

When comparing differences between baseline and flood conditions, more nuanced changes in interdependence can be capitalized on. Table 2.2.1 illustrates the nodes with the highest BC change from baseline to flood for the two case studies. For the Object-related Processes layer, the most profound change from HB to HF was "Provide food," which decreased in BC by 45%. Conversely, this had a positive increase in BC by 5% from NCB to NCF. Variation in the Generalized Functions layer reflect this, as "Ensure provision of food" had the greatest increase in BC—and hence interdependence—for Hulme, whereas it had a very small increase from NCB to NCF. This suggests that in Hulme, the function of "Ensure provision of food" becomes more central to the system after a flood to compensate for

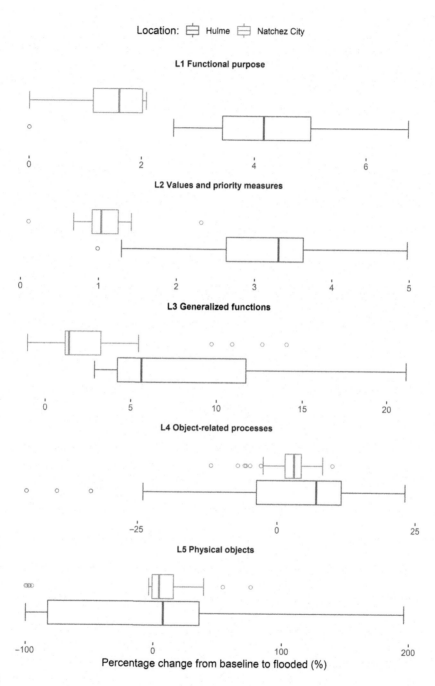

FIG. 2.2.4

Box plot illustrating the range of percentage change in BC from baseline to flood conditions for both case studies. Variable scales for each layer of the AH are used to illustrate the range of change in more detail.

Table 2.2.1 Nodes with greatest percentage increase and decrease in betweenness centrality from baseline to flood conditions for Natchez City and Hulme across all levels of the abstraction hierarchy (those marked with an asterisk denote an outlier from Fig. 2.2.4).

	Natchez City		Hulme	
	Greatest % increase		**Greatest % increase**	
Functional purpose	Freedom of movement and expression		Social opportunity	
	Social opportunity		Freedom of movement and expression	
	Economic opportunity		Urban ecosystem services	
Values and priority measures	Reliable communications and mobility*		Minimal human vulnerability	
	Sustainable economy		Comprehensive security and rule of law	
	Diverse livelihoods and employment		Effective leadership and management	
Generalized functions	Generate economic output*		Ensure provision of food	
	Generate tourism*		Generate tourism	
	Ensure provision of clothing*		Perform/support ceremonies …	
	Greatest % increase	**Greatest % decrease**	**Greatest % increase**	**Greatest % decrease**
Object-related processes	Make payments/wire money*	Produce goods*	Fertilize soil	Provide food*
	Provide drainage	Provide/maintain clothing*	Provide nutrient cycles	Provide place of worship*
	Facilitate and increase employment	Support health/fitness*	Provide water purification	Provide drink*
Physical objects	Shops (home and garden)*	Roads*	Food Banks	Roads
	Pet goods and services*	Railway lines*	Hotels (with restaurants)	Training providers
	Cemeteries	Money transfer services*	Sports grounds and facilities	Electricity providers

the reduced functionality of "Provide food." Moreover, for all four scenarios, "Minimal human vulnerability" was the most interdependent node for the Values and Priority Measures layer. This also had the greatest increase in BC from HB to HF. However, for NCB to NCF, Table 2.2.1 illustrates the greatest increase in interdependence was "Reliable communications and mobility,"—which was not identified as "high" BC. For the Functional Purpose layer, "Economic opportunity" is not within the most interdependent nodes, but is listed within Table 2.2.1 indicating it has a greater interdependence during flood conditions. Moreover, "Shared community resources" has the highest interdependence across all four scenarios, but does not appear among the nodes with the highest *increase* in interdependence during both NCF and HF (Table 2.2.1). These results highlight that *changes* to BC in different scenarios are as important as *high* BC in a single scenario, to identify critical system interdependencies, and prevent cascading impacts on overall city functioning.

2.2.4 **Discussion**

The AH is a tool that models how a city could function during a shock event such as flooding. It aims to capture critical interdependencies between social and technical assets and systems through interlinking spatial scales: from individual physical objects through to policy-level criteria, which evaluates whether a city meets its functional purposes. It allows city resilience planning to flexibly shift perspective from the physical properties through to the purposive properties of a city (Naikar, 2017). The overall aim of this study was to utilize the AH to capture exposure of a city system to flooding by identifying interdependencies between technical and social systems, and how the impacts of such an event propagate through the more abstract aspects of that system. The results presented in this chapter showed the most interdependent nodes during routine, everyday conditions, as well as during a period of disturbance, helping stakeholders to understand how the city functions under both conditions. Removal of these interdependent nodes is most likely to disconnect other parts of the system, reducing a city's ability to maintain function during a flood and hence its resilience. Therefore, the AH contributes to Priority Action 1 by allowing city planners to reduce the exposure of key nodes, which are crucial to city functioning through an improved understanding of interdependences.

When exploring baseline interdependence, the nodes with the highest interdependence were the same for the two case studies—although their relative ranking varied slightly. This confirms that the AH is able to identify subtle differences in baseline conditions for different urban layouts. However, the differences become more profound when considering the change in interdependence in different scenarios, allowing us to explore the role of density and redundancy on system exposure. It was expected that a higher population density would increase system exposure to flooding, with a greater number of system components impacted as a result. The AH was able to confirm this hypothesis, as the results showed Hulme had greater variation in BC from baseline to flood compared to Natchez City. Although the case studies have a similar population size, Natchez City covers a much greater area resulting in a lower population density and smaller flood extent. Therefore, Natchez City had more redundancies in the Physical Objects layer compared to Hulme. Greater redundancy in a system means that there are multiple ways to fulfill a given function. Redundancy is an important aspect of resilience, as having functionally similar components ensures that the system does not fail if one component fails. Natchez City therefore had alternative options in terms of functionality in the event of a flood. However, the role of density on resilience is complex. For example, on the one hand, the concentration of technical and social systems can make a highly dense city more exposed to a hazard such as flooding. On the other hand, a high population density can increase resilience through increased social interaction. The potential for innovation can be substantially higher in contexts where urban social networks have diverse properties, compared to a city with the same population but lower density (Martin-Breen and Anderies, 2011). Moreover, the source of a hazard can also determine the role of density on system resilience. For example, Phoenix's low population density has made it more vulnerable to droughts (Martin-Breen and Anderies, 2011). An area of future research is therefore to couple the outputs of the AH with an understanding of local context, in order to explore the interplay between density and redundancy for city resilience.

Furthermore, although Hulme demonstrated a greater overall change in interdependence between baseline and flood conditions, Natchez City had more outliers in the data, indicating that the impact of a flood event was more concentrated on a fewer number of nodes. A high degree of interdependence can reduce resilience, since a disruption in one sector cascades into impacts on other sectors (Cutter et al., 2008). Therefore, changes to BC in different scenarios are just as important as BC in a single scenario, to identify critical interdependencies and prevent cascading impacts on overall city functioning.

It is through this analysis of interdependence between technical and social systems that the AH can be used as a tool to operationalize tripartite resilience. By understanding system structure in this way, the method develops key aspects associated with Systems Resilience, such as the *ability to maintain functionality and cope with* a flood event. A Complex Adaptive Systems approach accounts for interdependencies and feedbacks between spatial scales, which allows for new dynamics to emerge and elicits the transformational aspect of resilience to become operational (McClymont et al., 2019). The AH can capture this emergent behavior by supporting adaptive problem solving through system design, which is specified in terms of constraints that shape behaviors, which are possible in any situation (Naikar, 2017). It models how a system *could* function under a range of flood scenarios, leading to an increased understanding of how the city as a whole will function during the impact, thereby accelerating the learning cycle for coping with hazards and building the *capacity to learn and transform* (Complex Adaptive Systems Resilience). This method can also test engineering strategies at a lower spatial scale by utilizing the change in BC as a rough proxy for risk. Input data can be adjusted to evaluate the impact of different types of hazards (e.g., cyber-attacks), as well as potential interventions designed to protect certain objects from those hazards. The propagation of the effect of each design solution can be traced throughout the city system from lower to higher levels of abstraction (Naikar, 2017). This develops the *capacity to withstand* aspect of resilience (Engineering Resilience). This multihazard aspect of the methodology is a key contribution to Priority Action 1, which recognizes the need for "long-term, multi-hazard and solution driven research in disaster risk reduction" (UNDRR, 2019).

However, there is ample opportunity for future work. Without answering the question of "resilience of what, but more importantly, resilience for whom," the resilience concept risks maintaining the status quo, leading to inequitable outcomes (Cutter, 2016, p. 112). In order to tackle stubborn disadvantage and truly enable transformation, future work should be done to link AH findings to community outcomes through Amaryta Sen's capability approach. This will add to understanding exposure in all its dimensions—in communities, assets, cities, and hazards—as stated in Priority Action 1. In order to better reflect local context, future work should also adjust the link weighting of Values and Priority Measures using the 100 Resilient Cities indicators. This will explore the role of latent conditions thought to be fundamental to achieving resilience (Gunderson and Holling, 2002; Cheshire, 2015; Martin-Breen and Anderies, 2011). These slower variables of resilience occur during baseline conditions over longer time scales, where their importance manifests themselves during faster shock conditions, such as a flood event. By adjusting the weight of these goals in the network metrics, it will be possible to measure their impact during a flood event in light of different policy or adaptation strategies. For example, "Minimal human vulnerability" was identified as becoming more interdependent to city functioning during flood conditions for both case studies. If this goal was strengthened by city council development strategies, the impact of this during flood conditions can be explored. Finally, future climate change scenarios and the associated uncertainty of extreme flows will be incorporated into the AH weightings.

2.2.5 Conclusion

This chapter demonstrates the crucial first steps in understanding true resilience in urban systems. The AH could be a powerful resilience tool, adding a structured technique to past approaches that aim to understand urban interdependencies and outcomes. It can be used to answer questions such as: are more interactions better for the city system to enhance resilience? Or is the key to foster smaller interactions? Would additional redundancies added into the system enhance overall resilience? By contributing to a deeper exploration of exposure and resilience, these questions will lead to an improved understanding of disaster risk, which will strengthen resilience and ease the negative effects of hazards—the central goal of the Sendai Framework. Though there are many opportunities for future work, the novel approach outlined in this chapter captures sociotechnical interactions at difference scales, filling a significant gap for research and practice.

Appendix: Full abstraction hierarchy
A.1 Physical objects

Category	Physical object
Public Accommodation	People
	Assisted living
	Estate agents
	Hotels (with restaurant/meals)
	Hotels (self-catering)
	Private accommodation
Education and professional services	Creches
	Job centers
	Driving schools
	Music services
	Schools
	Training providers
	Further education (university, colleges)
Finance	Accountancy firms
	ATMs
	Banks
	Credit unions
	Currency exchanges
	Debt collection companies
	Insurance providers
	Investment/financial advisors
	Money transfer services
	Pawn shops

Continued

Category	Physical object
Food and drink	Bars/nightclubs
	Eateries
	Food markets
Government	Armed forces
	Council offices
	Courthouses
	Fire departments
	Libraries/archives
	Parliamentary buildings
	Police departments
	Politician offices
	Procurator fiscal
	Registrar service offices
	Social work offices
Healthcare	Doctors' offices
	Hospitals
	Healthcare (other)
	Pharmacies
Industry	Construction and heavy industry
	Farms
	Tradesmen businesses
	Lawyers/solicitors
	Media outlets
	Security services
	Industry (other)
	Telephone exchanges
	Network access points
	Derelict land
Infrastructure	Broadband providers
	Electricity providers
	Gas providers
	Waste collection
	Wastewater treatment works
	Water pumping stations
	Reservoirs
	Dykes (or other structural defense)
	Water tanks
	Water distribution network
Cultural, leisure, recreation, and tourism	Community centers
	Gyms
	Attractions (includes museums, concert halls, cinemas)
	Pet goods and services
	Shops (clothing)
	Shops (home and garden)
	Shops (nonessential)
	Sports grounds and facilities

Category	Physical object
Religion and major life events	Cemeteries
	Funeral directors
	Religious buildings
Social support	Business advice/support organizations
	Charity organizations
	Citizens Advice Bureau
	Day centers
	Employment services
	Food banks
	Homeless shelters
Transport and logistics	Airports
	Bus stations
	Couriers
	Bicycle parking points
	Distribution centers
	Petrol stations
	Rail stations
	Railway lines
	Post offices
	Roads
	Vehicle dealerships
	Vehicle repair services
	Cycle/pedestrian paths
Ecosystems	Parks
	Urban forests
	Rivers/canals/streams
	Storm water retention ponds

A.2 Object-related processes

Category	Object-related process
Accommodation	Act as main base of living
	Provide temporary or permanent shelter/protection from weather/events
	Provide assisted living (for elderly, disabled)
Education and professional services	Provide education services
	Provide professional/other qualification services
	Provide employment
	Provide youth/older childcare
	Undertake research

Continued

Category	Object-related process
Finance	Provide access to cash
	Make loans
	Provide other financial services/advice (e.g., mortgages, business)
	Make payments/wire money
	Provide access to consumer advice
Food and drink	Provide food
	Provide drink
Government	Investigate crimes
	Undertake legal action and proceedings
	Provide information on crime prevention
	Facilitate and increase employment
	Provide general legal advice
	Facilitate and process subsidence grants and/or benefits
	Facilitate and process taxes (council/income/business)
	Manage council-let housing
	Facilitate traveler sites
	Facilitate and process early childcare
	Facilitate and process community project grants
	Improve road conditions and safety
	Provide information on local resources (library services, archives, family history resources)
	Provide information on local conservation areas
	Facilitate waste collection and recycling
	Provide access to planning resources (guidance on process, existing local development plans, past building warrants/completion certificates/plans/specifications, current regulations)
	Provide street care (lighting, cleaning, dog fouling, stray dogs)
	Provide information on local road issues
	Provide information on listed building and applications for consent
	Enforce building standards
	Enforce food standards (cleanliness, food labeling, food hygiene, etc.)
Healthcare	Respond to medical emergencies
	Provide healthcare services
	Provide rehabilitation centers
	Provide medicines
Industry	Maintain/repair homes/shelter
	Keep roads in adequate condition (roadworks/gritting/etc.)
	Produce goods
	Environmental protection
	Environmental monitoring
Infrastructure	Transmit and receive communications
	Store and distribute fuel and energy
	Provide clean water
	Manage and dispose of sewage/wastewater
	Provide drainage

Category	Object-related process
Cultural, leisure, recreation, and tourism	Act as facility for leisure/recreation
	Provide luxury/recreational/nonessential goods/services
	Act as community/meeting space
	Support health/fitness
	Provide/maintain clothing
	Act as place of historical/cultural importance
Religion and major life events	Provide place of worship
	Facilitate use of/provide information on cemeteries, crematoriums, and memorial options
	Provide death-related services
	Facilitate registrar services
Social support	Provide help/support for vulnerable people/groups
	Assess risk level/needs of vulnerable people
	Advise and support carers
	Advise on relevant community activities
	Provide domestic violence advice/support
Transport and logistics	Enable transport
	Maintain/repair vehicles
Ecosystems	Provide microclimate regulation (reduce urban heat effect)
	Provide air filtration
	Provide natural hazard regulation
	Provide water purification
	Manage pest control
	Provide nutrient cycles
	Provide sources of energy
	Fertilize soil
	Provide esthetic value
	Provide species habitat

A.3 Generalized functions

Generalized function
Maintain public health
Maintain law and order
Support community activities and engagement
Ensure provision of food
Ensure provision of housing
Ensure provision of clothing
Provide financial services

Continued

Generalized function
Generate tourism
Perform planning activities
Support travel (people, not goods)
Enable people to work
Provide emergency services
Regulate government activities
Provide communications systems
Support observance of religion
Support distribution of goods (logistics)
Manage and dispose of waste
Supply energy
Supply of/access to clean water
Provision of sanitation
Generate economic output
Preserve and protect historical/cultural features
Preserve and protect natural features
Promote learning and education
Perform/support ceremonies, services, etc. for major life events
Conservation of environmental assets
Access to clean air
Provide physical security (personnel and property)
Enable social interaction

A.4 Values and priority measures

Value and priority measure	Definition
Minimal human vulnerability	The extent to which everyone's basic needs are met.
Diverse livelihoods and employment	This is facilitated by access to finance, ability to accrue savings, skills training, business support, and social welfare.
Effective safeguards to human health and life	This relies on integrated health facilities and services, and responsive emergency services.
Collective identity and community support	This is observed as active community engagement, strong social networks, and social integration.
Comprehensive security and rule of law	This includes law enforcement, crime prevention, justice, and emergency management.
Sustainable economy	This is observed in sound management of city finances, diverse revenue streams, and the ability to attract business investment, allocate capital, and build emergency funds.
Reduced exposure and fragility	This relies on environmental stewardship, appropriate infrastructure, effective use planning, and enforcement of planning regulations.

Value and priority measure	Definition
Effective provision of critical services	This results from diversity of provision, redundancy, active management and maintenance of ecosystem services and infrastructure, and contingency planning.
Reliable communications and mobility	This is enabled by diverse and affordable multimodal transport systems and information and communications technology networks, and contingency planning.
Effective leadership and management	This relates to government, business and civil society and is recognizable in trusted individuals, multistakeholder consultation, and evidence-based decision-making.
Empowered stakeholders	This is underpinned by education for all, and relies on access to up-to-date information and knowledge to enable people and organizations to take appropriate action.
Integrated development and management	This is indicated by the presence of a vision, an integrated development strategy, and plans that are regularly reviewed and updated by crossdepartmental groups.

A.5 Functional purposes

Functional purpose	Definition
Economic opportunity	This includes economic security and diversity of jobs as well as trade, investment, job creation, and economic growth.
Cultural heritage and sense of place	Cities should provide a "sense of place," which is defined as the degree of meaning people ascribe to a particular location, as well as a strong identity that is felt by all.
Urban ecosystem services	Direct and indirect contributions of ecosystems to human well-being. It includes all green and blue spaces within an urban area, which generate a habitat that deliver services to city inhabitants. These services include climate change mitigation and adaptation, improved air quality, esthetic values, reduced stress, and improved social cohesion.
Shared community resources	These are collectively built at a lower cost to the individual (economies of scale). Includes payment of public services, emergency response, rule of law, etc.
Freedom of movement and expression	This includes ideas, communications, people and goods, as well as the degree of anonymity, which enables people to freely choose who to associate with and express themselves.
Social opportunity	Cities provide higher quantities and diversity of people, which drives knowledge, innovation, and social relations.
Globalized connectivity	This includes global connectivity with travel and trade. It includes a city becoming an international player, but also building social relations on a larger scale and give a voice to local interest.
Physical settlement	Cities should provide adequate housing and accommodation, which provides homes for people as well as driving tourism.

Acknowledgments

The work presented was carried out as part of the EPSRC EP/NE30419/1 project "Water Resilient Cities: Climate Uncertainty & Urban Vulnerability to Hydrohazards."

References

Arup, 2015. City Resilience Framework. Available from: https://www.rockefellerfoundation.org/report/city-resilience-framework/. (Accessed 3 August 2018).

Bedinger, M., Beevers, L., Collet, L., Visser, A., 2019. Are we doing 'systems' research? An assessment of methods for climate change adaptation to hydrohazards in a complex world. Sustainability 11(4). https://doi.org/10.3390/su11041163.

Bedinger, M., Beevers, L., Walker, G.H., Visser-Quinn, A., McClymont, K., 2020. Urban systems: mapping interdependencies and outcomes to support systems thinking. Earth's Future 8. https://doi.org/10.1029/2019EF001389.

Beevers, L., Walker, G., Strathie, A., 2016. A systems approach to flood vulnerability. Civ. Eng. Environ. Syst. 33 (3), 199–213.

Census Bureau, 2019. Available from: https://msc.fema.gov/portal/home. (Accessed 7 March 2019).

Cheshire, L., 2015. 'Know your neighbours': disaster resilience and the normative practices of neighbouring in an urban context. Environ. Plan. A 47 (5), 1081–1099.

Cutter, S.L., 2016. Commentary resilience to what? Resilience for whom? Geogr. J. 182 (2), 110–113.

Cutter, S.L., Barnes, L., Berry, M., Burton, C., Evans, E., Tate, E., Webb, J., 2008. A place-based model for understanding community resilience to natural disasters. Global Environ. Change 18, 598–606.

Cutter, S.L., Emrich, C.T., Webb, J.J., Morath, D., 2009. Social vulnerability to climate variability hazards: a review of the literature. Final Report to Oxfam, Hazards and Vulnerability Research Institute Department of Geography, University of South Carolina.

Environment Agency, 2018. Environment Agency copyright and/or database right 2016. All rights reserved. Contains public sector information licensed under the Open Government Licence. Available from: http://apps.environment-agency.gov.uk/wiyby/37837.aspx. (Accessed 6 June 2018).

European Commission, 2019. Available from: https://ec.europa.eu/echo/partnerships/relations/european-and-international-cooperation/sendai-framework-disaster-risk-reduction_en. (Accessed 3 October 2019).

Fekete, A., Tzavella, K., Baumhauer, R., 2017. Spatial exposure aspects contributing to vulnerability and resilience assessments of urban critical infrastructure in a flood and blackout context. Nat. Hazards 86 (1), 151–176.

Fisher, L., 2015. Disaster responses: more than 70 ways to show resilience. Nature 518 (7537), 35.

Folke, C., 2006. Resilience: the emergence of a perspective for social-ecological systems analyses. Global Environ. Change 16, 253–267.

Gunderson, L.H., Holling, C.S., 2002. Panarchy: Understanding Transformation in Human and Natural Systems. Island Press.

Handayani, W., Fisher, M.R., Rudiarto, I., Setyono, J.S., Foley, D., 2019. Operationalizing resilience: a content analysis of flood disaster planning in two coastal cities in Central Java, Indonesia. Int. J. Disast. Risk Reduct. 35, 1–11.

Kermanshan, A., Derrible, S., 2017. Robustness of road systems to extreme flooding: using elements of GIS, travel demand, and network science. Nat. Hazards 86, 151–164.

Martin, S.A., 2015. A framework to understand the relationship between social factors that reduce resilience in cities: Application to the city of Boston. Int. J. Disast. Risk Reduct. 12, 53–80.

Martin-Breen, P., Anderies, J.M., 2011. Resilience: a literature review. In: Bellagio Initiative. IDS, Brighton.

McClymont, K., Morrison, D., Beevers, L., Carmen, E., 2019. Flood resilience: a systematic review. J. Environ. Plan. Manage. https://doi.org/10.1080/09640568.2019.1641474.

Microsoft, 2019. Copyright (c) Microsoft Corporation. Available from: https://github.com/Microsoft/USBuildingFootprints. (Accessed 7 March 2019).

Miller, F., Osbahr, H., Boyd, E., Thomalla, F., Bharwani, S., Ziervogel, G., Walker, B., Birkmann, J., Van der Leeuw, S., Rockström, J., Hinkel, J., Downing, T., Folke, C., Nelson, D., 2010. Resilience and vulnerability: complementary or conflicting concepts? Ecol. Soc. 15 (3), 11.

Naikar, N., 2017. Cognitive work analysis: an influential legacy extending beyond human factors and engineering. Appl. Ergon. 59, 528–540.

Nelson, K., Gillespie-Marthaler, L., Baroud, H., Abkowitz, M., Kosson, D., 2019. An integrated and dynamic framework for assessing sustainable resilience in complex adaptive systems. Sustain. Resilient Infrastruct. https://doi.org/10.1080/23789689.2019.1578165.

O'Hare, P., White, I., 2018. Beyond 'just' flood risk management: the potential for—and limits to—alleviating flood disadvantage. Reg. Environ. Change 18, 385–396. https://doi.org/10.1007/s10113-017-1216-3.

Opsahl, T., Agneessens, F., Skvoretz, J., 2010. Node centrality in weighted networks: generalising degree and shortest paths. Soc. Networks 33 (3), 245–251.

Ordnance Survey, 2018. OS data © Crown copyright and database right. Available from: https://www.ordnancesurvey.co.uk/opendatadownload/products.html. (Accessed 7 November 2018).

Patorniti, N.P., Stevens, N.J., Salmon, P.M., 2018. A sociotechnical systems approach to understand complex urban systems: a global transdisciplinary perspective. Hum. Factors Ergon. Manuf. Serv. Ind. 28 (6), 281–296.

Rasmussen, J., 1986. Information Processing and Human-Machine-Interaction – An Approach to Cogitative Engineering. North Holland, Amsterdam.

Restemeyer, B., Woltjer, J., van den Brink, M., 2015. A strategy-based framework for assessing the flood resilience of cities – a Hamburg case study. Plan. Theory Pract. 16 (1), 45–62.

Sayers, P.B., Horritt, M., Penning Rowsell, E., Fieth, J., 2017. Present and future flood vulnerability, risk and disadvantage: a UK scale assessment. A report for the Joseph Rowntree Foundation published by Sayers and Partners LLP.

UKPOL, 2018. Available from: http://www.ukpol.co.uk/james-bevan-2018-speech-on-flood-prevention/. (Accessed 18 October 2018).

UNDRR, 2019. Available from: https://www.unisdr.org/we/coordinate/sendai-framework. (Accessed 3 October 2019).

Whittle, H.J., Palar, K., Hufstedler, L.L., Seligman, H.K., Frongillo, E.A., Weiser, S.D., 2015. Food insecurity, chronic illness, and gentrification in the San Francisco Bay Area: An example of structural violence in United States public policy. Soc. Sci. Med. (143), 154–161. https://doi.org/10.1016/j.socscimed.2015.08.027.

MITIGATION

Disaster risk reduction beyond command and control: Mapping an Australian wildfire from a complex adaptive systems' perspective

3.1

Leonardo Nogueira de Moraes and Alan March

University of Melbourne, Melbourne, VIC, Australia Bushfire and Natural Hazards Cooperative Research Centre, Melbourne, VIC, Australia

CHAPTER OUTLINE

Understanding Disaster Risk. https://doi.org/10.1016/B978-0-12-819047-0.00003-2

3.1.1 **Introduction**

Coastal tourist destinations are fundamentally connected with the natural phenomena and resources that provide their inherent tourism values, but which are often also potential sources of harm. Wye River (Victoria, Australia) is located in a forested and highly wildfire-prone area, with limited evacuation options, having experienced a devastating wildfire in December 2015 during a peak tourist season (IGEM, 2016). Victoria is one of the most wildfire-prone areas in the world (2009 VBRC, 2010c, p. viii) while also being highly economically dependent on tourism—6% of GDP and 7.2% of employment (Tourism Victoria, 2016, p. 4), the Great Ocean Road (where Wye River is located) being a Priority Victorian Destination for tourists (Ruzzene and Dunn, 2012).

This chapter is a discrete output of a research program that explores the effects of regional tourism development on local resilience to natural hazards. It contributes to the Sendai Framework for Disaster Risk Reduction—SFDRR Priority 1: Understanding Disaster Risk by interrogating DRR using a social-ecological resilience lens.

3.1.2 **Methodology**

The research presented in this chapter aims to interrogate the practice of disaster risk reduction from a resilience-building perspective and has as its primary question:

- What are key challenges and opportunities to examining disaster risk reduction from a resilience-building perspective in the context of wildfires?

And as its secondary questions:

- How does the concept of resilience evolve since the 1970s and reach the domains of disaster risk reduction internationally and of natural hazard mitigation through urban planning and emergency management in Australia, particularly in the State of Victoria?
- How does resilience building relate to natural hazard mitigation, preparedness, response, and recovery in a real case?

A qualitative case study approach informed data collection and analysis allowing in-depth investigation of Australian, Victorian and local disaster risk reduction arrangements and the way they translate into local resilience to natural hazards. It constituted an embedded case study (Yin, 2014, p. 50) in which natural hazard mitigation, preparedness, response, and recovery were analyzed in a local tourist destination (Wye River) in the context of global, national (Australia), state (Victoria), regional (Great Ocean Road), and local (Colac-Otway Shire) frameworks for emergency management and urban planning for natural hazard mitigation. Data were analyzed using a Grounded Theory approach. NVivo 12 Plus supported the transcription of interviews and the coding of collected data to allow moving from substantive evidence to theoretical understandings.

Fieldwork involved direct observation of local conditions and audio recorded semistructured interviews in English with 10 representatives from key organizations who were selected by a combined purposive and snowball approach to sampling. In addition, data collection also targeted academic texts, government documents, meeting minutes, legislation, news articles, newsletters, and institutional social media[a] relating to the 2015 Wye River-Jamieson Track Fires. The collection of interviews, direct observation, and digital sources allowed the identification of key elements of the social structure operating locally. The generation of interactive timelines allowed visualization of how these elements relate temporally from a complex adaptive systems perspective. The framework for mapping social structure and agency derives from that developed by Nogueira de Moraes (2014), using network theory to identify institutional, relational, and embodied social structures (López and Scott, 2000). Widely adopted in Australia since the 1990s as a comprehensive approach to emergency management arrangements, PPRR[b] framed the way the results and discussion sections are presented in this chapter.

3.1.3 Resilience, disaster risk reduction, natural hazard mitigation, and emergency management

This section explores resilience as an evolving concept that makes way into disaster risk reduction and natural hazard mitigation in the international and Australian contexts, particularly in the State of Victoria.

3.1.3.1 From social-ecological resilience to natural hazards and disaster resilience

From a scientific perspective, resilience—as the term and concept being currently used by academics, governments, NGOs, and the private sector—can trace its origins to research and theory emerging from the study of systems, particularly as it incorporated complexity theory and gave rise to complex adaptive systems theory (Alexander, 2013). By the 1970s, two complementary avenues of inquiry were being developed in parallel: one focusing on the study of social-ecological systems and another on the psychology of individuals—both targeting different levels of nested systems. Despite their different foci, both streams recognize the intrinsic connection between individuals and communities and how these affect individual and system resilience, especially when individuals are also understood as systems themselves.

More recently, these two streams of scientific developments of resilience have been intersected by specific applications of the term. Relevant to this chapter are the concepts of *urban resilience* and *hazard/disaster resilience*. Meerow et al.'s definition of *urban resilience* captures the social-ecological essence of the term and adds a socio-technical dimension to it, proposing that: "[u]rban resilience refers to the ability of an urban system—and all its constituent socio-ecological and socio-technical networks across temporal and spatial scales—to maintain or rapidly return to desired functions in the face of disturbance, to adapt to change, and to quickly transform systems that limit current or future adaptive capacity" (Meerow et al., 2016, p. 39). In contrast, definitions of disaster

[a]Institutional tweets containing #wyeriver were searched and those identified as relating to Wye River, Colac-Otway Shire Council, VIC, Australia, were classified as the social media data corpus for this research.
[b]Prevention/Mitigation, Preparedness, Response, and Recovery.

resilience (also used interchangeably with *resilience to [natural] hazards*) have focused on the capacity of social systems and communities to prepare for, respond to, and recover from disasters.[c]

Absent from both definitions are direct references to the concepts of natural hazard mitigation and that of avoidance or decrease of exposure to hazards, reinforcing the idea of resilience being the flipside of vulnerability (Folke, 2006, p. 262)—although it is well established that traditional risk assessment and treatment processes do pay heed to these risk factors. This is particularly important when we account for Disaster Risk Assessment rationales that understand risk as a function of hazard, exposure, and vulnerability (Crichton, 1999).

In contrast, present in the earlier definitions of urban and disaster resilience is the concept of adaptive capacity, something that gained strength with the development of the field of Climate Change Adaptation—CCA, including the idea of adaptation pathways for change and response as a strategy to dealing with climate change associated risks (Wise et al., 2014). Interestingly, when certain climate change adaptation strategies—such as retreating settlements from areas at risk of flooding because of sea-level rise (Hino et al., 2017)—are considered, one could argue that they target the avoidance or reduction of exposure to a hazard. Similarly, adaptation strategies aiming to reduce urban heat-island effects (Hoverter, 2012) would be contributing to hazard mitigation, meaning it is hard to consider resilience as being only connected to preparedness, response, and recovery, but actually also incorporating natural hazard mitigation through its adaptation component.

3.1.3.2 Disaster risk reduction, the national strategy for disaster resilience, and the Sendai Framework

As humans' intensive reshaping of the global environment delineates a new geological epoch—the Anthropocene (Zalasiewicz et al., 2011) with human-induced climate change and increased frequency of disasters (IPCC, 2012) stemming from natural hazards and vulnerable social structures, those involved with emergency management and disaster risk reduction start to integrate resilience into their rationale and operations. This integration bears weight in academic, governmental, nongovernmental, and private sector domains.

From the adoption of measures targeting severe disasters in the 1960s and the creation of the United Nations Disaster Relief Office in the early 1970s, the UN shifted its focus to disaster preparedness and prevention in the 1980s with the declaration of the 1990s as the International Decade of Natural Disaster Reduction. Through conferences, resolutions, and the adoption of strategies and frameworks, the UN built its structural apparatus to promote concerted action in relation to disaster risk reduction (UNISDR, 2018)[d]. Ultimately these efforts translate as a "call for action," with the 1994 *Yokohama Strategy and Plan of Action for a Safer World* (IDNDR, 1994) highlighting the need for **risk** assessment for disaster reduction as its first principle, each country bearing "the primary **responsibility** for protecting its people" (ibid, p. 8), there being a "strong need to strengthen the **resilience** and self-confidence of local communities to cope with natural disasters" (ibid, p. 10).

---

[c]The United Nations Office for Disaster Risk Reduction proposes that "[disaster] resilience [is] [t]he ability of a system, community or society exposed to hazards to resist, absorb, accommodate, adapt to, transform and recover from the effects of a hazard in a timely and efficient manner, including through the preservation and restoration of its essential basic structures and functions through risk management" (UNDRR, 2017).
[d]UNISDR is now known as UNDRR.

In line with the UN's International Decade for Natural Disaster Reduction (1990–2000), Australian emergency management agencies started to focus on disaster mitigation in addition to response, recovery, and preparedness (EMA, 1999) from the mid-1990s onwards, adopting a comprehensive approach or PPRR—Prevention/Mitigation, Preparedness, Response and Recovery (EMA, 2004, p. 4). The month after the Yokohama Strategy was launched at the 1994 *World Conference on Natural Disaster Reduction*, the Australian *Senate Standing Committee on Industry, Science, Technology, Transport, Communications and Infrastructure* tabled a report on *Disaster Management* that called for the development of the Australian/New Zealand Standard for Risk Management (Standards Australia and Standards New Zealand, 1995) highlighting **the role of risk treatments in risk management**, disaster mitigation being a "key treatment strategy for reducing communities' disaster risk" (EMA, 1998). Australia actively engaged with the decade's program, incorporating risk assessment into its emergency management arrangements. As part of the decade's national program, the Commonwealth of Australia funded diverse initiatives, including a research project on the *Development of a Measure for Community Resilience to Disasters* (Pooley and O'Connor, 1999).

By the early 2000s, Australia sought to develop land-use planning guidelines for natural hazards as part of a National Framework for Mitigation (EMA, 2000), although many prior examples exist dating back to the early 1970s. Subsequently, a Disaster Mitigation Research Working Group (BTRE, 2001) helped inform a 2002 report (DoTaRS, 2004) to the Council of Australian Governments—COAG on reforming mitigation, relief, and recovery arrangements in Australia, with disaster resilience, infrastructure resilience, community resilience, and the resilience of the built environment being widely employed terms.

In 2005, the Yokohama Strategy was reviewed, and the *Hyogo Framework for Action 2005–2015: Building the Resilience of Nations and Communities to Disasters* (ISDR, 2005) was adopted, with **building resilience to hazards** as one of its three strategic goals. By the end of 2008, the Council of Australian Governments—COAG's Ministerial Council for Police and Emergency Management—Emergency Management (MCPEM-EM) proposed a National Disaster Resilience Framework. Two months later, the 2009 Victorian Bushfires[e] resulted in the largest loss of human lives by wildfire in the history of Victoria and the most devastating for Australia (2009 VBRC, 2010a; National Museum of Australia, 2019), triggering the appointment of a Royal Commission, whose report is explicitly supportive of the concept of shared responsibility, presenting 67 recommendations that target not only emergency and incident management, but also planning and building arrangements.

At the national level, COAG issued a *National Disaster Resilience Statement* endorsing a *National Strategy for Disaster Resilience* in 2011 (Australia, 2011). Acknowledging different vulnerabilities among community members and emphasizing the importance of risk assessment and treatments, the strategy cites the 2010 VBRC's report, highlighting the idea of **shared responsibility** between communities and governments when it comes to Disaster Risk Reduction and recognizing the need to better balance *prevention and recovery* with *preparedness and response* in emergency management. In parallel, a *Risk Assessment Framework* led by the National Emergency Management Committee resulted in the development of the *National Emergency Risk Assessment Guidelines*—NERAG in 2010 (Tasmanian SES, 2010), later reviewed in 2015 (EMA, 2015).

Starting in 2012, reviewing of the *Hyogo Framework* led to the adoption of the *Sendai Framework for Disaster Risk Reduction 2015–2030* in March 2015 (UNISDR, 2015). With a strong focus on

---

[e]Wildfires are usually referred to as bushfires in the Australian context.

disaster risk management and governance instead of disaster management, the *Sendai Framework* resonates with the *Australian National Strategy for Disaster Resilience* in its highlighting of mitigation and recovery, linking the two through the concept of "preparedness to build-back better." In May 2017, Australian and New Zealand Ministers responsible for law and justice, police, and emergency management agreed that the implementation of the Sendai Framework should be pursued as a national priority through the *Law, Crime and Community Safety Council—LCCSC* (2017). As a result, the *Australia-New Zealand Emergency Management Committee—ANZEMC* established an interjurisdictional *Sendai Framework Working Group* to support the development of a national roadmap for the implementation of the framework in Australia (NSW-SEMC, 2017, p. 2) and the establishment of "a set of nationally appropriate indicators based on the United Nations' 38 global indicators to measure progress against the Sendai Framework's seven global targets" (Merrin-Davies, 2018, p. 2).

In December 2017, following the creation of the *Department of Home Affairs* and its incorporation of *Emergency Management Australia—EMA*, a *National Resilience Task Force* was established to lead the development of a *National Disaster Risk Reduction Framework* in collaboration with representatives of all levels of government, industry and communities to "translate[…] many components of Sendai into action that is appropriate to Australia" (Department of Home Affairs, 2018a, p. 4). In October 2018, a Draft Framework was presented to the Ministerial Council for Police and Emergency Management which supported the continued development of the framework "and noted, subject to negotiations with the States and Territories, that in principle disaster resilience funding be aligned with the Framework" (MCPEM, 2018, p. 1).

In April 2019, the Department of Home Affairs published the final version of the framework (Department of Home Affairs, 2019), which was later endorsed by the Ministerial Council for Police and Emergency Management in June 2019 (MCPEM, 2019, p. 1). This version highlighted not only the Sendai Framework for Disaster Risk Reduction 2015–2030 as a guiding contextual driver for the development and implementation of the National Disaster Risk Reduction—DRR Framework, but also the 2030 Agenda for Sustainable Development—SD and the Paris Agreement, establishing a vision, goals, and priorities for 2030 that align DRR and Climate Change Adaptation—CCA to support Sustainable Development Goals—SDGs (Department of Home Affairs, 2018b). A set of documents on Guidance for Strategic Decisions on Climate and Disaster Risk were launched in 2019 to complement the Framework (AIDR, 2019b).

Following the record-breaking Australian wildfires of 2019–2020, which by the 18th of February 2020 had "burned more than 10 million hectares of land in southern Australia, greater than the combined area burned in the Black Saturday 2009 and Ash Wednesday 1983 bushfires" (CSIRO, 2020), the Australian Commonwealth Government appointed a Royal Commission into National Natural Disaster Arrangements, which will produce recommendations on land use planning, zoning and development approvals, among others (Australia, 2020).

3.1.3.3 Wildfires, emergency management, and resilience building in Victoria, Australia

The State of Victoria comprises some of the most wildfire-prone lands in the world (EMV, 2014). The 1851 Black Thursday, the 1939 Black Friday, the 1983 Ash Wednesday, and more recently the 2009 Black Saturday wildfires have all considerably shaped how the State of Victoria—and the Commonwealth of Australia—have structured current emergency management arrangements (2009 VBRC,

2010b; CFA, 2009; FFM Victoria, 2014). Presenting the largest human death toll, the 2009 Black Saturday triggered the appointment of the 2009 Victorian Bushfires Royal Commission. As a reflection of their recommendations (2009 VBRC, 2010a), a comprehensive set of reforms were implemented, including the creation of an all-hazards management agency—Emergency Management Victoria (DPC, 2012), changes in land-use planning controls—most importantly the Bushfire Management Overlay—BMO (Victorian Planning Provisions, 2011), and a greater focus on resilience building and the concept of shared responsibility. This process was developed in interaction with the formulation of the National Strategy for Disaster Resilience (Australia, 2011).

When it comes to resilience building, three key streams of action can be observed in Victoria: one led by the Commonwealth level through the provision of grant funding administered by Emergency Management Victoria—EMV on behalf of the Victorian State Government—National Disaster Resilience Grant Scheme—NDRGS (EMV, 2015); one led by the State Government of Victoria focusing on the development of social cohesion in local communities (Victoria, 2015); and a third led by the City of Melbourne and supported by the University of Melbourne as part of the Rockefeller Foundation's 100 Resilient Cities Programme (Melbourne, 2016).

Cofinanced by the Victorian Government, 10 Councils within the Barwon South West Region and various partners, including the Australian Government and Water Authorities, the Climate Resilient Communities of the Barwon South West project was established in 2012 "with the aim of helping communities throughout the region to understand what risks and opportunities might be presented by future extreme weather events" (South West Climate Change PCG, 2019). The project led to the development of the Colac-Otway Shire Climate Change Adaptation Plan published in early 2017, about a year after the Wye River Fires of December 2015, helping to contextualize elements of natural hazard mitigation and disaster risk reduction within a climate change adaptation strategy. In its Action A.1.3, the plan proposes "working with tourism operators and holiday rental owners to raise awareness and build capacity of the sector to understand bushfire risk" (ARUP, 2017, p. 16), therefore contributing to the SFDDR Priority 1.

3.1.4 The Wye River-Jamieson Track Wildfire

Wye River is an Australian regional town in the Colac-Otway Shire in the State of Victoria. In 2016, its permanent resident population summed 63 people, while a total of 199 private dwellings were accounted in the last census (ABS, 2017), a number that far exceeds its permanent population. Wye River is a popular coastal tourist destination, explaining this disparity resulting from the great number of holiday houses built by metropolitan Melbourne dwellers seeking frequent retreats from life in the city.

Known as the Wye River-Jamieson Track Fire or the 2015 Christmas Fire, this devastating wildfire was ignited by lightning strikes on December 19, 2015. It hit Wye River and Separation Creek on December 25 and was fully contained only on January 21, 2016. It resulted in no human casualties but a total of 98 houses were destroyed in Wye River and 18 in its neighboring settlement Separation Creek (IGEM, 2016). Situated in a historically wildfire-prone region, Wye River has experienced severe wildfires of state and regional significance, including the 1939 Black Friday Wildfires (220,212.91 ha), the 1962 (2588.47 ha), and the 1968 Wildfires (1776.33 ha). Wye River's neighboring suburbs have also experienced other significant events, highlighting the region's proneness to wildfires—Separation Creek: 1959 (608.66 ha), 1964 (669.97 ha), and 1984 Wildfires, and Kennett River: 1967 Wildfires (754.96 ha) (VICMap, 2017).

3.1.4.1 Mitigation[f]

Mitigation measures applicable to the site include those stemming from institutional reform triggered by the 2009 Victorian Bushfires Royal Commission recommendations. Responding to some of the planning and building recommendations, amendment VC83 of the Victorian Planning Provisions (2011)—VPPs included a shift from targeting simply the minimization of disaster risk to an approach that addresses risk as one of the strategies to strengthen "community resilience to bushfire." Clause 44.06 of the VPPs was also amended to reflect the call for "a comprehensive Bushfire-prone Overlay provision, resulting in the creation of the BMO, in place of the former Wildfire Management Overlay—WMO." As a supplement, Clause 52.47 (Planning for Bushfire) was introduced to provide specific guidelines for the design and construction of buildings in areas subject to the overlay, including a reference to AS3959-2009. Despite advancements, the reworked overlay has proven to need further adjustments to become more comprehensive. According to CSIRO (Leonard et al., 2016), out of seven dwellings that were built in accordance to AS3959-2009, only three survived, in comparison to a total of 11 out of 14 that were built in accordance to previous standards. Key planning control issues pointed out by CSIRO include "the separation distances between buildings [...] to limit structure-to-structure spread" and "the materials used and location of retaining walls proximal to buildings." In addition, there were instances in which combustible external cladding had been used and/or flammable material had been stored underneath buildings in ground-level exposed subfloors, a common building design for the area, due to its topographic characteristics—steep terrains (Leonard et al., 2016, pp. 7, 45–46).

3.1.4.2 Preparedness

Evident in the community newsletters that date back to 1988 (WRSCPA, 1988), there is a long history of community engagement with the Country Fire Authority, which has a specific branch for Wye River with local-serving volunteer firefighters. The Wye River Surf and Lifesaving Club—S&LSC also played a significant role in promoting community safety through its life-saving and education activities, but also by its role in promoting local social cohesion and youth empowerment. Despite not affecting Wye River, the 2009 Black Saturday event had triggered greater concern about wildfire safety, manifested in the S&LSC intent to become the safest place in Wye River (a town with a single road for escape) for anyone unable to evacuate in case of an emergency (Bordignon, 2016). Additionally, there was the Bushfire Scenario event for the local community promoted by the Wye River Fire Brigade and the Wye River Community Volunteers on November 7, 2015 at the S&LSC to increase understanding of local risk for that wildfire season. It included the demonstration of possible scenarios of wildfire impacting the area by use of computer-generated wildfire modeling through *Phoenix RapidFire*, support in understanding the Community Emergency Management Plan and for the development of individual Bushfire Plans (WRSCPA, 2015). While the

---

[f]"The lessening or minimising of the adverse impacts of a hazardous event" (UNDRR, 2017) or "measures taken in advance of a disaster aimed at decreasing or eliminating its impact on society and environment" (AIDR, 2019a).

topography of Wye River results in a settlement that is split into two pockets—one in each side of the Wye River valley, the settlement's overall scale and that of its beach and entertainment businesses cater for a community that is likely to coutilize key public and private spaces that can create opportunities for encounter and engagement. Despite being a tourist destination and, therefore, experiencing a mix of permanent and temporary populations, the fact that the tourist mix includes holiday home owners and frequent campers and caravanners could also result in greater social networking among its permanent and temporary populations, which may have facilitated efforts such as the Bushfire Scenario event, purposefully organized to happen on a weekend, so as to allow participation of holiday-home owners.

3.1.4.3 Response

Response to the wildfire involved direct attack, the creation of containment lines, aerial bombing, back-burning to reduce fuel load, early warning systems, community engagement throughout the process, and a timely call for evacuation (commencing at 11:57 a.m. on December 25, 2015) that included the national Emergency Alert telephone warning system, Twitter posts, door knocking, and the previous identification of vulnerable residents. Due to the steep terrain and narrow winding roads that prevail in most built areas of Wye River, fire trucks were not sent to most properties, for safety reasons. The response operation was successful in so far there were no human casualties and available resources were allocated as appropriate (IGEM, 2016). As part of the response stage, community engagement meetings were held in the Wye River Surf and Lifesaving Club (see Fig. 3.1.1), reinforcing its importance as the community hub for bringing together community self-organization and formal emergency management planning and deployment. The successful evacuation and protection of Wye River's human population are likely to have been facilitated by this bridge between community and emergency management services, including the local network of community member volunteers acting in various roles, including fire-fighting, as well as fundraising activities organized by the local CFA Auxiliary. In terms of community engagement and information, this included the advertising and running of a public meeting on the December 23 in Wye River, the opening of Colac-Otway Shire information centers on Christmas Day, and the conclusion and dissemination of an evacuation plan for Wye River by the evening of 23rd of December (IGEM, 2016, pp. 35–36). In its description of good practices and learning lessons from the initial response to the Wye River-Jamieson Track Fires, IGEM points to the "value of local knowledge" (IGEM, 2016, p. 41), reinforcing the importance of community engagement and participation in the process of response.

3.1.4.4 Recovery

The recovery process started before the wildfire was fully contained, with the first community recovery newsletter issued January 17, 2016 (Colac Otway Shire Council and EMV, 2016). A focal point for residents to assist them with recovery was first established in Apollo Bay and later in central Wye River, bordering the Great Ocean Road. This was coupled with dedicated communication channels that included a direct phone line, a website, newsletters, social media, and the establishment of community forums and community recovery groups, including the self-reinforcing *Community Connection and*

FIG. 3.1.1

Snapshot of the North side of the Wye River Valley showing the Wye River Surf and Life Saving Club on the bottom and, in the background, the steep terrain in which many of the affected properties are located.

Wellbeing Working Group and the *Work Group on Flora, Fauna and Beachscape* targeting landscape and mental recovery; and the *Business and Tourism Working Group* targeting local economy recovery.

A cascading effect of the Wye River-Jamieson Track Fires was the subsequent clearing of the very steep terrain that is predominant in most of Wye River. This led to severe landslides during the heavy rain that followed the wildfire, blocking sections of the Great Ocean Road and further affecting regional tourism (Steeth, 2016). As part of the recovery package, resources were allocated to fix and upgrade sections of the road (Colac Otway Shire Council and Emergency Management Victoria, 2016). As part of the National Disaster Resilience Grant Scheme—NDRGS Scheme, a grant of AUD 250,000.00 made available to *Volunteering Victoria* supported fund the *HelpOut Emergency Volunteering Program* to "improve the coordination and management of spontaneous emergency volunteers," which included, among other initiatives, the deployment of bird boxes (habitat kits) in Wye River—in partnership with *Southern Otway Landcare Network*, to support native wildlife renewal post the 2015 wildfires (Volunteering Victoria, 2018, p. 10, 29).

To guide the process of revegetation of individual properties to make them more resilient to future wildfires, a partnership between the *WyeSep Vegetation Restoration Committee*, the *Flora, Fauna and Beachscape workgroup*, *Parks Victoria*, the *Southern Otway Landcare Network*, the *Department of Environment, Land, Water and Planning—DELWP*, the *Country Fire Authority—CFA* and the shire councils of *Colac-Otway* and *Surf Coast* has resulted in the publication and distribution of the Land-scaping your Coastal Garden for Bushfire guide. If well promoted by existing community networks that are part of Wye River's social structure, the wide employment of this guide can leverage the overall settlements resilience to future wildfires, by decreasing individual vulnerability across the settlement and containing the potential for fire spread during a wildfire event.

3.1.5 **Discussion**

Analysis of mitigation, preparedness, response, and recovery in the context of the 2015 Wye River-Jamieson Track Fires has led to the following considerations regarding resilience building:

- **mitigation** was mostly concentrated on statutory planning measures (such as the Bushfire Management Overlay and the Australian Standard AS 3959-2009) targeting individual levels of exposure and vulnerability of new buildings, not necessarily accounting for existing buildings and existing infrastructure—in other words, the construction of new buildings in accordance with current standards was not always an effective way to increase building resilience when surrounding buildings and infrastructure (including retaining walls) were not rebuilt to current standards;
- in contrast, **preparedness**, **response**, and **recovery** were marked by a greater deal of community engagement and participation and the establishment of bridges between government agencies and the local community through key organizations such as the local Wye River CFA and CFA Auxiliary, the Wye River Surf & Life Saving Club, the Wye River and Separation Creek Progress Association, the Otway Coast Tourism Association, and the Wye River Community Volunteers, among others. This is evident in the Bushfire Scenario event for preparedness, the successful settlement evacuation as part of response and the community steering of and active participation in the recovery process.

In summary, while community effort was successful for preparing, responding and recovering from the 2015 Wye River-Jamieson Track Fire, consideration of CSIRO's report—that argues that colocation with more vulnerable dwellings increases the vulnerability of dwellings built to higher standards—raises the need for the community to act together to increase the overall settlement's built-form resilience to wildfire. This brings to light a potential tension between development existing prior to and after the introduction of current building standards, raising the question of who should bear the costs of retro-fitting or even rebuilding older dwellings not yet destroyed by wildfires when these may be increasing the overall vulnerability of the settlement.

In contrast, the recovery process has brought an important contribution to improving overall settlement resilience by collaboratively producing a guide to *Landscaping Your Coastal Garden for Bushfire*. If well adopted by property owners, the garden design principles and landscape actions proposed in this guide can serve as a starting point to further increase overall settlement and community resilience, possibly leveraging similar processes for the settlement's existing buildings. Whether this will translate into independent and/or government/community supported redevelopment and/or retro-fitting of surviving/existing buildings and their surrounding infrastructure to current wildfire protection standards is yet to be seen.

3.1.6 **Conclusions**

Australia's long history in dealing with natural hazards has promoted the development of emergency management arrangements that are constantly evolving and starting to incorporate the concept of resilience as part of their rationale. However, despite advances to date, the Australian system can still improve the implementation of a comprehensive approach to natural hazard mitigation as part of its urban and regional planning arrangements. Overall, there is a need to move from a site/individual approach to a community/settlement approach when it comes to consideration of disaster risk and the management of existing and new development. This was evident in the Wye River case, which illustrates how the individual vulnerability of existing buildings built prior to the introduction of current standards can lead to greater overall settlement vulnerability that, in turn, can dampen the effects of individual efforts to increase the resilience of their buildings to hazards.

By portraying systems as nested, complex adaptive systems theory brings an **opportunity** to understand community/settlement-level vulnerability and resilience as interlinked with individual/site-level vulnerability and resilience. This, in itself, is an important contribution to the SFFDRR Priority 1: Understanding Disaster Risk, especially for its item (o), "to enhance collaboration among people at the local level to disseminate disaster risk information through the involvement of community-based organisations and non-governmental organisations" (UNISDR, 2015, p. 15).

An inherent element of complex adaptive systems, adaptive capacity is also intrinsically linked with the concept of resilience. In that respect, the extent to which individual/sites and collective/settlement capacity to adapt can be targeted as interlinked is likely to determine the future success of resilience building processes for communities such as that of Wye River.

If resilience is employed from a perspective that looks at systems as nested—as proposed by its roots in social-ecological systems' theory that looks at them as being complex adaptive systems (Folke, 2006), community resilience can be understood as related to social structure resilience. This perspective would account for the resilience of individual members of the community and of the institutional structures (Nogueira de Moraes, 2014) they need to support their process of resilience building. From this conceptual understanding, resilience could reinforce the idea of shared responsibility (2009 VBRC, 2010a) between institutional structures, communities, and individuals in adapting to reduce disaster risk rather than the transferring of responsibility from government to communities/individuals or vice versa.

Among the **challenges** of employing resilience theory in the context of disaster risk reduction is the varied use of the word *resilient*. Risk treatment is a key part of emergency-related risk management processes and comprises "consider[ing], [...] select[ing] and assess[ing] measures to reduce **risk levels**" (EMA, 2015, p. 20). These derive from the analysis of the **likelihood** of an "event happening given the current level of control" and the **potential** "consequences to people, the economy, the environment, public administration and the social setting" (EMA, 2015, p. 48). Likewise, resilience refers to the **degree** to which a system can withstand impacts and continue to carry out its essential functions. When terms like *disaster resilient infrastructure, resilient communities,* or *resilient cities* are employed, they contradict the primary foundation of risk and resilience conceptualization—their relativity to complex conditions and events yet to occur and not fully predictable and understood. As Meerow and Newell (2019) suggest, it is important to ask the question: resilience to what? In other words, what are the scenarios being used to describe something, someone or some group is resilient to?

Without reference to these scenarios, the use of the term "resilient" could be understood as a catchword implying settlements, infrastructure, and communities can be future-proofed, which is not the case.

While future-proofing human settlements may seem far from reality, the reduction of the likelihood and potential consequences of natural hazard events seems a more attainable endeavor. But this requires comprehensive approaches, concerted action, and effective communication. The better integration of different knowledge and practice domains, namely, disaster risk reduction and urban and regional planning, might cater for understandings of resilience that promote the better integration of different agency programs that deal with natural hazard mitigation (March et al., 2018).

Another challenge raised by the employment of resilience theory to examine the field of disaster risk reduction is the way in which the concept of resilience could be integrated with the idea that risk is a function of hazard, exposure, and vulnerability across a wider range of interdisciplinary fields beyond emergency management. As discussed previously, by incorporating the capacity to adapt as a core part of its concept, resilience can be understood not only as the flipside to vulnerability but also as potentially comprising avoidance or exposure to hazards and hazard mitigation. This leads us to the question as to whether vulnerability itself can be seen as a separate category to hazard and to exposure or whether it also refers to the capacity to mitigate hazards and to avoid or reduce exposure to them. If the first is the case, then we can conclude resilience is a more encompassing concept than just being an "antonym of vulnerability" (Miller et al., 2010). Otherwise, we can conclude the concept of resilience can be used to critically re-examine the concept of the risk triangle (Crichton, 1999) if it is to be applied beyond risk assessment for the financial insurance of physical infrastructure, but also to examine settlements risk to disasters from a social-ecological complex adaptive systems perspective.

Additionally, the importance of natural environment restoration to postdisaster community mental well-being opens a door to building social-ecological resilience as part of developing community resilience to disasters—even if from an anthropocentric perspective of the environment being the provider of ecosystem services to human needs. This could bring the concept of disaster resilience (ISDR, 2005) closer to that of urban resilience (Meerow et al., 2016), better aligning urban planning and emergency management approaches to natural hazard mitigation.

Finally, as observed in the 2019 National Disaster Risk Reduction Framework, its accompanying guidance documents and the 2012 Climate Resilient Communities of the Barwon South West project, Climate Change Adaptation and Disaster Risk Reduction are starting to be integrated in local, state and national levels in Australia. From these examples, resilience seems to be part of a common terminology and underpinning concept, whereas mitigation assumes different meanings in the two domains.[g]

Acknowledgments

The development of the research presented in this chapter and the development and presentation of an early version of this chapter in the 8th International Conference on Building Resilience were made possible by the award of a Conference Travel Grant and an Early Career Researcher Internal Grant by the Faculty of Architecture, Building and Planning of the University of Melbourne, Australia.

---

[g]When the UNDRR (2017) defines mitigation, it notes that "in climate change policy, 'mitigation' is defined differently, and is the term used for the reduction of greenhouse gas emissions that are the source of climate change."

References

2009 Victorian Bushfires Royal Commission Final Report – Summary (978-0-9807408-1-3). Retrieved from: http://www.royalcommission.vic.gov.au/finaldocuments/summary/HR/VBRC_Summary_HR.pdf.

2009 Victorian Bushfires Royal Commission Final Report – Volume I Appendix C – Fire in Victoria – A Summary (978-0-9807408-2-0). Retrieved from: http://www.royalcommission.vic.gov.au/Finaldocuments/volume-1/PF/VBRC_Vol1_AppendixC_PF.pdf.

2009 Victorian Bushfires Royal Commission Final Report – Volume III – Establishment and Operation of the Commission. Retrieved from: http://royalcommission.vic.gov.au/Commission-Reports/Final-Report/Volume-3/High-Resolution-Version.html.

ABS, 2017. 2016 Census QuickStats. (12/01/2017) Retrieved from: http://www.abs.gov.au/.

AIDR, 2019a. Australian Disaster Resilience Glossary. Retrieved from: https://knowledge.aidr.org.au/glossary/.

AIDR, 2019b. Guidance for Strategic Decisions on Climate and Disaster Risk. Retrieved from: https://knowledge.aidr.org.au/resources/strategic-disaster-risk-assessment-guidance/.

Alexander, D.E., 2013. Resilience and disaster risk reduction – an etymological journey. Nat. Hazards Earth Syst. Sci. 13 (11), 2707–2716. https://doi.org/10.5194/nhess-13-2707-2013.

ARUP, 2017. Colac Otway Shire Climate Change Adaptation Plan 2017–2027. Retrieved from: http://www.swclimatechange.com.au/resources/Colac%20Otway%20Shire%20Climate%20Change%20Adaptation%20Plan.pdf.

Australia, 2011. National Strategy for Disaster Resilience – Buiding the Resilience of Our Nation to Disasters (978-1-921725-42-5). Retrieved from: https://www.aidr.org.au/media/1313/nationalstrategyfordisasterresilience_2011.pdf.

Australia, 2020. Royal Commission into National Natural Disaster Arrangements, Retrieved from: https://naturaldisaster.royalcommission.gov.au/publications/commonwealth-letters-patent-20-february-2020.

Bordignon, M., 2016. 2016 Victorian Life Saving Conference Presentation Notes. Wye River SLSC, Wye River, p. 4.

BTRE, 2001. Economic Costs of Natural Disasters in Australia. Retrieved from: https://web.archive.org/web/20070901110223/. http://www.btre.gov.au/publications/99/Files/r103_lores.pdf.

CFA, 2009. Major Fires in Victoria. CFA—Country Fire Authority. Retrieved from: http://www.cfa.vic.gov.au/about/major-fires/.

Colac Otway Shire Council, Emergency Management Victoria, 2016. Tramways in the hills, Online News. Wye-Sep Connect News. (04/07/2016). Retrieved from: http://wyesepconnect.info/tramways-in-the-hills/.

Colac Otway Shire Council, EMV, 2016. Wye River and Separation Creek – Wye River Fire – Community Resilience Newsletter – 05 July 2016. In: Community Resilience. Colac Otway Shire, WyeSep Connect.

Crichton, D., 1999. The risk triangle. In: Ingleton, J. (Ed.), Natural Disaster Management – A Presentation to Commemorate the International Decade for Natural Disaster Reduction (IDNDR). Tudor House, Leicester, pp. 1990–2000.

CSIRO, 2020. The 2019–20 Bushfires: A CSIRO Explainer. Retrieved from: https://www.csiro.au/en/Research/Environment/Extreme-Events/Bushfire/preparing-for-climate-change/2019-20-bushfires-explainer.

Department of Home Affairs, 2018a. National Disaster Risk Reduction Framework Consultation Draft.

Department of Home Affairs, 2018b. National Disaster Risk Reduction Framework (978-1-920996-78-9). Retrieved from: https://www.homeaffairs.gov.au/emergency/files/national-disaster-risk-reduction-framework.pdf.

Department of Home Affairs, 2019. Emergency Management Resources – National Disaster Risk Reduction Framework. (05/04/2019). Retrieved from: https://www.homeaffairs.gov.au/about-us/our-portfolios/emergency-management/resources.

DoTaRS, 2004. Natural Disasters in Australia – Reforming Mitigation, Relief and Recovery Arrangements. Retrieved from: http://content.webarchive.nla.gov.au/gov/wayback/20050615003821/. http://www.dotars.gov.au/localgovt/ndr/nat_disaster_report/naturaldis.pdf.

DPC, 2012. Victorian Emergency Management Report – White Paper. Retrieved from: https://web.archive.org/web/20170119031517/. http://www.dpc.vic.gov.au/images/images/featured_dpc/victorian_emergency_management_reform_white_paper_dec2012_web.pdf.

EMA, 1998. Mt Macedon Paper Number 5/1997 – Record of the Disaster-Mitigation Workshop 27–30 April 1997 – Prepared for The Australian Emergency Management Institute Mt Macedon. Emergency Management Australia, Canberra.

EMA, 1999. Final Report of Australia's Coordination Committee for the International Decade for Disaster Reduction (IDNDR) 1990–2000. Emergency Management Australia, Canberra, Australia.

EMA, 2000. Developing a National Framework for Mitigation. (23/10/2000) Retrieved from: https://web.archive.org/web/20001209191400/. http://www.ema.gov.au:80/fs-education.html.

EMA, 2004. Emergency Management in Australia – Concepts and Principles. Retrieved from: https://knowledge.aidr.org.au/media/1972/manual-1-concepts-and-principles.pdf.

EMA, 2015. National Emergency Risk Assessment Guidelines, second ed. Retrieved from: https://www.aidr.org.au/media/1489/handbook-10-national-emergency-risk-assessment-guidelines.pdf.

EMV, 2014. Emergency Management Victoria State Bush Fire Plan 2014. Retrieved from: http://files.em.vic.gov.au/EMV-web/State-Bush-Fire-Plan-2014.pdf.

EMV, 2015. NDRGS Recipients 2015–2017. Emergency Management Victoria, Melbourne.

FFM Victoria, 2014. Past Bushfires – A Chronology of Major Bushfires in Victoria From 2013 Back to 1851. Retrieved from: https://www.ffm.vic.gov.au/history-and-incidents/past-bushfires.

Folke, C., 2006. Resilience – the emergence of a perspective for social-ecological systems analyses. Global Environ. Change 16 (3), 253–267. Retrieved from: http://www.sciencedirect.com/science/article/B6VFV-4KFV39T-1/2/21ccf91cced363dbd098af70b8eb525d.

Hino, M., Field, C.B., Mach, K.J., 2017. Managed retreat as a response to natural hazard risk. Nat. Clim. Change 7 (5), 364–370. https://doi.org/10.1038/nclimate3252.

Hoverter, S.P., 2012. Adapting to Urban Heat – A Tool Kit for Local Governments. Retrieved from: https://kresge.org/sites/default/files/climate-adaptation-urban-heat.pdf.

IDNDR, 1994. Yokohama Strategy and Plan of Action for a Safer World. Retrieved from: https://www.unisdr.org/files/8241_doc6841contenido1.pdf.

IGEM, 2016. Review of the Initial Response to the 2015 Wye River – Jamieson Track fire (978-0-9944237-6-4). Retrieved from: http://www.igem.vic.gov.au/documents/CD/16/86381.

IPCC, 2012. Managing the Risks of Extreme Events and Disasters to Advance Climate Change Adaptation – Special Report of the Intergovernmental Panel on Climate Change. Retrieved from: https://www.ipcc.ch/pdf/special-reports/srex/SREX_Full_Report.pdf.

ISDR, 2005. Hyogo Framework for Action 2005–2015 – Building the Resilience of Nations and Communities to Disasters. Retrieved: http://www.unisdr.org/files/1037_hyogoframeworkforactionenglish.pdf.

LCCSC, 2017. Law, Crime and Community Safety Council Meeting Communiqué – Melbourne, 19 May 2017. Retrieved from: https://www.ag.gov.au/About/CommitteesandCouncils/Law-Crime-and-Community-Safety-Council/Documents/19-May-LCCSC-Communique.pdf.

Leonard, J., Opie, K., Blanchi, R., Newnham, G., Holland, M., 2016. Wye River – Separation Creek Post-bushfire Building Survey Findings – Report to the Victorian Country Fire Authority (CSIRO Client Report EP16924). Retrieved from: http://wyesepconnect.info/wp-content/uploads/2016/05/Wye-River-Separation-Creek-final-V1-1.pdf.

López, J., Scott, J., 2000. Social Structure. Open University Press, Buckingham, Philadelphia.

March, A., Nogueira de Moraes, L., Riddell, G.A., Stanley, J., van Delden, H., Beilin, R., … Maier, H., 2018. Practical and Theoretical Issues – Integrating Urban Planning and Emergency Management (1). Retrieved from: https://www.bnhcrc.com.au/file/8951/download?token=u4pNzrhc.

MCPEM, 2018. Ministerial Council for Police and Emergency Management Meeting Communiqué – Canberra, 26 October 2018. Retrieved from: https://www.homeaffairs.gov.au/how-to-engage-us-subsite/files/mcpem-communique-20181026.pdf.

MCPEM, 2019. Ministerial Council for Police and Emergency Management Communiqué - Adelaide, 28 June 2019. Retrieved from: https://www.homeaffairs.gov.au/how-to-engage-us-subsite/files/mcpem-communique-20190628.pdf.

Meerow, S., Newell, J.P., 2019. Urban resilience for whom, what, when, where, and why? Urban Geogr. 40 (3), 309–329. https://doi.org/10.1080/02723638.2016.1206395.

Meerow, S., Newell, J.P., Stults, M., 2016. Defining urban resilience – a review. Landsc. Urban Plan. 147, 38–49. https://doi.org/10.1016/j.landurbplan.2015.11.011.

Melbourne, 2016. Resilient Melbourne – Viable, Sustainable, Liveable, Prosperous. Retrieved from: https://resilientmelbourne.com.au/wp-content/uploads/2016/05/COM_SERVICE_PROD-9860726-v1-Final_Resilient_Melbourne_strategy_for_web_180516.pdf.

Merrin-Davies, M., 2018. The review of the national principles for disaster recovery. Austral. J. Emerg. Manage. 33 (3), 16–17. Retrieved from: https://knowledge.aidr.org.au/media/5825/ajem_july2018_kh.pdf.

Miller, F., Osbahr, H., Boyd, E., Thomalla, F., Bharwani, S., Ziervogel, G., … Nelson, D., 2010. Resilience and vulnerability: complementary or conflicting concepts? Ecol. Soc. 15 (3) Retrieved from: http://www.ecologyandsociety.org/vol15/iss3/art11/.

National Museum of Australia, 2019. 'Black Saturday' bushfires. Retrieved from: https://www.nma.gov.au/defining-moments/resources/black-saturday-bushfires.

Nogueira de Moraes, L., 2014. Inheriting Sustainability: World Heritage Listing, the Design of Tourism Development and the Resilience of Social-Ecological Complex Adaptive Systems in Small Oceanic Islands: A Comparative Case Study of Lord Howe Island (Australia) and Fernando de Noronha (Brazil). (Doctoral Thesis)The University of Melbourne Digital Repository, Melbourne. Retrieved from: http://hdl.handle.net/11343/48400.

NSW-SEMC, 2017. New South Wales State Emergency Management Committee Meeting Communiqué – Wollongong, 14 September 2017. Retrieved from: https://www.emergency.nsw.gov.au/Documents/communiques/semc-communique-meeting-106-20170914.pdf.

Pooley, J.A., O'Connor, M., 1999. International Decade for Natural Disaster Reduction: Resilience of Disaster Communities. Edith Cowan University, Joondalup, WA.

Ruzzene, M., Dunn, F., 2012. Great Ocean Road Destination Management Plan. Retrieved from: http://www.greatoceanroadtourism.org.au/wp-content/uploads/2015/01/Otways-WEB-version-final-10-Feb.pdf.

South West Climate Change PCG, 2019. Climate Change Resilient Communities. Retrieved from: http://www.swclimatechange.com.au/cb_pages/climate_resilient_communities.php.

Standards Australia, Standards New Zealand, 1995. AS/NZS 4360:1995 Risk Management; 1995. Retrieved from: https://www.saiglobal.com/PDFTemp/Previews/OSH/AS/AS4000/4300/4360-1995(+A2).pdf.

Steeth, C., 2016. Fire tourism – Victorian Great Ocean Road fires cause more tourism in south-east South Australia, particularly for wine, Online News. ABC News. (13/01/2016). Retrieved from: http://www.abc.net.au/news/rural/2016-01-13/great-ocean-road-fires-causing-more-wine-tourism-in-sa/7085596.

Tasmanian SES, 2010. National Emergency Risk Assessment Guidelines, first ed. (978-0-9805965-1-9). Retrieved from: http://coastaladaptationresources.org/PDF-files/1438-National-Emergency-Risk-Assessment-Guidelines-Oct-2010.PDF.

Tourism Victoria, 2016. Tourism Victoria Annual Report 2015–2016 (2206-6381); 2016. Retrieved from: https://www.business.vic.gov.au/__data/assets/pdf_file/0003/1518528/DEDJTR_TV_Annual_Report_2015_2016-1.pdf.

UNDRR, 2017. Terminology on Disaster Risk Reduction 2017. (02/02/2017). Retrieved from: https://www.unisdr.org/we/inform/terminology.

UNISDR, 2015. Sendai Framework for Disaster Risk Reduction 2015–2030. Retrieved from: http://www.unisdr.org/files/43291_sendaiframeworkfordrren.pdf.

UNISDR, 2018. UNISDR – History. Retrieved from: https://www.unisdr.org/who-we-are/history#idndr.

VICMap, 2017. Colac Otway Shire Maps [Report Map]. Retrieved from: http://cos.cerdi.com.au/cos_map.php?view=1811_ac57c14.

Victoria, 2015. Strategic Framework to Strengthen Victoria's Social Cohesion and the Resilience of Its Communities (978-1-922222-70-1). Retrieved from: https://www.dpc.vic.gov.au/images/documents/about_dpc/Strategic_Framework_to_Strength_Victorias_Socoal_Cohesion.pdf.

Victorian Planning Provisions, 2011. Clause 44.06 – Bushfire Management Overlay – Consolidated With Amendment VC83; 2011. 14 July 2013.

Volunteering Victoria, 2018. Spontaneous Emergency Volunteering in Victoria – A Final Report on the HelpOUT and MSEV Programs. Retrieved from: http://volunteeringvictoria.org.au/wp-content/uploads/2018/08/Volunteering-Victoria-HelpOUT-Final-Report.pdf.

Wise, R.M., Fazey, I., Stafford Smith, M., Park, S.E., Eakin, H.C., Archer Van Garderen, E.R.M., Campbell, B., 2014. Reconceptualising adaptation to climate change as part of pathways of change and response. Global Environ. Change 28, 325–336. https://doi.org/10.1016/j.gloenvcha.2013.12.002.

WRSCPA, 1988. Wye River and Separation Creek Progress Association Newsletter – 1988. Wye River and Separation Creek Progress Association.

WRSCPA, 2015. Wye River and Separation Creek Progress Association Community Newsletter – November 2015. Wye River and Separation Creek Progress Association.

Yin, R.K., 2014. Case Study Research – Design and Methods (Applied Social Research Methods), fifth ed. SAGE, Thousand Oaks, Calif.

Zalasiewicz, J., Williams, M., Haywood, A., Ellis, M., 2011. The Anthropocene – a new epoch of geological time? Phil. Trans. R. Soc. A 369 (1938), 835–841. https://doi.org/10.1098/rsta.2010.0339.

PREPAREDNESS

Resilience work in Swedish local governance: Evidence from the areas of climate change adaptation, migration, and violent extremism

4.1

Evangelia Petridou, Jörgen Sparf, and Kari Pihl
Risk and Crisis Research Centre, Mid Sweden University, Östersund, Sweden

CHAPTER OUTLINE

4.1.1 Introduction

Scandinavian countries in general and Sweden in particular have historically rarely been plagued by disasters, and they generally have low-risk national profiles. With the exception of large EU-support in combatting forest fires in recent years, the Swedish country image is that of the donor rather than its recipient. The Swedish government contributed 11.1 million USD in unearmarked (and therefore flexible) funds to the United Nations Office for Disaster Risk Reduction (UNDRR) in 2016–2017, the latest period for which UNDRR has released donor information, making Sweden its top contributor. As a comparison, the U.S. contributed 2.9 million USD of earmarked funds (UNDRR, n.d.).

Recently, the emergent erosion of American soft power worldwide as well as Russian posturing has foregrounded the Nordic[a] model in politics, culture, and society. The "Nordic exceptionalism" comprises discourse that was developed in the 1960s to describe the "third way," the compromise adopted by the Nordic countries in a then-polarized world between two competing ideologies, liberal democracy, and command-and-control economies (Danbolt, 2016). The concepts of the "Nordic model" and "Nordicity" have gained exceptional status, historically standing for "progress, modernization and for being *better than* other models" in terms of socioeconomic organization as well as stability, social advancement, lack of international claims and colonial history, and rationality in the political sphere (Browning, 2007, p. 27, emphasis in the original; Danbolt, 2016). The bipolar world power alignment shifted to an American hegemony, which in turn gave way to several states striving to gain a surplus in the power differential. Despite the euphoria of the early 1990s giving way to various endisms (Fukuyama, 1989/1999), history is not over and the investigation of the Nordic model, exceptional or not, is perhaps now more relevant than ever before.

At the same time that these shifts highlight the salience of the Nordic soft power (see Andersson, 2017), new threats are emerging while older ones change in scope and intensity. For example, in addition to the Russian posturing creating anxiety in the Baltic-Nordic region, climate change poses an existential threat to Arctic and subarctic regions, one would argue that the manner the Nordic countries handled the increased influx of refugees poses an existential threat as well: for the first time ever in 2016, border control on the Öresund bridge between Denmark and Sweden was initiated. Domestically, Kangas and Kvist (2013) projected that increased unemployment combined with decreased efficacy of social assistance might result in higher poverty levels. Demographic changes (an aging population) and ethnicity issues (a large number of foreign-born individuals some of whom may find integration in the job market difficult[b]) challenge the universality principle of the Nordic model. In other words, social cleavages are exacerbated. (Kangas and Kvist, 2013; for a more scathing but poetic retrospective of the Swedish utopia, see Pred, 2000).

Various, if relatively rare, extraordinary events have recently played out even in Sweden. Recent events with national repercussions include the terrorist attack in central Stockholm in April 2017, and the forest fires in the summers of 2014 and 2018, for example. The cultural imagination of these realized disasters and many other, chimerical ones, influences the response to current challenges. More specifically, the "emergency imaginary" dictates the perception of disasters among professionals (Calhoun, 2008) and, in turn, this perception determines what risks the professionals communicate to the public and how they do so. In May 2018, the Swedish Civil Contingencies Agency (MSB) sent a pamphlet to every Swedish household detailing how to respond to a variety of risks, from extensive electricity blackouts to invasion. The presaging text explained that even though Sweden compared to many other countries is relatively safe, it nonetheless faces various potential threats and urges the citizenry to be prepared, to be resilient (Swedish Civil Contingencies Agency, 2018b; Petridou et al., 2019).

---

[a]Nordic countries include Iceland and Finland, whereas Scandinavian countries generally do not. We have to note here that the distinction not always sharp. See, for example Einhorn and Logue, 2003.
[b]The percentage of foreign-born people living in Sweden is 19.1% in 2018 numbers (Central Intelligence Agency, 2019).

The questions that emerge concern the ways an advanced western democracy, such as Sweden, conceptualizes and understands risk, and, in turn, resilience. Through the analysis of the former (conceptualizing and understanding risk), we seek to produce new knowledge about the latter, namely, resilience in Sweden. More specifically, the aim of this chapter is to problematize existing theory on resilience from the Swedish perspective and to produce new knowledge regarding the way the concept may be applied in the Swedish context. Our empirical focus is the municipality level in Sweden as it is this local governance level that has the mandate (as the first stage) to provide crisis management. The remainder of the chapter is structured as follows: the two sections that follow describe different aspects of the Swedish case, followed by an explanation of the data and method used in our research. We then present the results and wrap up the chapter with a concluding discussion and avenues for further research.

4.1.2 **Resilience and the Swedish context**

This study is part of a larger project investigating the concept of societal resilience in the Swedish context. The term is not uncontroversial, perhaps not unsurprisingly if one considers the pitfalls that normally befall concepts that make the journey from the natural to the social sciences (see Prindle, 2012, for a treatment on a different concept). Even though an exhaustive literature review on the concept of resilience is beyond the scope of this chapter, we largely concur with Adger (2000), who defines resilience in society as "[t]he ability of human communities to withstand external shocks or perturbations to their infrastructure, such as environmental variability or social, economic, or political upheaval, and to recover from such perturbations" (quoted in Olsson et al., 2015, p. 2). As Olsson et al. (2015) point out, despite its analytical utility, resilience is difficult to use as a universal concept because it combines resistance—which is a static term—with adaptation, which is a dynamic term. In other words, resilience is a way to think about change (Miles and Petridou, 2015; Walker and Cooper, 2011) as much as it promotes the status quo (Harris et al., 2018). What is more, we view resilience not as a "goal or consensus outcome", some kind of achievable stasis, but a set of negotiated societal arrangements, a "[...] process through which ideals, policies, and agendas are sought, pursued and at times forced" (Harris et al., 2018, p. 198).

In addition to the criticism that it privileges the status quo, other critical voices have noted that resilience legitimizes a retreating state abdicating its responsibility to keep its citizens safe by shifting the blame to vulnerable (i.e., nonresilient) individual, an argument that also sheds light on the tensions between resilience and vulnerability (Chandler, 2012; Evans and Reid, 2014; Schott, 2013). Additional criticisms include that resilience, when used as a blanket policy instrument, tends to obfuscate power differentials and politically sensitive decisions (Coaffee and Lee, 2016; Olsson et al., 2015); that it has a normative function by applying a one-size-fits-all norm to everyone (Schott, 2013), and that it contributes to distributive injustice by ignoring unequal distribution of benefits in society (Chu et al., 2017).

In line with Harris et al. (2018), we appreciate the criticism leveled against resilience, but we believe that the concept has analytical value in understanding, theorizing, and planning for, complex social-ecological change. More specifically, and given the criticism, it is our view that the Swedish corporatist governance system with firm roots in welfare ideals and redistributive policies would be a fruitful case for the study of the concept of resilience. In this chapter, we do this by investigating how local bureaucratic organizations use the concept of resilience in Sweden to understand and work with risk.

Empirically, resilience is not an established concept among the actors working with DRR in Sweden. Instead, the term often used is *emergency preparedness* [krisberedskap]. Even though the latter points to the potential handling of large emergencies, the two terms are similar in substance: the overarching aim of Swedish emergency preparedness is to protect the population's life and health as well as society's functionality and ability to maintain basic values such as democracy, rule of law, human rights, and freedom (Swedish Civil Contingencies Agency, 2018a). Rather, the connection between the Sendai framework's priority *Understanding Risk* and emergency preparedness in Sweden should instead be considered as an aspect of operationalizing emergency preparedness, which translates to collaboration, principles for organization and control in crisis management, and risk-and vulnerability-analyses. Scandinavian democracy in general and Swedish democracy, in particular, privileges consensus building and consultative processes in drafting legislation and policy making. Additionally, ministers have a less direct role in day-to-day administration while governmental agencies have considerable authority (Einhorn and Logue, 2003). The need for collaboration is a corollary of political autonomy on the one hand and consensus politics on the other is the need for collaboration. In practice, vertical and horizontal collaboration is necessary in order to solve collective problems at hand.

Especially in the case of crisis management, the Swedish Civil Contingencies Agency (MSB) supports the development of policy, routines, research, technology, and training in the broader theme of collaboration. A large part of this collaboration between agencies and private actors takes place in six collaboration areas: financial security; hazardous substances; geographic areas of responsibility; protection, rescue, and care; technical infrastructure, and transportation. Civil society or volunteer actors are not recognized as an integral part of the crisis management system. Additionally, various social questions regarding risk reduction and resilience are not compatible with the scope of these collaboration areas (Swedish Civil Contingencies Agency, 2018a).

The Swedish national platform for disaster risk reduction was established in 2007 and its work is focused on three main areas: cooperation and coordination between stakeholders such as organizations and authorities, effective provision of data, and research and development (Swedish Civil Contingencies Agency, n.d.). A heading in the report by the national agency "Sweden also suffers from severe natural events" indicates that even though rarely severe natural events make international headlines, they do occur in Sweden with attendant economic, social, and environmental consequences (Swedish Civil Contingencies Agency, n.d., p. 4). The disaster risk reduction of the national agency focuses on long-term planning for resilience, an integrated approach with the rest of Europe, and privileges collaboration.

4.1.3 The Swedish case: Municipalities[c]

Swedish public administration is characterized by considerable independence of local governments and national agencies. At the local level, public administration is largely the responsibility of the local authorities/municipalities. Despite the considerable autonomy the 290 Swedish municipalities enjoy, local governance is, to a certain extent, subject to implementation of national policies

---

[c]We use the terms local authority, local government and municipality interchangeably in this chapter.

(Larsson and Bäck, 2008). Political reforms in Sweden since the 1970s have resulted in consolidation under central control in the 1970s, to a decentralization in the 1980s and 1990s, partly as a way to deal with cutbacks and challenges to the national welfare state. The past two decades have witnessed a push toward regionalization and calls for collaboration, both with other local authorities, as well as with extramunicipal organizations, often in form of networked governance (Montin and Hedlund, 2009; Niklasson, 2016). The considerable autonomy of the municipalities combined with established processes for emergency preparedness can hinder the implementation and usage of the Sendai-framework terminology and working methods at the local level. Additionally, as we show in the remainder of this section, the municipalities generally work with a different set of terms, partly because the Sendai-framework terminology is not (yet) in the Swedish emergency preparedness vernacular.

Crisis management in Sweden is based on three principles: (i) the *principle of responsibility*, according to which, entities responsible for a function during normal times retain this responsibility during a crisis or war; (ii) the *principle of parity*, according to which, entities retain their existing structure and location during extraordinary times (in crisis or war), and (iii) the *principle of proximity*, according to which, extraordinary events are to be handled at the lowest possible level of government (SOU, 2001, p. 41). Municipalities have a number of mandates, which include rescue operations, environmental protection, and civil defense (Swedish Association of Local Authorities and Regions, 2019). Additionally, municipalities have the mandate to perform risk analyses, as well as coordinate activities and collaborate during extraordinary events (Ministry of Justice, 2006; see also Petridou and Sparf, 2018). Similar to other European countries, the local level bears the biggest responsibility for disaster risk reduction (DRR) activities (Sparf and Migliorini, 2019).

More specifically, local authorities have the responsibility to produce a detailed risk - and vulnerability - analysis every 4 years, which then may be followed up yearly by the county government (Swedish Civil Contingencies Agency, 2015). County governments then in turn send the analyses of the local authorities to MSB. These analyses must necessarily (i) describe the local authority and its geographical area; (ii) describe the process and method of the analysis; (iii) identify socially vital operations within their jurisdiction; (iv) identify critical dependencies for the socially vital operations; (v) identify and analyze risks for the local authority and its jurisdiction; (vi) describe identified vulnerabilities and gaps in emergency preparedness in the local authority's jurisdiction, and (vii) identify measures to rectify any issues that emerge from the analysis.

All local authorities, regardless of size, have the obligation to provide the mandated services to their citizens. Sweden's territory measures just over $450,000\,km^2$ with a population of about 10 m. More than 90% of the population inhabits the southern half of the country (SCB, 2019a), mainly clustered around the three biggest cities Stockholm, Gothenburg, and Malmö, while northern interior areas are sparsely populated (EFGS, n.d.). As a result, there is great differentiation among the 290 municipalities, classified in three main categories: (i) large cities and municipalities near large cities, with a population of at least 200,000 inhabitants in the largest urban area; (ii) medium-sized towns and municipalities near medium-sized towns, with a population of at least 50,000 and at least 40,000 in the largest urban area, and (iii) small towns-municipalities with a population of at least 15,000 in the largest urban area (Swedish Association of Local Authorities and Regions, 2017). One local authority per category (Malmö, Örebro, and Arboga, respectively) comprises the cases studied for this chapter.

More specifically, Malmö has a population of 340,802; Örebro's population is 154,152, while in Arboga live 14,128 people (SCB, 2019b). Örebro has a slightly lower unemployment rate at 7.7%, compared to Malmö's 8.2 and Arboga's 8.0. As of December 2018, the municipality of Malmö employed 22,048 while in 2019, Örebro reported approximately 12,000 employees. As of November 2018, Arboga employed 1,375 individuals (Ekonomifakta, n.d.).

All three municipalities are situated in the southern half of the country, but otherwise represent a wide range of local authorities. This was an important factor for the research design underpinning this study, as we describe in the following section.

4.1.4 Research design

4.1.4.1 Rationale

We use three empirical fields representing potential risks for Swedish local authorities through which to study resilience: climate change adaptation, migration, and violent extremism. Our level of analysis is the organization and concerns the way they understand and conceptualize these potential risks and the ways the organize themselves in order to address them. The choice of risks represents those which are currently most prominent in the national agenda (see Center mot våldsbejakande extremism [Centre against violent extremism], 2019; SOU, 2017, p. 12; Klimatanpassningsportalen [the Swedish Portal for Climate Change Adaptation], 2019) and issues that, to a different extent, concern all three local authorities. As mentioned elsewhere in this chapter, the choice of local authorities aimed at providing a wide range of municipalities. This allows us to understand how municipalities conceptualize possible risks nuanced by their size, the resources, and know-how at their disposal. Aiming at the widest possible variation will allow us to cautiously make generalizable knowledge claims for the entire country, claims that we may, at a minimum, test at the next stage of this research process.

Sweden, taken as a case study in which to study resilience, is a form of an "extreme case" (Gerring, 2008). The rationale underpinning the value of this case study for theory development is a high score on a variable of interest—in this case, political, social, and economic stability. Even though this study is qualitative and our vocabulary does not include values on variables, the aim of the knowledge claims produced is to tease out the concept of resilience, good practices as well as challenges, policies and practices, in a stable advanced western democracy.

4.1.4.2 Method and material

Group interviews with key informants comprised the data for this chapter. More specifically, we conducted 15 semistructured interviews with public servants and civil society workers in the municipalities of Malmö, Örebro, and Arboga in September 2018 and February 2019, in the fields of climate change adaptation, migration, and violent extremism. In aiming for dynamic interviews, our objective was to recruit as many different participants at each municipality and empirical field as possible. These participants were interviewed together, with the exception of the field of violent extremism in Malmö, where the municipal public servants and civil society were interviewed separately and additional interviews were conducted with civil society in the field of migration as well. Participating informants ranged from 2 to 6 per interview, with the discussion lasting an average of 2 h, conducted at the respective municipality's (or, alternatively, the civil society organization's) premises. All the interviews were recorded and subsequently transcribed. In addition to these data, we have also had frequent, periodic contact with key persons in the municipalities and the national agency as a conduit of result validation, quality control, and contextual relevancy.

We were aware of "the perils of strategic reconstruction" (Rathbun, 2008, p. 694) that sometimes occur in interviews as a means of data collection, and ensuring a diverse group of participants in each group interview was a means to combat this. We welcomed the participants' narratives that went beyond them answering the number of *a priori* questions comprising the interview guide.

The analysis for this chapter was inductive. We conducted a thematic analysis of the transcribed material on Atlas.ti (Friese, 2014). The coding team (the three authors) consisted of two senior researchers in political science and sociology and a research assistant. One author produced a draft code book based on a subset of the interviews, which was renegotiated and edited in ensuing author-team meetings (see Frisch-Aviram et al., 2020 and Jones et al., 2016 for a similar process). A first meeting to reconcile the coding yielded 30 code groups and 166 individual codes. Subsequent meetings yielded 168 codes and 29 groups, not all of which are used in the ensuing discussion. The discussions resulting in the development of codes into emerging categories included a dialogue of the team members' feelings and relationship to research in general, in addition to the material in particular (Hall et al., 2005). These dialogues reflected the different disciplinary backgrounds, biases, as well as experience levels of the team. Iterative communication not only brought these items to the surface, but also made the coding process transparent and negotiable, the analysis more creative, and enabled crossdisciplinary knowledge sharing (Phoenix et al., 2016). Coding became an integral part of the analysis of the material, while it shaped the research relation of the team. The collaborative process enriched the analysis through the dialectic process of the inductive analysis (Weston et al., 2001).

4.1.5 Results

In this section we unpack and analyze the narratives emerging from the group interviews, especially with a view to revealing similarities across municipalities and empirical areas when it comes to processes furthering resilience in emergency preparedness. The narratives we report here are tied to items identified in the Sendai framework as ways toward a better understanding of risk and center on collaboration and knowledge transfer. Additionally, even though this study is inductive, the connection between resilience and collaboration (especially in the form of collaborative governance) as policy knowledge transfer and resilience are not new (see for example, Hartley, 2018; Patton and Johnston, 2017; Tierney, 2003).

4.1.5.1 Articulations of resilience in local government bureaucracy

Few interviewees spoke of resilience per se. As noted elsewhere in this chapter, public servants did not directly work with a resilience-based conceptual or implementation framework. As evidenced by, *inter alia,* recent work toward a new governing document in Malmö, this is beginning to change. In Malmö specifically, the public sector worked with *sustainability*; public servants must delineate the two concepts, especially in terms of resilience being "a way of thinking of how to find alternative solutions"; indeed "being creative, finding new solutions, a new way to work" (R 1.1.1; R 1.3.3). Such need for flexibility is particularly relevant during extraordinary events when the system is stressed. Respondents report improvising and taking an elastic approach when it comes to established routines as strategies to handle extraordinary stressors. At the same time, the importance of working preventatively is emphasized, namely, that work toward resilience necessitates adequate preparedness. Related to this and at the individual level, public officers working under stress during their daily work—such as social workers—are thought to be more resilient in the sense that resilience translates to being more well prepared for an extraordinary stressor. Furthermore, social interactions and an emphasis on the individual level characterize the understanding and operationalizing of resilience: "so, talking about resilience, this is how we build resilience, we build resilience by giving information, by interacting with people" (R 1.4.1).

4.1.5.2 **Collaboration as a dimension of resilience: Conduits and hurdles**

A common theme cutting across empirical fields and municipalities was that of collaboration. Collaboration, beyond its being encouraged and even mandated as a top-down imperative way of organizing work, is at the center of any discussion of emergency preparedness regardless of the size of the local authority, whether the actor comes from the public or volunteer sector or the area in which they work. Broadly speaking, collaboration has normative value, but also substantive importance; it is multifaceted and not unproblematic. It is contingent on extant power relations and it also engenders and reproduces power relations. It may be formal or informal with outputs and impacts that are hard to measure. Finally, there exist interdependencies between collaboration, legitimacy, intergovernmental relations, and governance.

More specifically, collaboration that is, the act of working with others, sharing information, pooling resources, and coordinating actions, is a desirable state of affairs: "collaborate, coordinate, pool procurement resources: that's what it's all about if we are going to have any effect" (R 1.3.2). This includes collaboration among different departments of the local authority, vertical collaborative activities (with the county government and national agencies), horizontal collaboration among local authorities, as well as multiorganizational collaboration, including a mixture of public and third sector actors.

Tangible benefits include the building of trust during normal operations that facilitates the handling of extraordinary events as well as sharing information, including trends, in order to construct a joint situational awareness. Notably the latter is not a straightforward term either, and at least one respondent pointed out the need for defining what joint situational awareness is, despite the fact that collaborative activities within the local authorities are constructed around this term.

However, in practice, a lot of resilience work at the local level takes place in silos and collaboration (with the exception of intramunicipal collaborative networks and the police officers in the local authority of Arboga reporting increased information flow from the national security agency) seemingly takes on the qualities of a unicorn: everyone loves it and has an idea of what it looks like, but no one has actually seen it. Along this vein, respondents speak of collaboration in the subjunctive, expressing wishes of how things ought to be rather than how things. Collaboration is desirable and should take place: "we have to find a way to work together" (R 1.3.2); "we would like to have a long-term contact with civil society" (R 2.3.2).

However, there is a host of hurdles. Collaboration, especially with civil society, entails high transaction costs:

> It would entail quite a bit of work, just like if we are talking about having a local presence, then one must also be active there, having meetings, because one has to give also, not just take. We are talking about working together, and at least from my perspective, I find it difficult finding time for this part, to work like this (R 2.3.3).

These transaction costs are compounded by the (perceived) fluidity of civil society organizations, operationalized as high turnover of personnel. A specific issue expressed by local authority public servants working in the fields of migration and refugeehood and violent extremism is the legitimacy of civil society organizations, specifically these organizations that are established and run by people recently arrived in Sweden [in Swedish: nyanlända]—another way to refer to recent immigrants in Sweden. Swedish public servants cannot be seen working with organizations based on premises contravening the fundamental Swedish ideas of democracy and equality, especially gender equality: "[...] it's enough with a seed of suspicion, then it comes quite difficult to collaborate in an honest and good way" (R 2.3.2).

Conversely, civil society workers note the rigidity and the perceived high level of bureaucratization of the municipal public workers, as well as the politicization of issues and the question of turf:

> And the conversation has always been that 'yeah, we want to do that', but then nothing really happens and they wonder why nothing is happening. And you realize soon after that this is all politics, you know. They want to keep you out as much as possible. There in Stockholm, this is all about policies and it's all about … It's lots of money involved in it. It is a lot prestige involved (R1.4.1).

The term *prestige* in this quote refers to the fact that the local authority has competence in certain issues and the mandate to provide certain public goods and services. Asking for help from a civil society organization may at times be perceived as a weakness from the side of the public sector, a sign of a perceived inability to handle things, and thus creating the image that the public servants are not performing well in their jobs. In comparison, both local authorities and civil society organizations tout their collaboration with experts in local and international universities. These collaborative activities become a proxy for legitimacy, stemming from a reliance on expertise for the appropriate methodologies to use in order to provide high-quality emergency preparedness.

A considerable part of collaborative activities takes place in project form. Though the projectification of bureaucratic activities is seen as promoting creativity and may be a laboratory for the institutionalization of new routines (Sparf, 2019), it also creates project application fatigue among civil society organizations that have neither the time for, nor the expertise in, grant writing and administrative reporting, which, as respondents note, is crucial for the continuation of funding in the future. It may also result in fragmentation of activities and lack of continuity.

In addition to project financing, some civil society organizations receive funds from the municipality on a yearly basis. Control of the purse implies that the municipality is able to influence the work of the civil society organization and require evaluations. Respondents from both the local authorities and the civil society organizations concurred that prevention work is difficult to evaluate as it is long term and rarely shows immediate tangible results. They also pointed to the paradox that the better work they do, the fewer the incidents that happen and the bigger the potential for their budget to be cut. Public servants from all empirical areas and municipalities touched upon the issue of politicization. If the question they are working on, be it climate change adaptation, migration, or violent extremism, receives media attention, their budget is more likely to increase, but at the same time they are subject to more scrutiny; such politicization cuts both ways.

4.1.5.3 **The utility of science and learning as a dimension of resilience**

An evidence-based approach is apparent, at least in the discourse of our informants. This is in line with the Sendai framework guidelines on understanding risk through the use of science in decision-making processes. We must note here that Sweden is seen as a pioneer in the use of scientific knowledge in policymaking and the culture of evaluation in entrenched in bureaucratic functions (Knill and Tosun, 2012, see also Varone et al., 2005; Viñas, 2009). Informants from the two larger municipalities, both from the local government and civil society, explicitly referred to their work (methods and routines) being grounded on research, even reporting collaboration with universities in the area.

Having said that, practitioners expressed frustration regarding translation work: "it's difficult for researchers to talk to practitioners and for practitioners to talk to researchers" (R1.3.2). The language of academe is different to the one used by practitioners in scope, detail, and level of abstraction. Though researchers ought to distill the results in less abstract, concrete terms, an extreme reductionist approach would threaten the integrity of research results. Additionally, one informant lamented the research funding timing mismatch: in 2014, she would have liked to pick up the phone and ask researchers to observe the handling of forest fires, but the research funding system is not that agile.

4.1.6 **Conclusions and avenues for further research**

Increasing shifts in the geopolitical order, globalization, and an erosion of American soft power, have contributed to the relevance of the Nordic model in many societal arenas, including that of societal resilience. In this chapter, we aimed at problematizing existing theory on resilience from the Swedish perspective as well as produce new knowledge, through an empirical focus at the local governance level, on how risk is understood and managed in the context of emergency preparedness and in line to the Sendai framework for disaster risk reduction. In this chapter, we focused on issues relating to collaboration (especially with civil society) and knowledge transfer.

The inductive, qualitative analysis of our three case studies at the local governance in Sweden reveals a set of commonalities when it comes to resilience and emergency preparedness. The dominant emerging theme is that of collaboration, and this is not surprising for a few reasons. First, substantively, a certain pooling of resources is needed in order to handle extraordinary events, and this presupposes a level of collaboration during normal operations. Additionally, this is a theme that has permeated the national discourse for many years in the sense that it is an integral part of publications by the Civil Contingencies Agency and an imperative for local authorities.

However, public servants still work in silos. They collaborate within the boundaries of the municipality, but other forms of collaboration tend to be more difficult to achieve and maintain. There seems to be a mismatch between the organization of work that takes place at the municipality, centering on reporting, risk and vulnerability analyses and frequent intra-municipal network meetings, and that organization of work of outside organizations that tend to be less bureaucratic, less professional, at times unstable, and perhaps less permanent. Additionally, the local authorities often finance the operations of civil society organizations, which creates power differentials that are reproduced through the projectification of work and the constant need to apply for funding. Power differentials exist vertically as well, with municipalities in turn being dependent on national agencies for information and funding. There is a level of mistrust between the public and volunteer sector and the latter attempt to assert legitimacy through claims of long experience in the field or collaboration with academics. Local authorities also rely on expertise in an apparent turn toward evidence-based local policy making.

Having said all this, further research is needed to fully understand the organization of emergency preparedness on the one hand, and its effectiveness on the other, in Sweden. Such research could focus on the differences between local authorities as well as contextual factors related to the size of the local authority, the expertise available, and the occurrence of an extraordinary event and how they, in turn, affect resilience work at the local level. Crosscountry comparative work would also be of value in revealing nuances in local emergency management practices. Since, as mentioned elsewhere in this chapter, natural (as well as anthropogenic) events take place in Sweden as well, such further research will be of theoretical, but also of empirical and practical value.

References

Adger, W.N., 2000. Social and ecological resilience: are they related? Prog. Hum. Geogr. 24 (3), 347–364. https://doi.org/10.1191/030913200701540465.

Andersson, S., 2017. Sverigebilden håller på att bli mer nyanserad. Svenska Dagbladet. Editorial, 11 August (in Swedish).

Browning, C.S., 2007. Branding nordicity: models, identity and the decline of exceptionalism. Coop. Confl. 42 (1), 27–51. https://doi.org/10.1177/0010836707073475.

Calhoun, C., 2008. A world of emergencies: fear, intervention, and the limits of cosmopolitan order. Can. Rev. Sociol. 41 (4), 373–395. https://doi.org/10.1111/j.1755-618X.2004.tb00783.x.

Center mot våldsbejakande extremism, 2019. Available from: (in Swedish), (Accessed 19.06.19), https://www.cve.se/.

Central Intelligence Agency, 2019. The World Factbook, Sweden. (Accessed 19.06.19) Available from: https://www.cia.gov/library/publications/the-world-factbook/geos/sw.html.

Chandler, D., 2012. Resilience and human security: the post-interventionist paradigm. Secur. Dialogue 43 (3), 213–229. https://doi.org/10.1177/0967010612444151.

Chu, E., Anguelovski, I., Roberts, D., 2017. Climate adaptation as strategic urbanism: assessing opportunities and uncertainties for equity and inclusive development in cities. Cities 60, 378–387. https://doi.org/10.1016/j.cities.2016.10.016.

Coaffee, J., Lee, P., 2016. Urban Resilience. Palgrave Macmillan, London.

Danbolt, M., 2016. New Nordic exceptionalism. In: Kim, J.J., Einhorn, E. (Eds.), The United Nations of Norden and other realist utopias. J. Aesthet. Cult. 8(1). https://doi.org/10.3402/jac.v8.30902.

Einhorn, E.S., Logue, J., 2003. Modern Welfare States: Scandinavian Politics and Policy in the Global Age, second ed. Praeger, Westport, CT.

Ekonomifakta, n.d. Arbetslöshet nyckeltal [Unemployment figures]. (Accessed 19.06.19) (in Swedish) Available from: https://www.ekonomifakta.se/Fakta/Regional-statistik/Alla-lan/Vastmanlands-lan/Arboga/?var=17259.

European Forum for Geography and Statistics (EFGS), n.d. Available from: https://www.efgs.info/information-base/case-study/dissemination/visualisation-using-cartograms/.

Evans, B., Reid, J., 2014. Resilient Life: The Art of Living Dangerously. Polity, Malden, MA.

Friese, S., 2014. Qualitative Data Analysis With Atlas.ti, second ed. Sage, Thousand Oaks, CA.

Frisch-Aviram, N., Cohen, N., Beeri, I., 2020. Wind(ow) of change: a systematic review of policy entrepreneurship characteristics and strategies. Policy Stud. J. https://doi.org/10.1111/psj.12339.

Fukuyama, F., 1989/1999. The end of history? In: Foreign Affairs (Ed.), The New Shape of World Politics: Contending Paradigms in International Relations. Council of Foreign Relations, New York, pp. 1–25.

Gerring, J., 2008. Case selection for case-study analysis: qualitative and quantitative techniques. In: Box-Steffensmeier, J.M., Brady, H.E., Collier, D. (Eds.), The Oxford Handbook of Political Methodology. Oxford University Press, Oxford, pp. 645–684.

Hall, W.A., Long, B., Bermbach, N., Jordan, S., Patterson, K., 2005. Qualitative teamwork issues and strategies: coordination through mutual adjustment. Qual. Health Res. 15 (3), 394–410. https://doi.org/10.1177/1049732304272015.

Harris, L.M., Chu, E.K., Ziervogel, G., 2018. Negotiated resilience. Resilience 6 (3), 196–214. https://doi.org/10.1080/21693293.2017.1353196.

Hartley, K., 2018. Environmental resilience and intergovernmental collaboration in the Pearl River Delta. Int. J. Water Resour. Dev. 34 (4), 525–546. https://doi.org/10.1080/07900627.2017.1382334.

Jones, M.D., Peterson, H.L., Pierce, J.J., Herweg, N., Bernal, A., Lamberta Raney, H., Zahariadis, N., 2016. A river runs through it: a multiple streams meta-review. Policy Stud. J. 44 (1), 13–36. https://doi.org/10.1111/psj.12115.

Kangas, O., Kvist, J., 2013. Nordic welfare states. In: Greve, B. (Ed.), The Routledge Handbook of the Welfare State. Routledge, New York, pp. 148–160.

Klimatanpassningsportalen, 2019. The Swedish Portal for Climate Change Adaptation. (in Swedish) Available from: https://www.klimatanpassning.se/roller-och-ansvar/kommande-underlag/pa-gang-fran-myndigheterna-2019-1.146533.

Knill, C., Tosun, J., 2012. Public Policy: A New Introduction. Palgrave Macmillan, Basinstoke, Hampshire, UK.

Larsson, T., Bäck, H., 2008. Governing and Governance in Sweden. Studentlitteratur, Malmö.

Miles, L., Petridou, E., 2015. Entrepreneurial resilience: role of policy entrepreneurship in the political perspective of crisis management. In: Bhamra, R. (Ed.), Organisational Resilience: Concepts Integration and Practice. CRC Press, Boca Raton, FL, pp. 67–81.

Ministry of Justice, 2006. Lag [Swedish legislature] 2006:544, om kommuners och landstings åtgärder inför och vid extraordinära händelser i fredstid och höjd beredskap. https://www.riksdagen.se/sv/dokument-lagar/dokument/svensk-forfattningssamling/lag-2006544-om-kommuners-och-landstings_sfs-2006-544. Accessed 19.06.19.

Montin, S., Hedlund, G., 2009. Governance som interaktiv samhällsstyrning—gammalt eller nytt i forskning och politik? In: Hedlund, G., Montin, S. (Eds.), Governance på svenska. Santérus Academic Press Sweden, Stockholm, pp. 7–36 (in Swedish).

Niklasson, L., 2016. Challenges and reforms of local and regional governments. In: Piere, J. (Ed.), The Oxford Handbook of Swedish Politics. Oxford University Press, Oxford, pp. 399–413.

Olsson, L., Jerneck, A., Thoren, H., Persson, J., O'Byrne, D., 2015. Why resilience is unappealing to social science: theoretical and empirical investigations of the scientific use of resilience. Sci. Adv. 1(4), e1400217. https://doi.org/10.1126/sciadv.1400217.

Patton, D., Johnston, D., 2017. DIsaster Resilience, second ed. Charles Thomas, Springfield, IL.

Petridou, E., Sparf, J., 2018. For safety's sake: the strategies of institutional entrepreneurs and bureaucratic reforms in Swedish crisis management, 2001–2009. Polic. Soc. 1–19. https://doi.org/10.1080/14494035.2017.1369677.

Petridou, E., Danielsson, E., Olofsson, A., Lundgren, A., Grosse, C., 2019. If crisis or war comes: a study of risk communication of eight EU member states. Int. J. Crisis Risk Commun. 2 (2), 207–232. https://doi.org/10.30658/jicrcr.2.2.3.

Phoenix, A., Brannen, J., Elliott, H., Smithson, J., Morris, P., Smart, C., Barlow, A., Bauer, E., 2016. Group analysis in practice: narrative approaches. Forum Qual. Soc. Res. 17(2). https://doi.org/10.17169/fqs-17.2.2391.

Pred, A., 2000. Even in Sweden: Racisms, Racialized Spaces, and the Popular Geographical Imagination. University of California Press, Berkeley.

Prindle, D.F., 2012. Importing concepts from biology into political science: the case of punctuated equilibrium. Policy Stud. J. 40, 21–44. https://doi.org/10.1111/j.1541-0072.2011.00432.x.

Rathbun, B.C., 2008. Interviewing and qualitative field methods: pragmatism and practicalities. In: Box-Steffensmeier, J.M., Brady, H.E., Collier, D. (Eds.), The Oxford Handbook of Political Methodology. Oxford University Press, Oxford, pp. 685–701.

SCB, 2019a. Statistical Database 2019. Available from: http://www.statistikdatabasen.scb.se/pxweb/en/ssd/?rxid=f45f90b6-7345-4877-ba25-9b43e6c6e299. Accessed 10.06.19.

SCB, 2019b. Folkmängd i riket, län och kommuner 31 mars 2019 och befolkningsförändringar 1 januari–31 mars 2019. Available from: https://www.scb.se/hitta-statistik/statistik-efter-amne/befolkning/befolkningens-sammansattning/befolkningsstatistik/pong/tabell-och-diagram/kvartals–och-halvarsstatistik–kommun-lan-och-riket/kvartal-1-2019/. Accessed 19.06.19.

Schott, R.M., 2013. Resilience, normativity and vulnerability. Resilience 1 (3), 210–218. https://doi.org/10.1080/21693293.2013.842343.

SOU, 2001. Säkerhet i en ny tid [Security in a New Age]. Publication No.: 2001:41. Statens Offentliga Utredningar [Official Government Review Series]. 2001. Stockholm (in Swedish).

SOU, 2017. Att ta emot människor på flykt Sverige hösten 2015. Publication no.:2017:12. Statens Offentliga Utredningar [Official government review series]; 2017. Stockholm (in Swedish).

Sparf, J., 2019. Interfaces in temporary multi-organizations in routine emergency management: the case of Stockholm. Saf. Sci. 118, 702–708. https://doi.org/10.1016/j.ssci.2019.05.043.

Sparf, J., Migliorini, M. (Eds.), 2019. Socioeconomic and Data Challenges: Disaster Risk Reduction in Europe. United Nations Office for Disaster Risk Reduction. Available from: www.preventionweb.net/publications/view/65182.

Swedish Association of Local Authorities and Regions, 2017. Kommungruppsindelning 2017. Available from: https://skl.se/tjanster/kommunerochregioner/faktakommunerochregioner/kommunernasataganden.3683.html. Accessed 19.06.19 (in Swedish).

Swedish Association of Local Authorities and Regions, 2019. Kommunernas åtaganden. Available from: https://skl.se/tjanster/kommunerochregioner/faktakommunerochregioner/kommunernasataganden.3683.html. Accessed 19.06.19.

Swedish Civil Contingencies Agency, 2015. MSBFS 2015:5 föreskrifter och allmänna råd om kommuners risk- och sårbarhetsanalyser. (In Swedish). Available from: https://www.msb.se/externdata/rs/15e78831-767b-4714-9fa4-3b4fd0df92a8.pdf.

Swedish Civil Contingencies Agency, 2018a. Gemensamma grunder för samverkan och ledning vid samhällsstörningar. Publication no.: MSB777. https://www.msb.se/RibData/Filer/pdf/28738.pdf. (Accessed 17.06.19) (in Swedish).

Swedish Civil Contingencies Agency, 2018b. Om krisen eller kriget kommer. Publication no.: MSB1186. https://www.msb.se/RibData/Filer/pdf/28494.pdf. (Accessed 19.06.19) (in Swedish).

Swedish Civil Contingencies Agency, n.d. Swedish National Platform for Disaster Risk Reduction: Dealing With Natural Disasters—A Matter of Cooperation. Available from https://www.msb.se/siteassets/dokument/publikationer/english-publications/dealing-with-natural-disasters–swedish-national-platform.pdf (Accessed 20.06.19).

Tierney, K.J., 2003. Conceptualizing and Measuring Organizational and Community Resilience: Lessons Learned From the Emergency Response Following the September 11, 2001 Attack on the World Trade Center. DRC Preliminary Papers; 329. Available at http://udspace.udel.edu/handle/19716/735. Accessed 28.11.19.

UN Office for Disaster Risk Reduction, n.d. Donor Partnerships. https://www.unisdr.org/who-we-are/donors (Accessed 19.06.19).

Varone, F., Jacob, S., De Winter, L., 2005. Polity, politics and policy evaluation in Belgium. Evaluation 11 (3), 253–273. https://doi.org/10.1177/1356389005058475.

Viñas, V., 2009. The European Union's drive towards public policy evaluation: the case of Spain. Evaluation 15 (4), 459–472. https://doi.org/10.1177/1356389009341900.

Walker, J., Cooper, M., 2011. Genealogies of resilience: from systems ecology to the political economy of crisis adaptation. Secur. Dialogue 42 (2), 143–160. https://doi.org/10.1177/0967010611399616.

Weston, C., Gandell, T., Beauchamp, J., McAlpine, L., Wiseman, C., Beauchamp, C., 2001. Analyzing interview data: the development and evolution of a coding system. Qual. Sociol. 24 (3), 381–400.

Another take on reframing resilience as agency: The agency toward resilience (ATR) model

4.2

Eva Louise Posch[a], Karl Michael Höferl[a], Robert Steiger[b], and Rainer Bell[a]

Department of Geography, University of Innsbruck, Innsbruck, Austria[a] Department of Public Finance, University of Innsbruck, Innsbruck, Austria[b]

CHAPTER OUTLINE

Understanding Disaster Risk. https://doi.org/10.1016/B978-0-12-819047-0.00013-5

4.2.1 Introduction

Over the last 15 years, the concept of resilience entered the debate on disaster risk reduction (DRR) and has become an accepted—although controversial—element in local and global DRR policies such as the Hyogo and Sendai Framework (UNISDR, 2007, 2015)[a] and in contemporary research on disaster risk (e.g., Adger, 2000; Birkmann, 2006; Cannon, 2008; Cutter et al., 2008; Klein et al., 2003). In order to invest in disaster risk reduction for resilience as named in Priority 3 of the Sendai Framework, we first need to improve our conceptual understanding of the interlinkages between disaster risk and social resilience (Priority 1 of the Sendai Framework) (UNISDR, 2015).

Thus, we focus in this contribution on the ongoing scientific debate on understanding disaster risk and social resilience to natural hazards. Most definitions of social resilience refer to capacities, which are an important dimension of disaster risk (UNISDR, 2015). We argue that, on the one hand, it is not enough to see social resilience as a set of capacities that can be measured. On the other hand, studies paying attention to agency toward resilience focus mainly on the ability to access various assets, structural conditions, and given properties to build resilience (e.g., Cutter et al., 2008; Ifejika Speranza et al., 2014; Magis, 2010; Norris et al., 2008; Sharifi, 2016; Sherrieb et al., 2010). However, it has been empirically proven that a narrow, one-dimensional focus on the "structural ability" cannot convincingly explain why some people take actions toward resilience while others do not. In contrast, values, beliefs, and worldviews seem to play a more essential role for understanding the individual's rationality toward resilient action than assumed (Appleby-Arnold et al., 2018; Posch et al., 2019; Rawluk et al., 2017; Rawluk et al., 2018). However, existing frameworks and models do not include these aspects. Therefore, we propose a dualist, agency-sensitive model on social resilience, which can guide future research in this field. The model rests on empirical findings from a case study in Nepal and a literature review (Posch et al., 2019). Given the vast number of publications on resilience in the context of disaster risk, we do not make the claim to be exhaustive regarding all conceptual aspects but believe that we have included key issues central to the discussion.

To this end, we start with a brief review of how the discussion on social resilience adopted a perspective focused on social entities and their capacities instead of a systems theory approach found in early works on resilience (Section 4.2.2). Section 4.2.3 justifies why agency is a rather vague aspect of social resilience, often approximated with the mere ability to act. In Section 4.2.4, we present three findings from this brief stroll through social resilience's more recent history, covering our limited understanding of agency toward resilience, the role of individuals for this agency, and how values and beliefs influence actions toward resilience. These findings prepare the ground for Section 4.2.5, in which we propose a model of Agency Toward Resilience (ATR). Some critical reflections (Section 4.2.6) and remarks on this ATR model (Section 4.2.7) conclude this chapter.

4.2.2 From systems to social entities and their capacities

Whether resilience is a valid and worthwhile concept for the study of society remains one of the main controversial issues in resilience research (Keck and Sakdapolrak, 2013). Doubts about this suitability are often driven by the observation that the current resilience debate does not address social issues

---

[a]UNISDR is now known as UNDRR.

(Béné et al., 2012; Cote and Nightingale, 2012; Fabinyi et al., 2014). This failure is most often discussed by referring to aspects of justice, the distribution of power resulting in unequal access to resources, institutional constraints and limited individual capacities to cope, adapt, or transform (Keck and Etzold, 2013; Keck and Sakdapolrak, 2013; Obrist et al., 2010). A common reason for doubt is seen in the uncritical transfer of resilience's system theory approach—originating from the sphere of natural systems—to social phenomena (Cannon and Müller-Mahn, 2010; Davidson, 2010). Thus, Bohle et al. (2009) first expressed the need to refocus on people-centered approaches to better understand "social" resilience by moving actors as well as their activities, capacities, and strategies in the center of attention. In the resulting discussion on social resilience, research focuses on social entities—individuals, households, or communities (Keck and Sakdapolrak, 2013). Most definitions of social resilience incorporate the idea of capacity (Cannon, 2008; Cannon and Müller-Mahn, 2010), understood as the ability "to cope, adapt, and transform in response to different types of changes, including sudden shocks or crises" (Brown and Westaway, 2011, p. 322).

During the last decade, numerous approaches have been applied to classify these capacities, resulting in divergent interpretations of the latter. Coming from the psychological sciences, Norris et al. (2008, p. 135) suggest that "resilience emerges from a set of adaptive capacities," which they identify as economic development, social capital, information and communication, and community competence. Obrist et al. (2010) introduce a twofold reading of capacities, inspired by Alexander's (2005) disaster-management cycle. Using adverse events as temporal anchors, they see social resilience as "the capacity of actors to access capitals in order to – not only cope with and adjust to adverse conditions (that is, reactive capacity) – but also search for and create options (that is, proactive capacity) and thus develop increased competence (that is, positive outcomes) in dealing with a threat" (Obrist et al., 2010, p. 289). Based on previous discussions (Béné et al., 2012; Norris et al., 2008; Obrist et al., 2010; Voss, 2009), Keck and Sakdapolrak (2013) advocate a threefold reading of capacities, including coping, adaptive, and transformative capacities. These three capacities are distinguished using the type of response to a threat, the temporal scope of the response, and the degree of change associated with this response as classifying variables. While coping capacity covers the ability to absorb, persist, or withstand immediate shocks and stressors caused by various kinds of hazardous events (Adger, 2000; Turner, 2003), adaptive and transformative capacities address the anticipatory capacities "that facilitate the ability of the social system to re-organize, change, and learn" (Cutter et al., 2008, p. 599) to deal with change and an uncertain future. Transformative capacities encompass the ability to engage in decision-making and governance processes aiming at progressive, paradigmatic change (Darnhofer, 2014). In a recent contribution, Hutter and Lorenz (2018) provide another perspective on capacities relevant for social resilience. Based on ideas outlined by Voss (2009), they argue that coping and participative capacities are best suited to explain social resilience, as they highlight the embeddedness of social actors in a "social fabric" and acknowledge different radii of agency.

Whether resilience is understood as tripartite combination of buffering, adapting, and transforming capacities, as a dual combination of re- and proactive capacities or any other combination, the capacities of social entities are an integral part of the ongoing discussion of social resilience.

On a more pragmatic level, the question arose, what constitutes these capacities. Most analyses put an emphasis on people's ability to act (Brown and Westaway, 2011) determined by the access to economic, social, natural, physical, or human assets. Analogue to the Sustainable Livelihood Framework (SLF—Scoones, 1998), this access to assets as well as given properties such as gender or age are seen as main determinants of community or household resilience (Adger, 2003; DFID, 1999; Norris et al., 2008).

4.2.3 **Why asset-based approaches might not be enough**

More recently, a significant number of scholars (e.g., Béné et al., 2012; Bohle et al., 2009; Hutter and Lorenz, 2018; Keck and Sakdapolrak, 2013; Lorenz and Dittmer, 2016; Obrist et al., 2010; Obrist, 2016; Voss, 2009) have addressed human agency as an important—and often forgotten—aspect of social resilience. Already in 1996, Brown and Kulig (1996, p. 30) claim that resilience "[…] needs to be grounded in a notion of human agency, understood in the sense of the capacity for meaningful, intentional action." In 2009, Bohle et al. suggested an "agency-based resilience framework," which recognized agency as "a vital issue in the conceptualization of social resilience and a theme that resilience discourse has, so far, failed to satisfactorily address" (Bohle et al., 2009, p. 8). This view is supported by Pain and Levine (2012, p. 3) who argue that a "focus on people's agency, their ability to make and follow through on their own plans in relation to socioeconomic security, might constitute a core ingredient of resilience." Despite these claims, agency remains a rather vague element in the discussion on social resilience (Obrist, 2016).

In livelihood and development studies, agency is commonly understood as the human capacity, ability, or capability to act (Ifejika Speranza et al., 2014; Keck and Sakdapolrak, 2013; Obrist, 2016). Kuhlicke et al. (2011, p. 807) provide a more nuanced definition of a social entity's agency as "the context-related ability to decide and to behave successfully in a certain situation in order to anticipate, respond to, cope with, recover from or adapt to the negative impacts of an external stressor (e.g., a hazardous event) as well as to employ the necessary resources." In line with this thinking, Coulthard (2012, para. 13) sees agency as "the capability of a person to act and make a difference to a pre-existing state of affairs or events." Inspired by sociological action theory (Emirbayer and Mische, 1998), several scholars (e.g., Obrist, 2016; Lorenz and Dittmer, 2016; Hutter and Lorenz, 2018) develop time-sensitive approaches to investigate the connection between human agency and social resilience. Here, human agency encompasses iteration (reactivating past patterns of action), projectivity (envisioning possible future courses of action), and practical evaluation (normative judgments). According to Emirbayer and Mische (1998, p. 970), human agency can be defined as "[…] the temporally constructed engagement by actors of different structural environments – the temporal-relational contexts of action – which […] both reproduces and transforms those structures in interactive response to the problems posed by changing historical situations."

The question how we can understand human agency in the context of social resilience is far from being settled (Hutter and Lorenz, 2018). Despite this open-ended discussion, an agency-based perspective on social resilience recognizes that individuals, households, or communities are not just passive owners of assets but actors with the ability to make their own choices in the face of adverse circumstances (Béné et al., 2012; Brown and Westaway, 2011; Pain and Levine, 2012).

4.2.4 **Three findings from the debate on social resilience**

Based on the briefly described discussion on capacities, access to assets and agency in the field of social resilience, we like to draw three conclusions:

4.2.4.1 **Our understanding of a social entity's agency toward resilience is limited**

Many approaches to social resilience refer to resilience as the general capacity to deal with change and surprises, influenced by the agency of a social entity and its ability to access critical assets. Often, definitions of these elements and their relationship to each other remain unclear. Beside this, the agency of social entities is often represented in a reductionist way, since most empiric analyses put an emphasis on a social entity's mere ability—not willingness—to act (Brown and Westaway, 2011). In this reductionist perspective, the access to economic, social, natural, physical, or human assets is seen as main determinant of an entity's resilience (Adger, 2003; DFID, 1999; Norris et al., 2008). Such asset-based approaches have been widely applied in disaster risk research and reduction initiatives to develop metrics for measuring (social) resilience and vulnerability (e.g., Bakkensen et al., 2017; Bruneau et al., 2003; Cutter et al., 2008, 2010; Fekete, 2009; Keating et al., 2017; Mayunga, 2007). By aggregating sociodemographic and -economic indicators such metrics—for an overview, see Cai et al. (2018)—highlight vulnerable and resilient households, communities, cities, etc. However, measuring resilience has proven to be like aiming at a moving target (Darnhofer, 2014) and often ends up in resilience benchmarking (Quinlan et al., 2016). Due to the focus on assets and not agency, the logic of benchmarking implies that the social entities under investigation behave like uniform and rational agents (Darnhofer et al., 2016). An improved access to assets would therefore lead to an inevitable increase in the entity's resilience (Ifejika Speranza et al., 2014).

In contrast to this assumed uniform rationality, human agency cannot be reduced to whether or not social entities have access to assets and resources, nor can we equate it with their capabilities and skills in doing so (Pain and Levine, 2012). Thus, a better understanding is needed of why social entities behave or act in a certain way, why they make certain choices, and why they have certain priorities and goals (Bristow and Healy, 2014). Otherwise, the entity under investigation remains a "black box for which some combined utility function is assumed" (Baghchi et al., 1998, p. 457).

4.2.4.2 **An agency-based perspective on resilience has to start with the individual actor**

If we try to unpack these "black box" social entities, we should have a clear picture of whose agency toward resilience we are interested in. Looking at how the term social entity has been interpreted so far, we see an insufficient consideration and reflection on the role of the individual in the current debate on social resilience. This neglect of the individual is remarkable, since the discussion on agency toward resilience is often based on the implicit assumption that the most basal carrier of agency is the individual.

Most studies focus on collective entities such as the region or the community (e.g., Cutter et al., 2008; Norris et al., 2008; Sharifi, 2016). The community—understood as a place-based entity with individuals sharing same interests and acting collectively (Sharifi, 2016)—is presented as particularly integral to resilience and thus is most often used as unit of analysis (e.g., Cutter et al., 2008; Magis, 2010; Sherrieb et al., 2010). Conceptualizing resilience at the level of communities leaves no room for analyzing tensions and constraints—driven by highly divergent aims and priorities—that might exist within a community (Pain and Levine, 2012; Ruszczyk, 2017). Hence, many scholars have highlighted the need to problematize "the community" as particularly elusive concept without clear definition (Cannon, 2008; Cannon and Schipper, 2014; Keating et al., 2017). Similarly, the household as unit of analysis can be criticized for the same reasons. In studies on livelihood strategies and disaster risk, we often define the household as "a single decision-making unit maximizing its welfare subject to a range of income-earning opportunities and a set of resource constraints" (Ellis, 1998, p. 12). Here again, using the household as unit of analysis may conceal issues as intrahousehold differences and power inequalities, which can constrain individual as well as collective action (de Haan and Zooners, 2005).

4.2.4.3 An actor-oriented, agency-based perspective on resilience is a culturally biased one

As consequence of our first two conclusions, we follow Bohle et al. (2009) who call for an actor-oriented, agency-based resilience framework, which consequently puts "actors, […] their creativity and their capabilities at the center of analysis" (ibid 2009:9). We further assume, that these individual actors are neither homogeneous, rational agents nor are their actions toward resilience completely random. Thus, we need to gain a deeper understanding of the rationalities behind their (in-)actions. So far, these rationalities of (in-)action are poorly understood and rarely accounted for in studies on disaster risk (Appleby-Arnold et al., 2018; Rawluk et al., 2017, 2018), as is the importance of normative and cognitive factors, which are crucial to understand people's willingness to cope, adapt, and transform in response to different types of shocks.

Trying to unravel these rationalities of (in-)action toward resilience, we can learn much from approaches developed in the fields of environmental psychology (Ajzen, 1991; Dietz et al., 2005; de Groot and Steg, 2007; Schwartz, 1977; Stern et al., 1998, 1999; Stern, 2000) as well as climate change and sustainability research (Leiserowitz, 2006; Leiserowitz et al., 2010; O'Brien and Sygna, 2013; O'Brien and Wolf, 2010). As for environmental psychology, there is an extensive body of work on how and why individuals take certain actions and respond to different types of environmental stressors. Also, many theoretical models—such as the Theory of Planned Behavior (Ajzen, 1991), the Norm-Activation Model (Schwartz, 1977), the Protection-Motivation Theory (Rogers, 1975), or the Value-Belief-Norm Theory (Stern et al., 1999)—have been developed and employed to evaluate (pro-)environmental behavior and the effects antecedent variables such as beliefs, values, worldviews, and norms can have on such behavior (Cordano et al., 2011). In the field of climate change research, such antecedent variables—often subsumed as "cultural bias"—have been used to better understand the attitudes of actors and their willingness to engage in practical responses to climate change (O'Brien and Sygna, 2013; O'Brien and Wolf, 2010). Based on these insights, O'Brien and Sygna (2013, p. 6) argue that transformations of beliefs, values, and worldviews have the most powerful effect on actions and strategies and can bring up different "action logics." To better understand these actor-specific rationalities on (in-)action, we argue for grounding an agency-based perspective on social resilience in these antecedent values, beliefs, and worldviews.

4.2.5 The best of both worlds—Reframing resilience as agency

Despite the explained reasons for establishing an actor-oriented and agency-based perspective, a transparent conceptual model for such an understanding of social resilience is still missing. To encourage this conceptual discussion, we propose such a model based on the idea to see resilience rather as process than an entity's property. Doing so, we follow Bohle et al.'s (2009, p. 12) idea of focusing on "social actors and their agency, arenas and respective agendas in the transformation of livelihoods in a resilient way." This perspective builds on the existing (asset-based) discussion on the ability of social entities to improve their resilience toward natural hazards. Additionally, we introduce the entity's willingness as counterpart to his or her ability to take actions toward resilience. Based on this combined ability and willingness to act, we propose a wider and more transparent definition of Agency Toward Resilience (ATR): ATR describes the nexus between practical (in-)actions, the willingness, and the ability of social entities to not only reactively respond and cope, but also proactively prepare, prevent, and adapt to impacts of natural hazards. To provide a valid ontological model for ATR, we use the next sections to describe the model's main components: (a) intended (in-)actions as well as (b) the ability and willingness to act before (c) combining them in our ATR model.

4.2.5.1 Intended (in-)actions as building blocks for social resilience

(In-)Actions are at the heart of our understanding of resilience (see Fig. 4.2.1): We see resilience as an ever-emergent "relational feature [... that] is both held in and produced through social (inter)actions" (Kruse et al., 2017: 2322) being taken by individual actors over and over again. In context of DRR, actions toward resilience can either be re-active or pro-active (Obrist et al., 2010) and focus on responding, recovering, preventing, or preparing to natural hazards. At the same time, we also see inaction— the situation where someone is able to act but opts against it—as a relevant form of intended action, as actions and inactions have practical consequences for an individual's resilience (Ajzen and Sheikh, 2013).

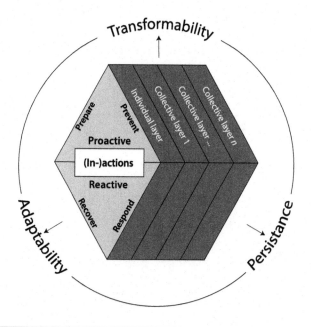

FIG. 4.2.1

Individual and collective (in-)actions toward resilience.

By drawing on the concept of multilayered resilience outlined by Glavovic et al. (2002), we acknowledge that there are different layers of action ranging from the individual to exclusively collective actions such as structural mitigation or spatial planning measures. Actions taken on different layers of action can therefore directly and indirectly improve the individual's persistence, adaptability, and transformability (Hutter and Lorenz, 2018). Hence, exploring resilience as manifestation of practical actions requires an investigation of the range of actions that individuals can and want to take at different layers of action.

4.2.5.2 The ability and willingness behind intended (in-)actions

Despite the complexity of (in-)actions toward persistence, adaptability, and transformability, interpreting ATR only as a sum of actor-specific actions is a too reductionist perspective. Looking also at what preceded and justified these actor-specific actions can reveal a more holistic understanding of ATR. Hence, we use a social entity's willingness and ability to justify individual (in-)actions toward resilience. The interrelation of ability and willingness to behave in a particular manner has only recently moved to the center of attention (Battilana, 2016; Chrisman et al., 2015; de Massis et al., 2014; Veider and Matzler, 2016). This interrelation implies that the likelihood to act "[…] is a function of [an individual's] willingness to act as such and of [his or her] ability to do so" (Battilana, 2016, p. 659). This ability-willingness-dualism acknowledges that an individual's ability to act is partly determined by the resources that they have access to, but also stresses the importance of an interest in doing so, which is determined by individual goals and motivations, as well as underlying values and beliefs (Battilana, 2016; Chrisman et al., 2015). de Massis et al. (2014, p. 347) refer to this dualism by arguing that "[…] ability without willingness, or vice versa, is logically and practically insufficient to produce a particular behaviour."

Regarding transparency, we see the need to further elaborate the ability and willingness to take actions toward resilience (see Fig. 4.2.2).

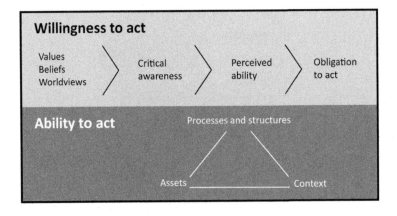

FIG. 4.2.2

The dualism between the ability and willingness to take actions toward resilience.

4.2.5.2.1 The ability to take actions toward resilience

Being able to act is facilitated or constrained by structural conditions and given circumstances, outlined for example in asset-based frameworks like the SLF (Scoones, 1998). Thus, we see the individual's range of actions determined by (a) the access to assets and embedded in (b) the broader context and (c) social structures that are historically, geographically, and culturally framed (Lorenz and Dittmer, 2016). Analogue to the SLF, the ability to act builds on five livelihood assets: human capital (such as education, skills, or health), social capital (such as social relationships, networks, group membership, or trust), natural capital (natural resources), physical capital (such as infrastructure, transport, or energy), and financial capital (such as savings or credits) (Scoones, 1998). These assets may be enhanced or constrained by broader structures and processes of society—such as formal or informal policies, institutions, politics, and power—and the general context, like exposure to natural hazards or local historic development paths.

4.2.5.2.2 The willingness to take actions toward resilience

We define the willingness to act as an individual's general openness to engage in a certain behavior driven by his or her values, beliefs, goals, and motivations (Gerrard et al., 2008). To further focus on this openness to engage, we build on two established theories from environmental and health psychology—the Value-Belief-Norm (VBN) theory (Stern et al., 1999) and the Protection Motivation Theory (PMT—Rogers, 1975). We chose these two theoretical approaches, since both connect basic values with actions and do this by providing a good balance between detail of explanation and the number of explaining factors. Comparing these two theoretical models to explain the willingness to take actions toward resilience, we conclude that four elements are most likely to inform an individual's openness to engage:

- *Beliefs, values, and worldviews*:
    The VBN theory suggests that human behavior is shaped by personal norms, which are linked to various beliefs such as the ascription of responsibility or general beliefs about the perceived relationship between humans and the environment. Basic values are at the heart of the model as stable cognitive structures that form early in life and act as guiding principles that precede beliefs and worldviews (Schwartz, 1977; Stern et al., 1998; Stern et al., 1999; Stern, 2000; Slimak and Dietz, 2006). These values inform and select general beliefs, which reflect among others people's worldviews regarding the environment and the effects of human activity on nature. Taken together, values, beliefs, and worldviews form the normative foundation for an individual's behavior.
- *Critical awareness of hazards*:
    This awareness—or "threat appraisal" in PMT—describes how an individual evaluates a certain risk based on (a) perceived probability and (b) perceived consequences. Originally developed to explain behavior in response to health threats, PMT has been repeatedly applied in natural hazards research (e.g., Bubeck et al., 2012; Corwin et al., 2017; Fox-Rogers et al., 2016; Grothmann and Reusswig, 2006).
- *Perceived ability to act*:
    The perceived ability to act—or "coping appraisal" in PMT—is subdivided into (a) perceived response efficacy (believing that the action is effective), (b) perceived self-efficacy (believing in one's own capacity to act), and (c) perceived response costs - capturing the perceived costs required for implementing an activity (Fox-Rogers et al., 2016). In analogy to the PMT, we assume that

individuals with a high awareness of hazards and the perceived ability to act are more likely to engage in protective behavior (Fox-Rogers et al., 2016).

- *Obligation to act*:

   We see the intrinsic motivation to take actions toward resilience not as a mere arithmetic product of being an aware and—according to their self-perception—able actor. In line with VBN theory, we understand this motivation as a moral obligation to act, which can result from reflecting one's own awareness and ability against personal values, beliefs, and worldviews. Thus, this obligation to act constitutes an important precondition for intentional (in-)actions toward resilience.

4.2.5.3 A model of agency toward resilience (ATR)

In a nutshell, our approach is in line with the central principle of the SLF that the ability to persist, adapt, and transform is shaped by the interplay of access to assets, context, and transforming structures and processes (Scoones, 1998). Building on PMT and VBN theory, we argue that the willingness to act is informed by critical awareness, perceived ability, and an obligation to act according to a social entity's beliefs and worldviews that are rooted in basic value orientations.

Combing these strands, we developed the model of Agency Toward Resilience (ATR—see Fig. 4.2.3), which grounds practical actions in the willingness and ability of social entities. To achieve this grounding, we introduced a linking element between the ability-willingness-dualism (see Fig. 4.2.2) and (in-)actions toward resilience (see Fig. 4.2.1). Reason for doing so is the simple observation that an obligation to act might not necessarily result in practical actions. We refer to this linking element as "intention," since only obligations, which transcend from the sphere of wishful thinking via a conscious decision, can build up the commitment of carrying out an action in the future. Besides this filtering, intention also includes goal-based premeditations that underlie concrete actions (Pomery et al., 2009).

Using "intent" to connect the ability-willingness-dualism (see Fig. 4.2.2) with practical actions toward resilience (see Fig. 4.2.1) also offers the chance for evaluating possible actions from an individual's perspective. While some practical actions lie within the willingness and ability range of an individual, other actions might overstrain an individual's ability or willingness to act. For the latter case it seems especially important to keep in mind that—despite many implementation guides for disaster resilience—only if someone perceives processes and actions as relevant, he or she will intentionally act on it—even if it's the "wrong" risk or action (Pain and Levine, 2012). Or in other words: Recommending "optimal" actions and activities that may fall out of an individual's ATR are neither effective nor appropriate.

Knowing the ATR of social entities might prove useful to explore "local optima," by working on ways to empower actors to fully exploit the range of actions they are able and willing to do. This leaves us in line with Pain and Levine (2012, p. 13) who argue that "in many situations, the best that people can strive for is a very conservative interpretation of resilience, with no possibility of transforming their lives, because they have limited choices and short-term horizons." Still, we might identify actions people are willing to implement but not able to do so. Outside support can encourage and facilitate these actions and activities and may serve as entry point to enhance resilience. However, to reach legitimate and lasting change, "social processes [...] can only come from within—where local people and local actors are the drivers of change" (Oxley, 2013, p. 4). Since the ATR model also includes values, beliefs, and worldviews, their change—taking place over time or by intentional interference (O'Brien and

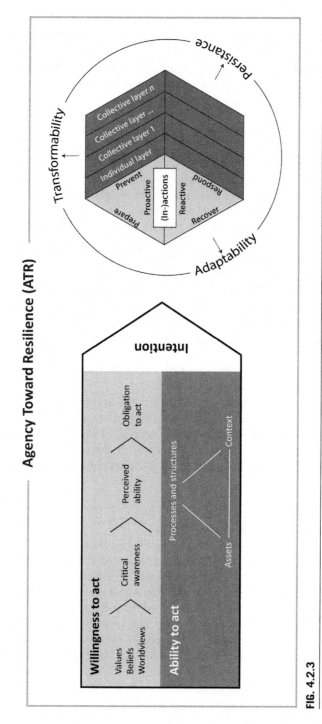

FIG. 4.2.3

An ontological model of Agency Toward Resilience (ATR).

Wolf, 2010)—can also be addressed. Turning the argument that values inform actions toward resilience upside down, one could ask if it is possible to influence values and beliefs to achieve certain actions toward resilience. Such "value engineering" might be justified with more effective and targeted interventions of policy stakeholders and institutions. Thus, we think that finding an ethically acceptable position on communication campaigns to influence values of social entities, or policy interventions based on value orientations is the most profound practical implication of our ATR model.

4.2.6 Reflecting the ATR model

Since we see our ATR model as actor-oriented and agency-based, one could conclude that we want to replace traditional, asset-based approaches to resilience with an agency-based one. This is not the case, since we acknowledge that an isolated focus on psychological and cognitive processes leads to an abstract and decontextualized understanding of (in-)actions (Dreier, 2008). Actions do not take place free-floatingly in a social vacuum nor are they detached from the social context (ibid.). There are long-standing discussions about how to conceptualize the relation of agency and structure in general and how to study the degree to which social structures affect the capacity of individuals to act independently in particular (Bourdieu, 2006; Giddens, 1984). Since we base our ATR model on the dualism between ability and willingness—which we see as functional equivalent of the structure-agency-dualism—it advocates an understanding of ATR, which reconciles ability and willingness.

Our model neither aims at explaining generalized causal chains between its key elements, nor at measuring the strength and effect of associations. Although some parts of our ATR model originate from psychometric works, we see the model as heuristic to better understand how ATR unfolds in given geographic, historic, and socioeconomic circumstances.

Further, we acknowledge that the type of practical action—according to the disaster management cycle—can strongly influence the importance of the willingness to act. In case of immediate actions following a disaster for example, structural conditions such as access to assets may play a more important role than the willingness to act. It could be worthwhile studying how this shift in the balance between willingness and ability manifests itself along the disaster management cycle.

4.2.7 Conclusions

In this chapter, we aimed to improve our understanding of the interlinkages between disaster risk and social resilience as claimed in Priority 1 of the Sendai Framework (UNISDR, 2015). While social resilience is often put on a level with a set of capacities, we argue that agency is crucial for the understanding of social resilience and thus understanding disaster risk. Therefore, we proposed a model for Agency Toward Resilience (ATR) that emphasizes the interrelations between a social entity's practical actions at different social levels with his or her ability and willingness to do so. An actor-oriented, agency-based approach toward resilience recognizes people as active agents who not only have the ability but also the willingness to take actions. The proposed ATR model is informed by elements of PMT and VBN theory, namely, basic values, general beliefs and worldviews, critical awareness, the perceived ability to act, and the feeling of obligation to act. Additionally, we included ideas of the SLF, recognizing that the wider context, access to assets and societal structures can facilitate or constrain a social entity's actions toward resilience. The proposed model further assumes that any practical action toward persistence, adaptability, and transformability is the outcome of a goal-based intention, which precedes an entity's willingness and ability for action.

We see our model as ontological approach to realize Bohle et al.'s (2009) claim of "reframing resilience as agency." This reframing enables us to focus on the willingness and ability of social entities to (re-)produce their resilience against natural hazards by practical actions. In the end, this refocus might push our knowledge on resilient development beyond the uniform normative claim of "be resilient."

References

Adger, W.N., 2000. Social and ecological resilience: are they related? Prog. Hum. Geogr. 24 (3), 347–364.

Adger, W.N., 2003. Social capital, collective action, and adaptation to climate change. Econ. Geogr. 79 (4), 387–404.

Ajzen, I., 1991. The theory of planned behavior. Organ. Behav. Hum. Decis. Process. 50 (2), 179–211. https://doi.org/10.1016/0749-5978(91)90020-T.

Ajzen, I., Sheikh, S., 2013. Action versus inaction: anticipated affect in the theory of planned behavior. J. Appl. Soc. Psychol. 43 (1), 155–162. https://doi.org/10.1111/j.1559-1816.2012.00989.x.

Alexander, D., 2005. Towards the development of a standard in emergency planning. Disaster Prev Manag 14 (2), 158–175. https://doi.org/10.1108/09653560510595164.

Appleby-Arnold, S., et al., 2018. Applying cultural values to encourage disaster preparedness: lessons from a low-hazard country. Int. J. Disaster Risk Reduct. 31, 37–44. https://doi.org/10.1016/j.ijdrr.2018.04.015.

Baghchi, D.K., et al., 1998. Conceptual and methodological challenges in the study of livelihood trajectories: case-studies in eastern India and Western Nepal. J. Int. Dev. 10, 453–468.

Bakkensen, L.A., et al., 2017. Validating resilience and vulnerability indices in the context of natural disasters. Risk Anal. 37 (5), 982–1004. https://doi.org/10.1111/risa.12677.

Battilana, J., 2016. Agency and institutions: the enabling role of individuals' social position. Organization 13 (5), 653–676. https://doi.org/10.1177/1350508406067008.

Béné, C., et al., 2012. Resilience: New Utopia or New Tyranny?; Reflection About the Potentials and Limits of the Concept of Resilience in Relation to Vulnerability Reduction Programmes. Brighton (IDS working paper 405.

Birkmann, J., 2006. Measuring Vulnerability to Natural Hazards: Towards Disaster Resilient Societies. United Nations University, New York.

Bohle, H.-G., Etzold, B., Keck, M., 2009. Resilience as agency: bro. IHDP Update 2.

Bourdieu, P., 2006. The Logic of Practice. Stanford University Press, Stanford, CA.

Bristow, G., Healy, A., 2014. Regional resilience: an agency perspective. Reg. Stud. 48 (5), 923–935. https://doi.org/10.1080/00343404.2013.854879.

Brown, D.D., Kulig, J.C., 1996. The concepts of resiliency: theoretical lessons from community research. Health Can. Soc. 4 (1), 29–52.

Brown, K., Westaway, E., 2011. Agency, capacity, and resilience to environmental change: lessons from human development, well-being, and disasters. Annu. Rev. Environ. Resour. 36 (1), 321–342. https://doi.org/10.1146/annurev-environ-052610-092905.

Bruneau, M., et al., 2003. A framework to quantitatively assess and enhance the seismic resilience of communities. Earthquake Spectra 19 (4), 733–752. https://doi.org/10.1193/1.1623497.

Bubeck, P., Botzen, W.J.W., Aerts, J.C.J.H., 2012. A review of risk perceptions and other factors that influence flood mitigation behavior. Risk Anal. 32 (9), 1481–1495. https://doi.org/10.1111/j.1539-6924.2011.01783.x.

Cai, H., et al., 2018. A synthesis of disaster resilience measurement methods and indices. Int. J. Disaster Risk Reduct. 31, 844–855. https://doi.org/10.1016/j.ijdrr.2018.07.015.

Cannon, T., 2008. Reducing People's Vulnerability to Natural Hazards Communities and Resilience. Helsinki (Research paper/UNU-WIDER 34).

Cannon, T., Müller-Mahn, D., 2010. Vulnerability, resilience and development discourses in context of climate change. Nat. Hazards 55 (3), 621–635. https://doi.org/10.1007/s11069-010-9499-4.

Cannon, T., Schipper, L., 2014. World Disasters Report 2014: Focus on Culture and Risk. Geneva.

Chrisman, J.J., et al., 2015. The ability and willingness paradox in family firm innovation. J. Prod. Innov. Manag. 32 (3), 310–318. https://doi.org/10.1111/jpim.12207.

Cordano, M., et al., 2011. A cross-cultural assessment of three theories of pro-environmental behavior: a comparison between business students of Chile and the United States. Environ. Behav. 43 (5), 634–657. https://doi.org/10.1177/0013916510378528.

Corwin, K.A., et al., 2017. Household preparedness motivation in lahar hazard zones: assessing the adoption of preparedness behaviors among laypeople and response professionals in communities downstream from Mount Baker and Glacier Peak (USA) volcanoes. J. Appl. Volcanol. 6 (1), 218. https://doi.org/10.1186/s13617-017-0055-8.

Cote, M., Nightingale, A.J., 2012. Resilience thinking meets social theory: situating social change in socio-ecological systems (SES) research. Prog. Hum. Geogr. 36 (4), 475–489. https://doi.org/10.1177/0309132511425708.

Coulthard, S., 2012. Can we be both resilient and well, and what choices do people have? Incorporating agency into the resilience debate from a fisheries perspective. Ecol. Soc. 17 (1), 13. https://doi.org/10.5751/ES-04483-170104.

Cutter, S.L., et al., 2008. A place-based model for understanding community resilience to natural disasters. Glob. Environ. Chang. 18 (4), 598–606. https://doi.org/10.1016/j.gloenvcha.2008.07.013.

Cutter, S.L., Burton, C.G., Emrich, C.T., 2010. Disaster resilience indicators for benchmarking baseline conditions. J. Homel. Secur. Emerg. Manag. 7 (1), 24. https://doi.org/10.2202/1547-7355.1732.

Darnhofer, I., 2014. Resilience and why it matters for farm management. Eur. Rev. Agric. Econ. 41 (3), 461–484. https://doi.org/10.1093/erae/jbu012.

Darnhofer, I., et al., 2016. The resilience of family farms: towards a relational approach. J. Rural. Stud. 44, 111–122. https://doi.org/10.1016/j.jrurstud.2016.01.013.

Davidson, D.J., 2010. The applicability of the concept of resilience to social systems: some sources of optimism and nagging doubts. Soc. Nat. Resour. 23 (12), 1135–1149. https://doi.org/10.1080/08941921003652940.

de Groot, J.I.M., Steg, L., 2007. Value orientations and environmental beliefs in five countries: validity of an instrument to measure egoistic, altruistic and biospheric value orientations. J. Cross-Cult. Psychol. 38 (3), 318–332. https://doi.org/10.1177/0022022107300278.

de Haan, L., Zooners, A., 2005. Exploring the frontier of livelihoods research. Dev. Chang. 36 (1), 27–47.

de Massis, A., et al., 2014. Ability and willingness as sufficiency conditions for family-oriented particularistic behavior: implications for theory and empirical studies. J. Small Bus. Manag. 52 (2), 344–364. https://doi.org/10.1111/jsbm.12102.

DFID, 1999. Sustainable Livelihoods Guidance Sheets.

Dietz, T., Fitzgerald, A., Shwom, R., 2005. Environmental values. Annu. Rev. Environ. Resour. 30 (1), 335–372. https://doi.org/10.1146/annurev.energy.30.050504.144444.

Dreier, O., 2008. Psychotherapy in Everyday Life. (Learning in Doing). Cambridge University Press, Cambridge.

Ellis, F., 1998. Household strategies and rural livelihood diversification. J. Dev. Stud. 35 (1), 1–38. https://doi.org/10.1080/00220389808422553.

Emirbayer, M., Mische, A., 1998. What is agency? Am. J. Sociol. 103 (4), 962–1023. https://doi.org/10.1086/231294.

Fabinyi, M., Evans, L., Foale, S.J., 2014. Social-ecological systems, social diversity, and power: insights from anthropology and political ecology. Ecol. Soc. 19 (4), 12. https://doi.org/10.5751/ES-07029-190428.

Fekete, A., 2009. Validation of a social vulnerability index in context to river-floods in Germany. Nat. Hazards Earth Syst. Sci. 9 (2), 393–403. https://doi.org/10.5194/nhess-9-393-2009.

Fox-Rogers, L., et al., 2016. Is there really "nothing you can do"? Pathways to enhanced flood-risk preparedness. J. Hydrol. 543, 330–343. https://doi.org/10.1016/j.jhydrol.2016.10.009.

Gerrard, M., et al., 2008. A dual-process approach to health risk decision making: the prototype willingness model. Dev. Rev. 28 (1), 29–61. https://doi.org/10.1016/j.dr.2007.10.001.

Giddens, A., 1984. The Constitution of Society. University of California Press, Berkley.

Glavovic, B., Scheyvens, R., Overton, J., 2002. Waves of adversity, layers of resilience: exploring the sustainable livelihoods approach. In: Contesting Development: Pathways to Better Practice: Proceedings of the Third Biennial Conference of the Aotearoa New Zealand International Development Studies Network. Massey University, Institute of Development Studies. December 5–7.

Grothmann, T., Reusswig, F., 2006. People at risk of flooding: why some residents take precautionary action while others do not. Nat. Hazards 38 (1–2), 101–120. https://doi.org/10.1007/s11069-005-8604-6.

Hutter, G., Lorenz, D.F., 2018. Social resilience. In: Fuchs, S., Thaler, T. (Eds.), Vulnerability and Resilience to Natural Hazards. Cambridge University Press, pp. 190–213.

Ifejika Speranza, C., Wiesmann, U., Rist, S., 2014. An indicator framework for assessing livelihood resilience in the context of social–ecological dynamics. Glob. Environ. Chang. 28, 109–119. https://doi.org/10.1016/j.gloenvcha.2014.06.005.

Keating, A., et al., 2017. Development and testing of a community flood resilience measurement tool. Nat. Hazards Earth Syst. Sci. 17 (1), 77–101. https://doi.org/10.5194/nhess-17-77-2017.

Keck, M., Etzold, B., 2013. Resilience refused. Wasted potentials for improving food security in Dhaka. Erdkunde 67 (1), 75–91. https://doi.org/10.3112/erdkunde.2013.01.07.

Keck, M., Sakdapolrak, P., 2013. What is social resilience? Lessons learned and ways forward. Erdkunde 67 (1), 5–19. https://doi.org/10.3112/erdkunde.2013.01.02.

Klein, R.J.T., Nicholls, R.J., Thomalla, F., 2003. Resilience to natural hazards: how useful is this concept? Environ. Hazard. 5 (1), 35–45. https://doi.org/10.1016/j.hazards.2004.02.001.

Kruse, S., et al., 2017. Conceptualizing community resilience to natural hazards—the emBRACE framework. Nat. Hazards Earth Syst. Sci. 17 (12), 2321–2333. https://doi.org/10.5194/nhess-17-2321-2017.

Kuhlicke, C., et al., 2011. Perspectives on social capacity building for natural hazards: outlining an emerging field of research and practice in Europe. Environ. Sci. Pol. 14 (7), 804–814. https://doi.org/10.1016/j.envsci.2011.05.001.

Leiserowitz, A., 2006. Climate change risk perception and policy preferences: the role of affect, imagery, and values. Clim. Chang. 77 (1), 45–72. https://doi.org/10.1007/s10584-006-9059-9.

Leiserowitz, A., et al., 2010. Global Warming's Six Americas. June 2010.

Lorenz, D.F., Dittmer, C., 2016. Resilience in catastrophes, disasters and emergencies. In: Maurer, A. (Ed.), New Perspectives on Resilience in Socio-Economic Spheres. Springer Fachmedien Wiesbaden, Wiesbaden, pp. 25–59.

Magis, K., 2010. Community resilience: an indicator of social sustainability. Soc. Nat. Resour. 23 (5), 401–416. https://doi.org/10.1080/08941920903305674.

Mayunga, J.S., 2007. Understanding and applying the concept of community disaster resilience: a capital-based approach. In: A Draft Working Paper Prepared for the Summer Academy for Social Vulnerability and Resilience Building, Munich.

Norris, F.H., et al., 2008. Community resilience as a metaphor, theory, set of capacities, and strategy for disaster readiness. Am. J. Community Psychol. 41 (1–2), 127–150. https://doi.org/10.1007/s10464-007-9156-6.

O'Brien, K., Sygna, L., 2013. Responding to climate change: the three spheres of transformation. In: Proceedings of Transformation in a Changing Climate. June 19–21, 2013, Oslo, Norway. University of Oslo. ISBN: 978-82-570-2000-1.

O'Brien, K., Wolf, J., 2010. A values-based approach to vulnerability and adaptation to climate change. Wiley Interdiscip. Rev. Clim. Chang. 1 (2), 232–242. https://doi.org/10.1002/wcc.30.

Obrist, B., 2016. Social resilience and agency. Perspectives on ageing and health from Tanzania. Erde 147 (4), 266–274.

Obrist, B., et al., 2010. Livelihood, malaria and resilience. Prog. Dev. Stud. 10 (4), 325–343. https://doi.org/10.1177/146499340901000405.

Oxley, M.C., 2013. A "people-centred principles-based" post-Hyogo framework to strengthen the resilience of nations and communities. Int. J. Disaster Risk Reduct. 4, 1–9. https://doi.org/10.1016/j.ijdrr.2013.03.004.

Pain, A., Levine, S., 2012. A Conceptual Analysis of Livelihoods and Resilience: Addressing the 'Insecurity of Agency'. HPG Working Paper.

Pomery, E.A., et al., 2009. From willingness to intention: experience moderates the shift from reactive to reasoned behavior. Personal. Soc. Psychol. Bull. 35 (7), 894–908. https://doi.org/10.1177/0146167209335166.

Posch, E. et al. (2019) 'Ke garne?—how values and worldviews influence resilience to natural hazards: a case study from Mustang, Nepal', Mt. Res. Dev., 39(4).

Quinlan, A.E., et al., 2016. Measuring and assessing resilience: broadening understanding through multiple disciplinary perspectives. J. Appl. Ecol. 53 (3), 677–687. https://doi.org/10.1111/1365-2664.12550.

Rawluk, A., et al., 2017. Public values for integration in natural disaster management and planning: a case study from Victoria, Australia. J. Environ. Manag. 185, 11–20. https://doi.org/10.1016/j.jenvman.2016.10.052.

Rawluk, A., Ford, R.M., Williams, K.J.H., 2018. Value-based scenario planning: exploring multifaceted values in natural disaster planning and management. Ecol. Soc. 23 (4), 15. https://doi.org/10.5751/ES-10447-230402.

Rogers, R.W., 1975. A protection motivation theory of fear appeals and attitude change. J. Psychol. 91 (1), 93–114. https://doi.org/10.1080/00223980.1975.9915803.

Ruszczyk, H., 2017. The Everyday and Events: Understanding Risk Perceptions and Resilience in Urban Nepal. Dissertation, Durham University.

Schwartz, S., 1977. Normative influences on altruism. In: Berkowitz, L. (Ed.), Advances in Experimental Social Psychology. Academic Press, Boston, MA.

Scoones, I., 1998. Sustainable Rural Livelihoods: A Framework for Analysis. IDS working paper 72.

Sharifi, A., 2016. A critical review of selected tools for assessing community resilience. Ecol. Indic. 69, 629–647. https://doi.org/10.1016/j.ecolind.2016.05.023.

Sherrieb, K., Norris, F.H., Galea, S., 2010. Measuring capacities for community resilience. Soc. Indic. Res. 99 (2), 227–247. https://doi.org/10.1007/s11205-010-9576-9.

Slimak, M.W., Dietz, T., 2006. Personal values, beliefs, and ecological risk perception. Risk Anal 26 (6), 1689–1705. https://doi.org/10.1111/j.1539-6924.2006.00832.x.

Stern, P., 2000. Toward a coherent theory of environmentally significant behavior. Soc. Psychol. Study Soc. Issues 56 (3), 407–424.

Stern, P.C., Dietz, T., Guagnano, G.A., 1998. A brief inventory of values. Educ. Psychol. Meas. 58 (6), 984–1001.

Stern, P.C., et al., 1999. A value-belief-norm theory of support for social movements: the case of environmentalism. Res. Hum. Ecol. 6 (2), 81–92.

Turner, B.L., 2003. A framework for vulnerability analysis in sustainability science. Proc. Natl. Acad. Sci. U. S. A. 100 (14), 8074–8079.

United Nations International Strategy for Disaster Reduction (UNISDR), 2007. Hyogo Framework for Action 2005–2015: Building the Resilience of the Nations and Communities to Disasters. Extract from the Final Report of the World Conference on Disaster Reduction, Geneva, Switzerland.

United Nations International Strategy for Disaster Risk Reduction (UNISDR), 2015. Sendai Framework for Disaster Risk Reduction 2015–2030. Geneva, Switzerland.

Veider, V., Matzler, K., 2016. The ability and willingness of family-controlled firms to arrive at organizational ambidexterity. J. Fam. Bus. Strat. 7 (2), 105–116. https://doi.org/10.1016/j.jfbs.2015.10.001.

Voss, M., 2009. Vulnerabilität. In: Hammerl, C. (Ed.), Naturkatastrophen: Rezeption—Bewältigung—Verarbeitung. StudienVerl., Innsbruck, pp. 103–121. Konzepte und Kontroversen, Bd. 7. (in German).

Risk management for forest fires at a world heritage site: Vulnerability and response capacity by Rapa Nui indigenous community

4.3

Constanza Espinoza-Valenzuela and Marcela Hurtado

Department of Architecture, Technical University Federico Santa María, Valparaiso, Chile

CHAPTER OUTLINE

4.3.1 Introduction

The World Heritage Site (WHS) called Rapa Nui National Park (RNNP) is located on the island of the same name, also known as Easter Island. It is the most remote island on the planet, located 3700 km off the mainland coast of Chile. It has a total area of 16,600 ha. The RNNP corresponds to 40% of the total (Fig. 4.3.1). In 1935, the park was declared a national monument by Chilean law, and in 1995, it was inscribed on the UNESCO world heritage list under criteria (i), (iii), and (v) (http://whc.unesco.org/en/list/715).

Rapa Nui Island is part of the Polynesian archipelago, made up of many islands where different forms of adaptation to the environment, manifestations, and cultural developments that originated

Understanding Disaster Risk. https://doi.org/10.1016/B978-0-12-819047-0.00014-7

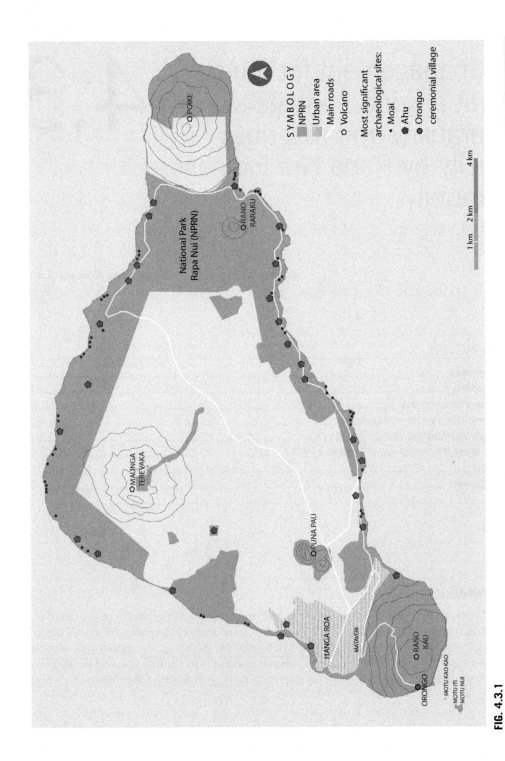

FIG. 4.3.1

Most significant archaeological sites in Rapa Nui island.

from the Ancestral Polynesian Culture have been developed for 3000 years. Although Rapa Nui Island is part of this system, the Rapa Nui developed their own culture in complete isolation, without foreign contact, until the 18th century. This is one of the most remarkable aspects of the Rapa Nui culture, together with their complex and fascinating worldview that is embodied in exceptional manifestations (Ramírez, 2008).

The Rapanui cultural heritage is varied with tangible and intangible expressions of great significance for their community. Among the first—tangible—highlights are archaeological sites and architectural remains. Some of the most outstanding sites are the moai, impressive stone sculptures that refer to the ancestors; the ahu, ceremonial altars arranged according to astronomical orientations; the reflection of the complex worldview of the Rapanui people; and the painting and rock art scattered around the island. One of the architectural vestiges is the ceremonial village of Orongo, composed of a series of stone houses in the shape of an ellipse, also strategically located in accordance with ancestral traditions (Corporación Nacional Forestal, Consejo de Monumentos Nacionales, 2010; Gobierno de Chile, 1995). As for intangible manifestations, there are various traditions and customs that have been passed on from generation to generation that are practiced today (Consejo Nacional de las Culturas y las Artes, 2012). However, the population is aware of the need to foster education that is focused on reaffirming the identity of younger generations in the face of globalization (Comisión Provincial de Monumentos Rapa Nui, Secretaría Técnica de Patrimonio Rapa Nui, 2014). There is therefore a close relationship between a community and its heritage. In fact one of the current challenges is to achieve sustainable development that integrates this cultural heritage. This is an urgent task because of the strong pressures that archaeological and environmental heritage suffer, as a result of multiple threats on the island, with the consequent social and economic impact.

According to ethnographic and archaeological research, the period of the greatest intensity of cultural development and production took place between the 11th and 16th centuries. This was interrupted in the 18th century by different events, such as contact with Western culture with the arrival of the Dutch expedition in 1722, as well as the depletion of the island's natural resources due to overexploitation and population increase, which produced a profound spiritual crisis among its inhabitants that resulted in a loss of power of the ancestors, whose spirits lived in the moai (Castro, 2006; Englert, 1983). As a result of this crisis, part of the moai was destroyed, and a new cult was installed: the birdman. In addition, the 19th century was especially critical for Rapanui culture due to the effects of colonization: diseases, other cults, and livestock were introduced and, most dramatically, the population was subjected to slavery. The result was an alarming demographic drop with consequent irreversible losses of traditions and cultural continuity (Ramírez, 2008; Gobierno de Chile, 1995). As a consequence of this historical process, the relationship between the different groups in the island today is difficult. The challenge to confront the threats to be exposed is to create a resilient community that must necessarily work in a coordinated manner. The social dimension will therefore be the focus of the present work.

Currently, hundreds of archaeological remain (more than 900 moai, 300 ahu, and 50 dwellings) scattered throughout the island, which are a testimony to the different historical moments and cultural processes in close relation with the landscape. Due to their exceptional nature, and as a way to safeguard their values and preserve them from disappearance, these assets are protected, first by Chilean law in 1935 (Law 17,288 of National Monuments) and later by UNESCO (World Heritage Convention) (Gobierno de Chile, 1995).

To date, the island has been affected by various threats, both natural and anthropic, that endanger the preservation of these valuable remains. Both the nomination dossier and periodic reports mention various threats, such as the impact of tourism, environmental pressures—increased by climate change—livestock (cows and horses) and fire (Comisión Provincial de Monumentos Rapa Nui, Secretaría Técnica de Patrimonio Rapa Nui, 2014; Gobierno de Chile, 2003, 2012). The nature of the property, represented by its physical characteristics and its location, becomes a factor that increases its exposure: Most of the sculptures are composed of porous volcanic stone (tufa), which suffers from the effects of humidity, wind, rain, and solar radiation. The location in coastal areas or on slopes also plays a role, as saline humidity and wind generate erosion and wear (Bahamondez and Isla de Pascua, 2000). In relation to the location of the archaeological remains, many of them are scattered on the island, being exposed to other threats such as fires or the passage of animals. However, the insular condition of the site appears to be a factor of vulnerability due to the potential limitation of resources to cope with the occurrence of disasters. The diversity and heterogeneity of groups present on the island, including the indigenous Rapanui group (Ma'u Henua) in charge of the site, is another distinctive feature to consider in the capacity to coordinate responses to disasters.

Among the threats mentioned, the occurrence of forest fires has alerted the local and national communities in recent times. The National Forestry Corporation (CONAF)—administrator of the site between 1995 and 2017—has registered 66 outbreaks of fires between 2011 and 2017 that have affected the RNNP and, with it, the archaeological cultural heritage of the site, leading to constant concerns for the damage to cultural and environmental heritage and its consequent social and economic impact for the island community (Corporación Nacional Forestal, n.d.; Corporación Nacional Forestal, Comunidad Indígenas Polinésica Ma'u Henua, Museo antropológico P. Sebastián Englert, Secretaría Técnica de Patrimonio Rapa Nui, 2017). It is worth mentioning that the site does not have a risk management plan; there is, however, a management plan from the year 1997 that is in the process of being updated where both natural and anthropic threats are identified (Comisión Provincial de Monumentos Rapa Nui, Secretaría Técnica de Patrimonio Rapa Nui, 2014). Furthermore, State of Conservation Reports submitted by the State Party to World Heritage Centre UNESCO (Gobierno de Chile, 2003, 2012) also have identified various threats affecting the Rapanui cultural heritage.

4.3.2 Methodology

The proposed methodology is based on recommendations issued by institutions specializing in risk management (UN Office for Disaster Risk Reduction, 2015) and in the preservation of world heritage (UNESCO, n.d.; UNESCO, 2012; Stovel, 1998). Fig. 4.3.2 synthesizes the process.

One of the fundamental axes is the principle of understanding risk, which is outlined in the Hyogo Framework for Action (UN Office for Disaster Risk Reduction, 2005) and stated in the Sendai Framework (UN Office for Disaster Risk Reduction, 2015). The case of forest fires on the island and their causes, associated vulnerability factors, impact on the community, and other pertinent components will be analyzed, thus seeking to determine the critical and differentiating aspects of these events to articulate with risk reduction strategies. Fire data were provided by CONAF, manager of the site between 1995 and 2017. Another axis also expressed in these reference texts is the importance of integrating the community in the risk management cycle as a fundamental aspect (Gottler and Ripp, 2017). In the case of the Rapa Nui Island, where a diversity of actors, different traditions, and customs coexist, it is a

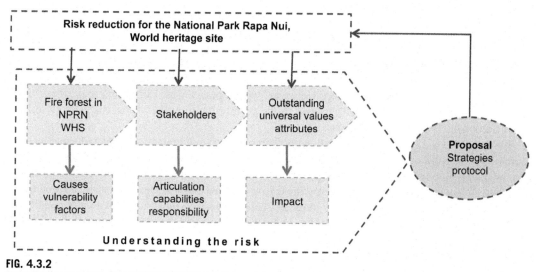

FIG. 4.3.2

Methodology.

critical point to evaluate; the groups, their link with the site, their capacities, the connections between the groups and their responsibility in risk management will be identified. One of the aspects to be evaluated is the management capacity of the new RNNP administration, the Ma'u Henua, as this is the first time that the administration is in the hands of an indigenous group. Finally, a central aspect is the WHS condition of the RNNP, which includes the life of the people, their assets, and the potential impact of this threat on cultural heritage and the outstanding universal values (OUV) that support the registration of a site on the UNESCO list. These topics are complemented by field work in gathering information, interviews with key actors, and bibliographic review.

The systematization and critical evaluation of this information is derived from the proposal of strategies and actions to improve the response capacity of the community residing on Rapa Nui Island in the face of forest fires, as a first step in the formulation of a comprehensive risk management plan for the site that considers other threats.

4.3.3 **Anthropologic context**

According to mythology, the first "settlers" from Hiva—the mythical ancestral land—were headed by the "royal family" with the Ariki Hotu A Matu'a as a spiritual leader. He led the colonization and distribution of the land for his children, thus initiating the tradition of the lineages, each of which occupied clearly delimited lands. These divisions persisted until the arrival of the Europeans in the 18th century. At the time of delimitation, the mana—the power of the ancestors represented by the moai—played an important role as a demarcation milestone. The most powerful families erected altars—ahu—to also demarcate their domains. Therefore, disputes over the island territory arose with the first occupation of the island (Ramírez, 2008; Campbell, 1987; Cornejo, 2005).

The legends about the origin of the settlement of the island give an account of a stratified society, with the Ariki family and the religious aristocracy in the upper part. Next, the specialists in diverse crafts (craftsmen, canoe builders, constructors of houses, agriculturists), servants, and so-called enemies were located. Part of this social order remained unaltered until the changes that began in the 17th century with the ecological crisis, a product of overpopulation, which impacted the beliefs that sustained this form of organization. However, when contact with Europeans took place, there was a clear leadership that survived, at least in the memory and myths of the islanders (Ramírez, 2008; Campbell, 1987; Cornejo, 2005).

The deepest crisis in social order began with claims by Holland, France, and Spain during the 18th century; but the most dramatic impact was the consequence of the transfer in 1860 of 2000 natives to Peru as slaves, among them the king and priests. The claims forced the return of the inhabitants to the island after a few years, but few survived and those who returned carried diseases that spread around the island. According to the records, in 1877, only 111 natives survived on an island with a population of 10,000 inhabitants (Gobierno de Chile, 1995). In the 20th century, with the annexation of the island to Chile, the situation did not change: The island was given to an English livestock company that continued the regime of subjugation of the islanders.

These facts, present in the memory of the islanders, are important to understand the feelings that the Rapanui community has today. It is understood that every culture is dynamic and changes over time. In the case of the Rapa Nui people, such changes have been a consequence of the relationship with the continent and the effects of globalization. Despite the strong roots of the group with its origins and its land, the Rapanui people wanted to leave the island three generations ago, for different reasons. The first generation (measured in the 20th century) wanted to escape the abuse imposed by the exploiting company Williamson & Baulfont and the representatives of the national army. The next generation emigrated in search of a partner, better education, or better job opportunities. Finally, many young people today study on the continent (especially in the capital, Santiago). Some of them are children of mixed marriages, were born on the continent and do not even know the island (Muñoz, 2007). Nevertheless, family relationships have remained—under the concept of extended family—as one of the great supports for the preservation of the roots of the traditional culture and the island itself. According to the last population census, only 45.3% of the inhabitants of the island are declared as part of the Rapanui ethnic group (Instituto Nacional de Estadísticas, n.d.).

In this context, recent studies have evaluated the vulnerability of the cultural heritage—both material and immaterial—of the island. The most recognized elements by diverse groups are the language, the moai and archaeological remains, and the oral traditions (Consejo Nacional de las Culturas y las Artes, 2012). There is a collective recognition of a critical situation on the island that affects the island's cultural heritage, among other factors. It is interesting to note that among the causes of this crisis are the lack of specific public policies for the island, the lack of resources for the safeguarding of the cultural heritage, social, and political problems within the island and between clans, the presence of many continentals on the island, and even the lack of interest from the people in the region (Consejo Nacional de las Culturas y las Artes, 2012; Comisión Provincial de Monumentos Rapa Nui, Secretaría Técnica de Patrimonio Rapa Nui, 2014). To date, there is a distrust from the Rapanui people toward the continentals that are institutionally represented by the public bodies present on the island, which should be integrated into risk management. Building spaces of trust is fundamental—as well as complex—but must be approached from similar experiences that, under the concept of coproduction and memoranda of understanding, have yielded good results (Shand, 2018).

These data, seen as opportunities and weaknesses, reflect a state of mind of the local community that should be considered in the definition of risk reduction strategies to deal with forest fires. Among these, the native language is seen as a key element in the preservation of identity and traditions and in the communication between generations.

4.3.4 Stakeholders for risk management

Next, the different groups present on the island that are related to risk management will be identified, from their capacities and responsibilities. Among the organizations specializing in risk management, there is a consensus on the importance of involving a broad community in the management cycle, that is, in the stages of prevention, control, and reconstruction as a key factor in reducing risk and developing resilient groups (UN Office for Disaster Risk Reduction, 2015; Gottler and Ripp, 2017). There has been integration between the original inhabitants and the so-called mestizos and continentals, carrying out commercial, productive, administrative activities, among others. There is, therefore, no segregation at the labor level, beyond the cultural differences that have been described.

In relation to the administration of the Rapa Nui National Park, when declared a protected wild area by the State of Chile in 1935, it remained under the administration of CONAF. It is a private law entity under the Ministry of Agriculture of Chile, whose main task is to administer Chile's forestry policies and promote the development of the sector. The scope of its action also includes the safeguard of the vegetation resources of the protected areas of the country, especially the fight against forest fires (Corporación Nacional Forestal, n.d.). After the inclusion of the RNNP in the UNESCO list, the administration remained in the hands of CONAF until 2017, when the administration was transferred to the community of the island. During the management of CONAF, five archaeological sites were kept open to the public, with 15 park rangers in charge who also worked to conserve and protect the island's ecosystems and fight against forest fires.

In the original community, there was always the desire to lead the administration of the site. This is how two of its most important organizations—Honui and the Easter Island Development Commission (CODEIPA)—promoted this idea (Comisión de Desarrollo de Isla de Pascua, 2015). In July 2016, the Ma'u Henua group was formed to comanage the site with CONAF. This is carried out through an agreement between the two groups, where a 1-year deadline is established for transferals and respective training. During this period, CONAF gave Ma'u Henua an operational plan that had to be carried out in one of the four zones of the park to verify the management capabilities.

In November 2017, the presidential decree was signed for the definitive delivery of the entire site to the Ma'u Henua group for 50 years, which generated great joy among the islanders. Among the efforts that have been made to date, the increase of park rangers and officials in charge of the park, as well as the increase of archaeological sites open to the public have been highlighted. The Ma'u Henua have also been involved in activities such as park maintenance, tourism management, and associated financial resources (https://www.mauhenua.com/). The maintenance includes conservation works of the archaeological heritage sites as well as the activities associated with the control of forest fires. Although the control of forest fires continues to be in the hands of CONAF, Ma'u Henua has also formed a brigade to deal with the fires with support from CONAF. Among the prevention actions for the Ma'u Henua forest fires, weeding, maintenance of the firebreaks, and training park rangers in contact with tourists and potential visitors during an emergency have been noted. Other groups within the island that are responsible for risk management are the local police, the army, and firemen. These last two groups are trained by CONAF to fight forest fires. Public institutions such as the provincial government, the representation of the national emergency office (ONEMI, an institution that is linked to the army), the Ministry of Public Works (MOP), the Council of National Monuments (CMN), the Anthropological Museum, and the municipality can also play an active role in case of an emergency. Finally, the media and the community organized in groups such as Honui, neighbors, and volunteers also constitute an interest group in risk management (Ninoska Huki Cuadros, 2017).

The management of forest fires in particular—the central theme of this article—is a critical issue for the site due to its insular condition and lack of both technical and human resources. This implies that both efficiency and coordination are necessary when dealing with an emergency. Currently, there are two protocols on the island to face forest fires from CONAF and the new administration, Ma'u Henua, that have well-defined different relationship schemes, involving the same actors.

The CONAF protocol is shown in Fig. 4.3.3. Upon receiving the fire alarm, a first call is made to the forest brigade, the firemen, and the Ma'u Henua community. The latter, in turn, summons its own brigades and park rangers, who alert tourists. If the fire fails to be fought by the brigades and firefighters, a second call is made to the ONEMI, who activates its support network with other institutions on the island: the army, the local police, and the municipality. The latter have the main task of facilitating the transit of emergency vehicles or providing water through a lorry available on the island owned by the municipality (Corporación Nacional Forestal, n.d.).

In addition, Ma'u Henua also has a protocol (Fig. 4.3.4). After receiving the fire alert, the members contact their brigades—who, in turn, contact the park rangers—and CONAF to support the forestry and army brigades. The ONEMI is also alerted along with the local police, the municipality, and the MOP (Rapu, 2017; Comunidad Indígena Polinésica Ma'u Henua, n.d.).

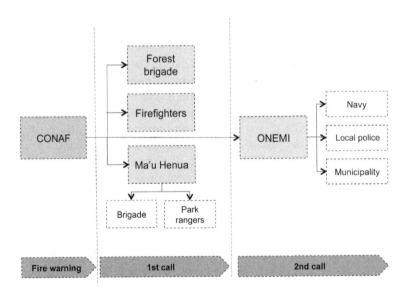

FIG. 4.3.3

Protocol again forest fire by CONAF.

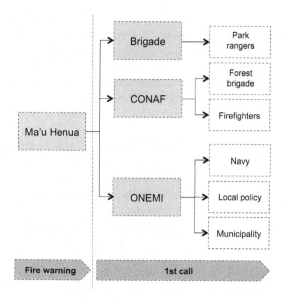

FIG. 4.3.4

Protocol against forest fire by Ma'u henua.

Both protocols involve various groups, especially the public bodies present on the island. However, as we will see later, the protocols do not explicitly include other relevant actors, such as community organizations or volunteers who can play an important role in the case of an emergency when properly trained and articulated with the other groups (Gottler and Ripp, 2017). This means integrating them into the process of prevention and preparedness of a disaster risk management plan.

4.3.5 **Threats affecting the site**

As indicated, the cultural heritage of the island is affected by various natural and man-induced threats, including the archaeological sites that are scattered throughout the island in a close relationship with the landscape and the activities of everyday life. This constitutes an intrinsic characteristic of the island's cultural heritage and, at the same time, a difficulty for conservation and management, due to the degree of exposure it supposes.

Natural hazards include those derived from environmental pressures, such as wind, rain, saline humidity, and UV radiation, and which especially affect archaeological assets built from volcanic stone with a soft, porous nature on the island (Gobierno de Chile, 2012; Bahamondez and Isla de Pascua, 2000; Corporación Nacional Forestal, Comunidad Indígenas Polinésica Ma'u Henua, Museo antropológico P. Sebastián Englert, Secretaría Técnica de Patrimonio Rapa Nui, 2017). Earthquakes are a minor but potential threat, as well as tsunamis. Along the same line, the effects of climate change, represented here by an increase in the number and intensity of the swells, also constitute a threat against which no mitigation actions have been taken. There are valuable archaeological sites located on the edge of the island, directly on the sea that are seriously threatened by these swells. However, there is no consensus regarding the relevance of moving them because of the significance of the location for these types of cultural assets. In terms of anthropic threats, the impact of not sufficiently controlled tourism stand out, including vandalism such as graffiti or highlighting rock art; the increase in the use of vehicles that impact the fragile soils of the island; cattle that roam uncontrolled in the RNNP; and the burning of grass, as part of the ancestral traditions of the Rapanui people, all of which result in large-scale forest fires (Gobierno de Chile, 2012; Corporación Nacional Forestal, Comunidad Indígenas Polinésica Ma'u Henua, Museo antropológico P. Sebastián Englert, Secretaría Técnica de Patrimonio Rapa Nui, 2017).

Vulnerability factors are related to contextual conditions, such as climate or soil type, and the intrinsic characteristics of those assets, especially the material composition (volcanic tuff) that makes archaeological assets highly vulnerable in the face of environmental pressures. Factors related to the usage or occupation can also increase the vulnerability of the cultural assets, including tourist usage, livestock practices, and customs rooted in the community, such as burning. In Table 4.3.1, the current and potential effects of these threats on the declared attributes of the island and the relationship with the vulnerability factors are indicated. The fragility of the cultural heritage of the island is explained due to various factors, such as its materiality, the location of the archaeological remains, the characteristics of the land, the forms of occupation or the management. All these aspects will be integrated in the evaluation of the forest fire as a threat.

Table 4.3.1 Main hazard affecting cultural heritage and vulnerability factors.

Cultural heritage	Main hazards	Impact (current or potential)	Vulnerability factors
Moai	– Environmental pressures – Tourism – Fire – Vandalism – Cattle	– Disintegration – Structural damage	– Materiality – Location – Management/ conservation – Social/ideological
Ahu	– Environmental pressures – Tourism – Fire – Vandalism – Cattle	– Disintegration – Structural damage – Integrity	– Materiality – Location – Management/ conservation – Social/ideological
Rock paintings	– Environmental pressures – Fire – Vandalism	– Disintegration – Authenticity	– Materiality – Location – Management/ conservation
Architectural remains	– Environmental pressures – Tourism – Fire – Vandalism – Cattle	– Structural damage – Integrity	– Materiality – Location – Management/ conservation

4.3.6 **Case study: The forest fire of September 2017**

Concerning the multiple threats listed, the case of forest fires will be analyzed in detail due to the increase in the number of hectares affected in recent years, as well as the relationship between the management of this threat—represented by the entire cycle of the risk management—and the different actors of the Rapanui community.

According to the information provided by CONAF between 2011 and 2017, 175 fires were registered on the island (Corporación Nacional Forestal, n.d.). Only 10 of these were located in the urban area (town of Hanga Roa), and the rest (165) affected the rural area of the island. Regarding the latter, 66 occurred within the limits of the RNNP, including a major event in September 2017, which will be analyzed in this study. Fig. 4.3.5 shows the location of the fires—inside and outside the boundaries of the RNNP—as well as the relationship with the most important archaeological sites on the island (Fig. 4.3.6). The data obtained indicate that the number of outbreaks has increased—from 6 in 2011 to 32 in 2017—as well as the number of hectares affected, which increased by more than 1000% (from 50 to 570 ha) in the period analyzed. It should be noted that the exceptional event from September 2017 that consumed nearly 300 ha is included in this calculation. However, if we exclude that fire, the number of hectares is still significantly increased (440%) (Fig. 4.3.7). According to CONAF records this increase in the number of forest fires on the island can be attributed to two factors related to climate change: the increase in temperature on the island and low rainfall. As a consequence, the reduction in environmental humidity. In 2017 there was a 44% deficit in rainfall, becoming the second driest year since there are records on the island (Corporación Nacional Forestal, n.d.; Dirección Meteorológica de Chile, 2019).

In relation to the origin of forest fires, it should be noted that the island does not offer environmental conditions that favor the spontaneous appearance of fire. These wildfires on the island are of anthropic origin, occasionally associated with vandalism. In addition, the inhabitants practice the burning of pastures, part of an ancestral ritual that favors the growth of new grass for the animals (López, 2017). Currently, only controlled burning of garbage is authorized within the island, under the supervision of CONAF and local police. However, this protocol is not respected, and pasture burns get out of control, causing most forest fires. Intentional burning caused by ideological and political conflicts between the clans and families of the resident indigenous community has also recurred. These conflicts are ancestral and difficult to solve. It should be taken into account that the control of the territory, seen as a scarce resource, has always represented a conflict within the Rapa Nui society (Ramírez, 2008).

Likewise, it should be noted certain conditions that favor the spread of fire, in addition to those already mentioned: the rapid combustion of grass—a dense type of vegetation introduced into the island (*Melinis minutiflora*)—the topography that facilitates the increase of wind, incrementing the intensity and progress of the fire, and the use of grass as cover material for certain traditional houses (Corporación Nacional Forestal, n.d.).

Analyzing the effectiveness of the response protocols, a case study of the fire on September 17, 2017, was analyzed (Fig. 4.3.8). On that occasion, residents saw a spotlight and instinctively notified firefighters, although they were mainly responsible for combating urban fires over forest fires. Firefighters, therefore, alerted CONAF, who activated its protocol, as well as Ma'u Henua, the site manager and link to the original, organized Rapanui community. Given the advance of the fire, CONAF alerted the ONEMI, who declared a red alert on the island. As a result, the army and local police

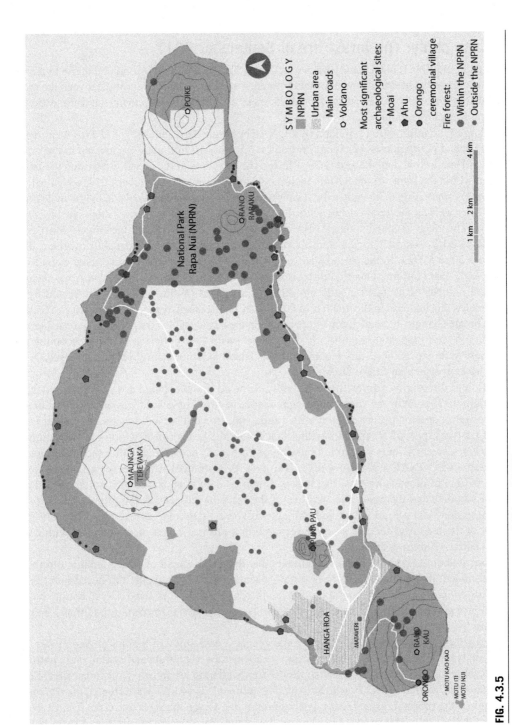

FIG. 4.3.5

Forest fire in Rapa Nui island between 2011 and 2017.

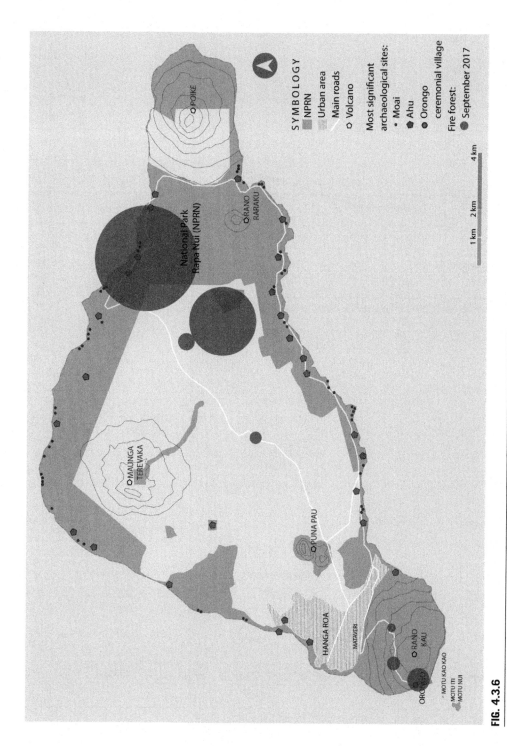

FIG. 4.3.6

Forest fire September 2017 in Rapa Nui island.

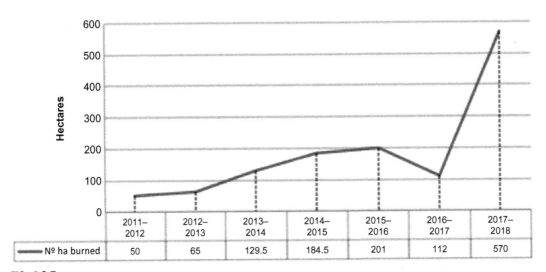

FIG. 4.3.7

Increase in hectares burned on Rapa Nui island between 2011 and 2018.

	2011–2012	2012–2013	2013–2014	2014–2015	2015–2016	2016–2017	2017–2018
N° ha burned	50	65	129.5	184.5	201	112	570

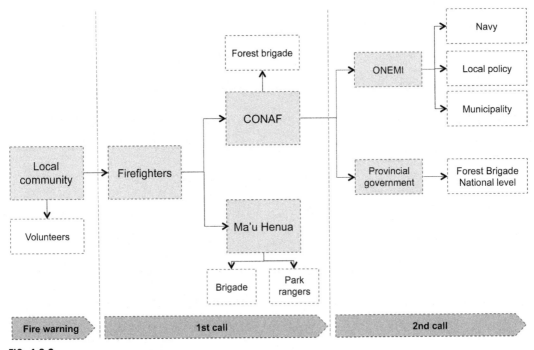

FIG. 4.3.8

Protocol activated in Rapa Nui island during fire forest September 2017.

supported the work, facilitating the circulation of emergency vehicles around the island and the provision of machinery and water through the tank truck. The municipality and the MOP also assisted this work. In parallel, the residents also joined as volunteers to fight the fire. After 48 h of activity, with more than six simultaneous outbreaks, the island was totally outnumbered in its ability to cope with the fire. It was necessary to ask for help from the continent, through the provincial government. Since it was a forest emergency, the forest brigade from the continent arrived on the island to aid in the extinction of the fire.

The effects of the fire were devastating, especially on the environment. Various institutions reported damage to the natural and cultural heritage sites, with pronounced effects on petroglyphs, which remain in a vulnerable condition, compared to other agents such as the aforementioned environmental pressures (Corporación Nacional Forestal, Comunidad Indígenas Polinésica Ma'u Henua, Museo antropológico P. Sebastián Englert, Secretaría Técnica de Patrimonio Rapa Nui, 2017).

This fire revealed the lack of coordination between the different entities, as well as ignorance from the neighbors about the accompanying protocols and the responsibilities of each group. This is a common feature, even at the level of competent public institutions.

4.3.7 **Results**

The analysis of the events that occurred on the island, especially the fire in September 2017, allows us to establish a diagnosis that identifies the main elements of the problem for an optimal risk management of forest fires. From this diagnosis, it is possible to define strategies and actions in the short and medium term to improve the management and resilience of this insular community. Regarding the origin of the fires, the causes are related to social and cultural aspects that are widely known by the population but poorly addressed. In this case, the major problem is not coming from the groups recognized as outsiders but between the clans themselves.

Table 4.3.2 shows the relationship between the forest fire and the series of factors that affect its occurrence in relation with the stages of prevention or control: (1) cultural and anthropological factors, related to the ancestral customs of the Rapanui people of burning pastures or disputes over territorial demarcations; within sociocultural factors is the distrust toward groups that do not belong to the

Table 4.3.2 Vulnerability factors in forest fire in relation with DRR cycle.

Vulnerability factors for forest fire		Prevention	Control
1. Cultural and anthropological	– Burning pastures	x	
	– Disputes between clans	x	
	– Distrust between groups in the island	x	x
2. Management	– No disaster risk management plan	x	x
	– No protocols and drills	x	x
3. Context	– Type of grass	x	
	– Temperature + humidity		x
	– Rainfall decrease		x
	– Topography + wind		x

Rapanui ethnic group, as well as toward the Chilean institutions in the island; (2) management factors, related with a lack of DRR plan resulting in lack of economic resources, lack of equipment for the different actors, and a lack of training to face forest emergencies; also the existence of different protocols with the same actors and no drills, which does not facilitate control of a forest fire; the diversity of actors, as they are not coordinated among themselves and do not have clear roles in managing emergencies; (3) context factors, related with geographical and environmental conditions of the island that have favored the spread of fire in recent years, such as increase of temperature. Taking account the origin of forest fire, and considering the strong relationship of the indigenous community with the site, they should have an active role in avoiding burning or restricting it to safe areas. Likewise, disputes between the clans should also be a topic of discussion, in that the councils of elders on the island can mediate conflicts, if not resolve them.

Regarding the protocols, the site manager should practice leadership in this matter. However, taking into account that the problem is the forest fire, this individual should work closely with professionals specialized in combat—like CONAF—to strengthen the capabilities of the brigade members of the Ma'u Henua. The protocols must also be known and tested, thus forming part of the prevention and control of the emergency stage. The idea of acting in an articulated manner, based on one's abilities and following a protocol, is essential. In addition to the site manager, this task should be supported by institutions at the national level (Fig. 4.3.9). The recovery stage, which will not be addressed in this study, should be included at this point. The proposal includes, in addition to the response stage, the subsequent stage prior to the reconstruction. This stage is critical because immediate measures must be taken depending on the degree of damage, the values and significance of the archaeological and architectural remains, and the available protection resources. All these measures must be taken by

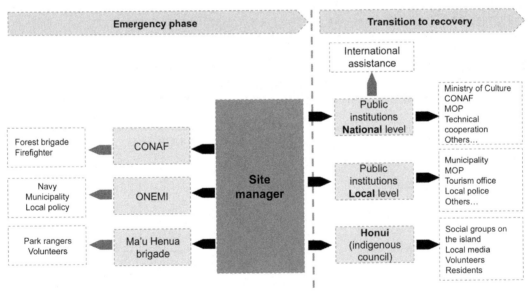

FIG. 4.3.9

Response protocol proposal.

specialized personnel with the purpose of both protecting the assets and applying procedures that do not cause further damage to the assets (International Council on Monuments and Sites, 2017; ICCORM, 2018).

In recent workshops carried out with the Rapanui community in the framework of the design of a site management plan, four areas emerged that were defined as priorities to guide the proper management of the island: education, identity, health, and environment (Comisión Provincial de Monumentos Rapa Nui, Secretaría Técnica de Patrimonio Rapa Nui, 2014). There is, therefore, evidence of the key areas to integrate into future initiatives for the island, linked to local culture and identity. All these areas can be considered in order to achieve a risk management culture through some concrete actions aimed at better prevention and control of forest fires (Table 4.3.3). Education is a key and central aspect of the process to relate the assessment and care of the vestiges of the ancestors with risk management. This also implies the issue of identity as an aspect that is reaffirmed from multiple manifestations, such as archaeological remains and traditions. The reference to health is interesting because it refers to mental and physical health. There is, therefore, an awareness of the social problems that are manifested in disputes or the aforementioned burnings. Finally, the environment is very relevant because the island presents critical environmental problems, such as garbage, deforestation, and lack of water, among others. Since forest fires are a real threat to the environment, there is also the opportunity to integrate measures that reduce risk into its conservation and management.

Table 4.3.3 Proposed actions to integrate in the reduction of risk from the traditions and areas that the islanders recognize.

Cultural sources/ areas to improve	Language	Extended family	Dances	Songs
Education	– Risk management plan, protocol and drills in native language	– Strengthen the transfer of traditional knowledge	– Integrate culture of prevention in dances	– Integrate culture of prevention in songs
Identity		– Importance form cultural heritage	– Working drills from dance groups	
Health (physical and mental)		– To improve relationships between clans	– Working drills from dance groups	
Environment	– Recover and disseminate the toponymy of the island	– Importance of environment in the history of the island – Strong relation of environment and cultural heritage	– Integrate the preservation and care of the environment to dances	– Integrate the preservation and care of the environment to songs

In addition, the diagnosis for cultural development carried out in 2012 (Consejo Nacional de las Culturas y las Artes, 2012) provides interesting and valuable results on the most valued cultural features, according to the Rapanui people themselves. The language is the most highly valued aspect, followed by extended family and the dances and songs expressed in diverse festivities. With these elements of immaterial culture present in the daily life of the islanders, there is an opportunity to integrate them into a culture of risk management. The subject of the language is fundamental and a constant demand: Part of the claims of the lack of recognition from Chile of the Rapanui identity is rooted in the lack of instruments emerged and expressed in their native language. Likewise, Rapanui native language, like many others, contains concepts and ideas based on a specific worldview, which are difficult to translate (Pedro Edmunds Paoa, 2017). The extended family also represents an opportunity to develop spaces for coordination, collaboration, and education around risk management considering, for example, the hierarchical structure maintained or the differentiation of roles. The extended family can also play an important role in improving relations between the clans, thereby reducing part of the problem of the origin of forest fires on the island and installing a more collaborative and supportive culture. The dances and songs are strongly present as part of the traditions associated with rites and customs. These can also contribute to installing good practices among the population in relation to the prevention of fires, the protocols, and the coordinated and articulated work.

Additionally, from the previous diagnosis, a series of actions have been proposed to implement a comprehensive risk management plan for the island, organized in four areas with their associated actions (Table 4.3.4). The institutional factors refer to the necessary coordination and joint work between the local, regional, and national institutions responsible for risk management, both inside and outside the island. In relation to this, especially at the local level, there is a need to have protocols that are unique and known by all the institutions and tested in simulacrum exercises.

It is fundamental to involve institutions and people in the actions proposed in the legal framework. As mentioned, the National Risk Management Policy is governed by the guidelines of the Sendai Framework. In this sense, for the first time, heritage management will be integrated with disaster risk reduction. This is an opportunity to relate culture to other legal instruments of different institutions. In the institutional sphere, the National Platform for Disaster Risk Management manages alerts from the ONEMI to the rest of the public institutions, being the instance to work in a collaborative and

Table 4.3.4 Proposal of risk reduction strategies.

Areas	Actions
Institutional framework	– Improve network between different institutions (national and local level) – Define protocol action between institutions – Strengthen site management office
Legal framework	– Integrate different levels and legislation related to risk preparedness – Integrate Disaster Risk Management for the World Heritage Site in urban planning
Human sources	– Identify different groups linked to the site, roles and responsibilities – Strengthen networks of community collaboration – Enhanced fire-fighter capacities for site conditions – Establish drills for residents and tourists
Financial sources	– Include allocation of national funds for risk management – Establish fund for disasters

coordinated way. Regarding urban planning, the recommendation is the same, as diagnoses are already available to identify other threats besides fires, such as those derived from climate change that currently affect areas with archaeological heritage. The WHS condition is an aspect that must always be kept in mind.

In the case of human resources, there are several tasks. Many of these duties should be led by the site manager because of his/her knowledge of the island's anthropological and cultural contexts. Several communal groups should be identified to review their relationship with the site and with the problem of forest fires in particular. These groups should be integrated into specific protocols to respond to emergencies, and these protocols should be disseminated with periodic drills, especially for neighbors and tourists. The role of community leaders and indigenous councils is very relevant because of the recognition they have within the Rapanui community. The value of the extended family on the island, the inheritance of the traditional clans, and the native language are elements that need to be strengthened and integrated in the cycle of risk management by designing training systems for the population in their language. It is also key to create response protocols that integrate the current family hierarchy within family groups, which can be framed in the type of collaborative, solidary and reciprocal, traditional activities of the island ("umanga"). In the specific area of forest fires, capacity building includes the Ma'u Henua brigades and the fire brigades. Here, training for firefighting should be led by CONAF, a group that addresses this issue and has ample experience working on the island.

Regarding financial resources, there should be funds to carry out the necessary actions of the entire risk management process: prevention, control, and reconstruction. These funds may come from the touristic activity in addition to funding provided at the regional and national levels.

4.3.8 Conclusions

The case of the NPRN raises a series of problems derived from its insular condition, the fragility of its ecosystems, the cultural heritage, and the diversity of the groups that inhabit it, as well. The group belonging to the original Rapanui ethnic group maintains a close relationship with its cultural heritage, despite the strong changes that have occurred on the island. This group has a certain authority over the management and conservation of these cultural remains, with ideological aspects prevailing over those that are exclusively technical or derived from the management of the assets (Comisión de Desarrollo de Isla de Pascua, 2015). However, a number of institutions that represent the Chilean state and provide important services on the island are not fully validated by the Rapanui community. This represents a problem because the island, as recognized by all its inhabitants, is absolutely dependent on the continent today.

In this context, the challenge of decreasing the vulnerability of this site while increasing the resilience of this community is the priority. The importance of analyzing specific contexts from both physical conditions and cultural processes is demonstrated. The case that was analyzed of the forest fires on the Rapa Nui island confirms this, since each of the stages of the risk management cycle appears to be strongly determined by conditions in this context, such as the causes derived from ancestral practices; the control protocols, with leadership problems among the groups of the island; and the management in general, with a new administration that needs to strengthen its capacities to efficiently manage and conserve the cultural heritage. Paradoxically, we are dealing with causes that might seem easy to handle, given that they are not natural, but they involve a relatively small group of people (7750 inhabitants, of whom 3512 correspond to members of the original Rapanui community) in a limited geographical space (16,628 ha). However, the sociopolitical reality on the island is complex, as in other world contexts. Nevertheless, there are several recent diagnoses that show the willingness of an important group of islanders to rescue their cultural and environmental heritage. The problems identified by the Rapanui people are common with other contexts (effects of climate change, environmental crisis, tourist overexploitation, etc.); therefore from this diagnosis and from the public information, available actions can be taken to initiate the process of formulating a comprehensive disaster risk management plan for the island. The importance of linking heritage with sustainable development in Rapa Nui, in both economic and social terms, is also understood. This can also be seen as an opportunity and a favorable scenario to implement measures and seriously address the threats that are affecting this valuable WHS.

References

Bahamondez, M., Isla de Pascua, M., 2000. conservación de su estatuaria: un proceso en desarrollo. Conserva 4, 57–70.

Campbell, R., 1987. La cultura de la isla de Pascua mito y realidad. Santiago, Andrés Bello.

Castro, N., 2006. Rapa Nui. El diablo, Dios y la profetisa, Rapa Nui. Rapanui Press.

Comisión de Desarrollo de Isla de Pascua, 2015. Ma'u Henua. Propuesta de nueva administración Parque Nacional Rapa Nui.

Comisión Provincial de Monumentos Rapa Nui, Secretaría Técnica de Patrimonio Rapa Nui, 2014. Compilado Resúmenes Mesas de Trabajo Seminario Umaŋa Haka Tika Mana'u Por un Plan Maestro Patrimonial Rapa Nui. Rapa Nui.

Comunidad Indígena Polinésica Ma'u Henua, n.d., Available from: https://www.comunidadmauhenua.com/ [Accessed 10 June 2018].

Consejo Nacional de las Culturas y las Artes. Estudio diagnóstico del desarrollo cultural del pueblo Rapa Nui. 2012. [Online] Available from: http://www.cultura.gob.cl/estudios/observatorio-cultural/estudiodiagnó sticodeldesarrolloculturaldelpueblorapanui.htm [Accessed 5 March 2019].

Cornejo, M., 2005. Isla de Pascua: Valparaíso. Valparaíso, Editorial Universidad de Valparaíso, Chile.

Corporación Nacional Forestal, n.d. Available from: http://www.conaf.cl/quienes-somos/ [Accessed 10 October 2018].

Corporación Nacional Forestal, Comunidad Indígenas Polinésica Ma'u Henua, Museo antropológico P. Sebastián Englert, Secretaría Técnica de Patrimonio Rapa Nui, 2017. Informe técnico pericial. Diagnóstico de conservación de rasgos arqueológicos afectados por incendio en la Isla de Pascua. Hanga Roa.

Corporación Nacional Forestal, Consejo de Monumentos Nacionales, 2010. Propuesta de declaración retrospectiva de valor universal excepcional. Parque Nacional Rapa Nui (N° 715, inscrito en la Lista del Patrimonio Mundial de 1995).

Dirección Meteorológica de Chile, 2019. Metadatos disponibles en el sistema SACLIM. [Online] Available from: https://climatologia.meteochile.gob.cl/application/informacion/ficha-de-estacion/270001, 2019 [Accessed 26 March 2019].

Englert, S., 1983. La tierra de Hotu Matu'a. Editorial Universitaria, Santiago.

Gobierno de Chile, 1995. Nominación a la lista de patrimonio mundial del Parque nacional Rapa Nui. Consejo de Monumentos Nacionales.

Gobierno de Chile, 2003. La presentación de informes periódicos sobre la aplicación de la convención del patrimonio mundial. Estado de conservación de bienes específicos del Patrimonio Mundial Estado Parte: Chile. Nombre del bien: Parque Nacional Rapa Nui.

Gobierno de Chile, 2012. La presentación de informes periódicos sobre la aplicación de la convención del patrimonio mundial. Estado de conservación de bienes específicos del Patrimonio Mundial Estado Parte: Chile. Nombre del bien: Parque Nacional Rapa Nui.

Gottler, M., Ripp, M., 2017. Community Involvement in Heritage Management Guidebook. Stadt Regenburg, City of Regensburg. [Online] Available from: http://openarchive.icomos.org/1812/1/FINAL_OWHC%20Guidebook%202017.pdf. [Accessed 22 March 2018].

ICCORM, 2018. First Aid to Cultural Heritage in Times of Crisis. Handbook 12. [Online] Available from: https://www.iccrom.org/sites/default/files/2018-10/fac_handbook_print_oct-2018_final.pdf. [Accessed 26 April 2019].

Instituto Nacional de Estadísticas n.d. Available from: https://ine.cl/bases-de-datos [Accessed 15 January 2019].

International Council on Monuments and Sites, 2017. ICOMOS Guidance on Post Trauma Recovery and Reconstruction for World Heritage Cultural Properties. Available from: http://openarchive.icomos.org/1763/. [Accessed 20 June 2018].

López, M.A., 2017. Archeologist. Interview. Secretaría Técnica de Patrimonio Rapa Nui Consejo de Monumentos Nacionales Servicio Nacional de Patrimonio Cultural November.

Muñoz, A., 2007. Rapa Nui Translocal: Estar Aquí, Estar Allá. VI Congreso Chileno de Antropología. Colegio de Antropólogos de Chile A. G, Valdivia. Accessible from: https://www.aacademica.org/vi.congreso.chileno.de.antropologia/42. [Accessed 26 April 2019].

Ninoska Huki Cuadros, 2017. Provincial Chief Rapa Nui Island. Interview. November.

Pedro Edmunds Paoa, 2017. Mayor Rapa Nui Island. Interview. November.

Ramírez, J., 2008. Rapa Nui. El ombligo del Mundo. Morgan Impresiones S.A., Santiago

Rapu, R., 2017. Archiolgist. Chief Department of Archeology National Park Rapa Nui. Interview. November.

Shand, W., 2018. Making Spaces for Co-production: Collaborative Action for Settlement Upgrading in Harare Zimbabwe; 2018. [Online] Accessible from: https://journals.sagepub.com/doi/10.1177/0956247818783962 [Accessed 4 March 2019].

Stovel, H., 1998. Risk Preparedness: A Management Manual for World Cultural Heritage. ICCROM, Rome. Available from: http://icorp.icomos.org/wp-content/uploads/2017/10/ICCROM_17_RiskPreparedness_en.pdf. [Accessed 4 May 2018].

UN Office for Disaster Risk Reduction, 2005. Hyogo Framework for Action 2005–2015: Building the Resilience of Nations and Communities to Disasters. [Online] Available from: https://www.unisdr.org/we/inform/publications/1037. [Accessed 25 March 2018].

UN Office for Disaster Risk Reduction, 2015. Sendai Framework for Disaster Risk Reduction 2015–2030. [Online] Available from: https://www.unisdr.org/we/inform/publications/43291. [Accessed 25 March 2018].

UNESCO, 2012. Risk Management at Heritage Sites: A Case Study of the Petra World Heritage Site. [Online] Available from: https://unesdoc.unesco.org/ark:/48223/pf0000217107. [Accessed 25 March 2018].

UNESCO. n.d. Managing Disaster Risks for World Heritage [Online] Available from: https://whc.unesco.org/en/managing-disaster-risks/ [Accessed 22 March 2018].

A people-centered approach to program design: Unpacking risks and vulnerabilities in Darfur, Sudan

4.4

Daniela Giardina[a] and Jessica Fullwood-Thomas[b]

Disaster Risk Reduction and Resilience Advisor, Humanitarian Theme Team, Oxfam America, Boston, MA, United States[a] Resilience and Fragility Advisor, International Programme Team, Oxfam GB, Oxford, United Kingdom[b]

CHAPTER OUTLINE

4.4.1 Introduction

The Sendai Global Framework for Action on DRR (2015–2030), a 15-year voluntary framework endorsed by the UN General Assembly and overseen by the United Nations Office for Disaster Risk Reduction (UNDRR), prioritized the need to better understand disaster risk as one of its four specific pillars (UNDRR, 2015). Priority 1 requires disaster risk management (DRM) policies to be based on a contextualized assessment of disaster risk, vulnerabilities, and capacities, which incorporates scientific information and traditional indigenous knowledge to enable prevention, mitigation,

Understanding Disaster Risk. https://doi.org/10.1016/B978-0-12-819047-0.00017-2

preparedness, and response interventions. Furthermore, the Sendai Framework's guiding principles promote empowerment, consultation, and inclusive knowledge production by emphasizing the requirement for an "all-of-society engagement and participation" strategy and the need for a "multi-hazard approach and inclusive risk-informed decision-making based on the open exchange and dissemination of disaggregated data" (UNDRR, 2015). The Sendai Framework therefore provides a vital bilateral platform to influence states on investment in vulnerability and risk assessment approaches, and it also helps generate avenues for diverse partnerships and alliances in support of more comprehensive, holistic, and multidisciplinary risk analysis.

This chapter will consider how understanding has evolved around vulnerability and risk analysis across a range of disciplines and then document process and findings of a contextualized locally driven qualitative methodology that Oxfam staff and partners have carried out in Darfur, Sudan, in November 2017.

The history of vulnerability analysis—the terminology that underpins it—has gone through several evolutionary stages from the natural sciences, geography, and human ecology disciplines to the social sciences and development practice. It has also taken on an increasingly localized and context-specific lens while maintaining a recognition of the global interdependencies and systems that shape risk across diverse socioecological landscapes. Vulnerability is experienced to some degree by everyone everywhere: it is never static and is shaped by intersecting and fluid dynamics that are informed by both our physical environment (location, access to key natural resources, weather and climate conditions) and our socially and culturally constructed norms (gender, age, religion, socioeconomic status). While there is no dominant or singular definition, there is some consensus that vulnerability is linked to the "susceptibility or fragility of communities, systems or elements at risk and their capacity to cope under hazardous conditions" (Birkmann et al., 2013). UNDRR notes that the conditions for vulnerability are informed by "physical, social, economic and environmental factors or processes, which increase the susceptibility of an individual, a community, asset or system to the impact of hazards" (UNDRR, 2015).

A distinct shift in understanding disasters occurred after the Cold War era; previously disasters were seen as the linear result of natural hazards/events and that human interventions could only respond through expert-led scientific monitoring, predictions, and preparedness (Knowles, 2011; Wisner, 2016), but a counternarrative emerged from human ecology and geography disciplines that began to highlight how people's understanding of risk and decisions around how they interacted with the natural environment, were factors in determining vulnerability—e.g., how land-use policy could influence the frequency and magnitude of flood disasters (White, 1945). That is to say, the natural hazard of flooding remained, but the pathway to a disaster could be influenced by human action or inaction. In this regard, there has been an awakening to the social constructs that inform and shape risk profiles and, in turn, the societal and behavior change shifts that are needed to reduce vulnerability and/or increase capacity. This thinking was taken further in some hazard perception testing in low-income countries, which highlighted the incompatibility of assumptions and perceptions when applied to different contexts. Furthermore, fieldwork results in these contexts pointed to the prominent role social and political issues like bad governance and unequal access to resources had, on turning a natural phenomenon, such as drought into a human-experienced disaster (Wisner, 2016). These findings situated the vulnerability of a person or asset not just within their physical environment, but also within preexisting social, economic and political conditions. This viewpoint has increasingly become the dominant narrative often referred to as the "vulnerability approach" (O'Keefe and Wisner, 1975; Susman et al., 1983; Chambers, 1983; Winchester, 1986). It has also given rise to a more nuanced exploration of the terminology around hazard, disaster, and risk that gives people greater agency and responsibility for how vulnerability is experienced (Anderson and Woodrow, 1998; Blaikie et al., 1994; Bohle et al., 1994; Cutter, 1996; Enarson and Morrow, 1997). Within development practitioners, there has been a growing complementary push to concede, or at least cohabit, scientific and quantitative data with indigenous practices and knowledge generated from the lived experience of people in highly vulnerable contexts (World Bank, 2001).

Recent evolutions in disaster risk reduction (DRR) and resilience research have shown that it is often not just the hazard that determines the scale or impact of a disaster but also the ability of a population to anticipate, respond to, and recover from its effects. This has forced a recognition that it is important to consider not just vulnerability but also capacity in order to understand how disasters occur. This shift in perception has facilitated a move from pure hazards identification, assessment, and ranking of vulnerabilities (including their unequal distribution in populations) (DFID, 2006) toward a greater emphasis on understanding existing capacities and channeling investment that supports the resilience of individuals, communities, and systems to manage change (Hilleboe et al., 2013). It has also facilitated a layering and packaging of interventions in response to analysis of risks that seeks simultaneously to build absorptive (protective actions that help people cope with known shocks and stresses), adaptive (strategies that change practices in the medium to longer term in anticipation of a disaster or response to change and in ways that create more flexibility), and transformative (intentional changes to stop or reduce the underlying causes of risk including poverty, inequality, exclusion and lack of rights) capacities (Pelling et al., 2014; Jeans et al., 2016; Jeans, 2017). This thinking is housed within the broader emergence of resilience narratives that Oxfam is adopting, by defining resilience as the ability of women and men to realize their rights and improve their wellbeing despite shocks, stresses, and uncertainty (Jeans, Thomas and Castillo, 2016). This future-looking, flexible approach seeks a step-change in capacity rather than only supporting people and systems to prepare and respond to inevitable shocks and stresses.

Given the current unpredictable and changing environment linked to the impact of climate change (Blunden et al., 2018; IPCC, 2018; Herring et al., 2019), which is also accompanied by shifting social norms, issues of governance, conflict, and greater fragility and instability (OECD, 2016; World Bank, 2018; Howard and Stark, 2018), it is increasingly important to find ways to assess and monitor the vulnerability of communities and key livelihood groups. Processes for mapping vulnerability and risk have incorporated scientific and climate modeling, analysis of loss and damage as well as community-dialogue processes that ensure those affected are able to share their experiences, priorities, and adaptive solution ideas (Birkmann et al., 2013). Initial thinking was housed within the natural science communities who primarily focused on quantifiable categories linked to physical vulnerability and/or economic or actual damages (Papathoma-Köhle et al., 2011). Social science approaches tended to consider broader context factors that shape who is most vulnerable and took a more person or community-centered approach in mapping harmful consequences (DFID, 1999; Wisner et al., 2004). In this regard, there is a strong acknowledgment that vulnerability is driven by "social inequality and historic patterns of social relations that manifest as deeply embedded in social structural barriers that are often resistant to change" (Phillips and Fordham, 2009, p. 12). As well as promoting improved partnerships between the science, technology, and private sectors for generating and sharing comprehensive multihazard disaster risk data, nonprofit organizations emphasize peer-to-peer collaboration through the involvement of community-based organizations, NGOs, and local groups. Delivering on priority 1 of the Sendai framework and finding a balance between scientific and nonscientific knowledge require a locally driven participatory methodology for vulnerability and risk analysis, which creates a platform for knowledge exchange, discussion, and exploration by a range of stakeholders and duty bearers.

Community-based participatory tools for vulnerability and risk assessments have therefore come to the fore (Chiwaka and Yates, 2000; IFRC, 2006) as has a greater realization within international frameworks and agreements that improvements are needed in understanding disaster risk in all its dimensions (UNDRR, 2015).

Oxfam has previously developed a Participatory Capacity and Vulnerability Assessment (PCVA) (Turnbull and Turvill, 2012) methodology, which identified strengths and weaknesses of households, communities, and institutions in the face of, primarily, environmental and climate hazards. These types of tools often focus on singular units of analysis at village level in which they aim to map risks within a small area and come up with community-based solutions. While these have proved effective and useful for localized interventions and building intracommunity activism, a critical limitation is the potential risk transfer on vulnerable community to be responsible for their own recovery and resilience building (as opposed to states and other duty-bearers). It can also limit solidarity with other stakeholders for building bigger transformative changes, such as advocacy on national policy decisions about infra-structure constructions or collective action for example against private sector unregulated logging and the environmental impacts of the extractives industry.

Given climate change consequences and extreme weather events affect large geographical areas and/or any changes in one location have knock-on effects (e.g., diverting rivers for reservoirs impacts downstream water flows and use), it is also important to find ways to support intercommunity dialogue. To better understand the complexity around vulnerabilities and take an intersectional and crossbound-ary approach and based on methodologies and tools available in the literature, Oxfam has in recent years developed a vulnerability and risk analysis (VRA) methodology. This tool takes a landscape-wide perspective of vulnerability, explores both natural hazards and social issues as causes of risk, and links stakeholders across various levels of governance, not only at community level, to coidentify risk priorities and solutions (Morchain and Kelsey, 2016; Fullwood-Thomas, 2017). Furthermore, this methodology is part of wider shifts in resilient development theory within Oxfam, including the per-ception of people in receipt of aid as not just victims or beneficiaries but as agents of change who have the skills and capacities to be part of their own development journey (Jeans et al., 2016).

Until recently, the VRA has been predominately used in developmental contexts but in this instance carrying it out in a protracted crisis setting like Darfur (Sudan) was an opportunity to explore how it can validate, reshape, and/or challenge thinking in more stressed environments that are primarily in receipt of humanitarian assistance but increasingly want to move toward recovery, rehabilitation, and resil-ience building interventions. The intention to explore both short and longer-term solutions in a joined-up, holistic way is in tune with wider reframing that is occurring within the aid sector on "new ways of working" under a nexus approach and the adoption of "local humanitarian leadership" principles (Slim, 2017; ICVA, 2017; Kittaneh and Stolk, 2018; OCHA, 2016). Conducting a VRA in such a protracted crisis setting where there is an inevitable convergence of risks also represented an opportunity to explore how underlying inequalities like gender, marginalization, and exclusion shape the more visible and commonly addressed natural hazard vulnerabilities within key livelihood groups in this case pastoralists, agropastoralists, and farmers. Moreover, we were also interested in investigat-ing how issues of land-rights and land-use have become entwined with the tensions and power dynam-ics associated with the conflict that has raged throughout the region for many decades (Osman and Cohen, 2014; Morchain et al., 2015).

4.4.2 **Background on the context**
4.4.2.1 **Livelihood and environment systems**

Darfur is a region in Western Sudan where communities are predominantly practicing a combination of farming and pastoralist activities as their major livelihoods systems. In this agropastoral production system, communities adopt farming as their primary livelihood and then complement this with animal husbandry, especially cattle and goats. South Darfur's agropastoralists have relatively higher and more stable crop production when compared to their North Darfur counterparts attributable to the higher annual rainfall in the area. Millet and sorghum are the major staple cereal crops produced in South Darfur often intercropped with okra and groundnut, which are cultivated as cash crops. Gum Arabic and other forest products are secondary sources of income for many households.

Pastoralism is a major—natural resource-based—livelihood system both in North and South Darfur. It relies on strategic seasonal migration where cattle herders trek North toward the wet-season pastures at the start of the rainy season (June/July) and return to the South in the dry season (November through April). This pastoral production system is associated with the *Baggara* (the cattle herders) and practiced by *Rizeigat* groups, nomadic Bedouin communities (Fitzpatrick and Young, 2015).

In North Darfur, the major pastoral production system is associated with the *Abbala* (the camel herders). Camel herding has always been characterized by nomadic mobility as the whole community moved seasonally tracking water and rangelands. While camel herds continue to move seasonally, a significant change has been noticed in recent years with regards to cattle- and goat-oriented pastoralists who have shifted from nomadism toward settled living (Fitzpatrick and Young, 2015).

Darfur has been the major exporter of livestock to other markets in Sudan, which reflects the relative quality of its livestock products (Fitzpatrick and Young, 2015). Its market was negatively affected by the conflict: according to the Chamber of Commerce in Khartoum, Darfur, it used to account for around 30% of Sudan's livestock exports before the conflict; this has now fallen by at least 15% (Buchanan-Smith et al., 2012).

Being considered part of the Sahel region, both North and South Darfur regions are highly exposed to climatic variabilities shaped by uneven distribution of rainfall spatially and temporally. Although rainfall data from across Darfur region show a declining trend, the most noticeable decrease is recorded in North Darfur. Since the beginning of rainfall data recording, the 10-year average in El Fasher (capital of North Darfur) dwindled from 300 mm to about 200 mm (UNEP, 2007). This annual rainfall variability is believed to have a significant impact on the livelihoods of North Darfur and other northern states of Sudan as they are based on rain fed agriculture or dependent on access to green pastures and water sources for livestock (UNEP, 2007).

According to the UN Convention to Combat Desertification, desertification is an outcome of the combined effect of climate change and human activities that degrade lands in arid, semiarid, and dry subhumid areas. UNEP identified three interlinked desertification processes that are a common challenge across Sudan: (1) over use of semidesert areas through deforestation; (2) overgrazing and (3) cultivation that converts a habitat into desert. In North Darfur and other northern states, there has been a trend in climate-based conversion of land types from semidesert to desert as a result of the aforementioned processes coupled with the overall reduction in annual rainfall, which led to the failure of less drought-resistant vegetation.

North Darfur's vegetation cover is characterized by variation in terms of density and diversity of plant species. The northern part of the state is desert sand dunes with no vegetation cover apart from in the surroundings of the major *Wadis* (seasonal rivers). Toward central parts of North Darfur, limited vegetation begins to emerge in the form of scattered and patchy growth of short grasses, good pasture for camel herders during the winter season. The vegetation cover in South Darfur is mostly woodland savannah as the area receives an average annual rainfall of 444 mm (Climate Data, 2019). The wood-land savannah of South Darfur is composed of a mixture of acacia trees, short shrubs, and perennial herbaceous species. In some areas, the vegetation cover takes the shape of grasslands, while in others it consists of dense tree growth.

4.4.2.2 Conflict

In 2003, the region of Darfur witnessed an unprecedented violent conflict that had highly disruptive consequences for farming and pastoralist communities, as well as causing the injury, death, and displacement of community members. While this period represented a peak in recent conflict dynamics, Darfur has a long history of human-made fragility linked to intercommunity fighting, conflict over resources, and wider political violence linked to the regional dimensions of the Sudan/South Sudan war and the attempts to annex Darfur/claim independence from Khartoum (BBC, 2019). The settled farming communities were significantly affected by this recent wave of conflict as they experienced loss of livelihood assets, especially animals and other accumulated wealth, and had very restricted access to their farming lands. This meant they were unable to plant, tend, and harvest crops. Pastoralist communities in different localities in North Darfur were also targeted by rebel forces for competing access to resources. According to UNOCHA, there are still about 1.76 million internally displaced persons (IDPs) in Darfur out of 2.1 million IDPs in Sudan as a whole (UNOCHA, 2018).

Following improvements in the security situation and access to conflict-affected areas, many households have been able to restore their livelihoods and UN estimates suggest that recently there have been about 386,000 returnees in Sudan, mainly in Darfur (UNOCHA, 2018). However, it remains very complex to identify who is a returnee and who is displaced, especially for those who are long-term IDPs who may have been displaced multiple times over a protracted period or are voluntarily moving between IDP centers and their villages of origin, e.g., during the cultivation season when they return to plant and harvest crops on their original land. Therefore, figures of IDPs at any given moment can only be representative estimates, but what they show is that conflict and instability continue to play a defining role in when, if, and how communities in Darfur are forced to move.

4.4.2.3 International assistance

Darfur has received extensive humanitarian assistance since 2003 in response to a complex, constantly evolving context. With recent increased stability, both the government and the international community are looking to move from a prolonged era of emergency response to one of recovery, rehabilitation, and sustainable development. The inability of all residents of Darfur to achieve sustainable livelihood strategies suited to the extremely variable climate has been at the heart of the on-going crisis. Any durable solutions must include a detailed understanding of the livelihood systems in Darfur, how they complement each other, and how they can be structured to not just meet the basic needs of the population, but also to allow them to adapt and thrive in the face of future climate vulnerability (Fitzpatrick and Young, 2015).

4.4.3 **Methodology**

The vulnerability risk analysis (VRA) assessment tool (Morchain and Kelsey, 2016; Fullwood-Thomas, 2017) was applied in North and South Darfur in November 2017 in partnership with two local organizations—Kabkabya Smallholders Charitable Society and Jabal Marra Charity Organization. The participatory and locally driven approach of this tool places particular importance on preparatory steps, which, depending on how they are completed, can play a defining role in both the success of the exercise and the usefulness and relevance of the data generated. One such critical task for the success of a VRA is the stakeholder mapping and associated power analysis, which is used to guide the subsequent selection of Knowledge Group participants (20–30 people representing the community and livelihood groups being analyzed as well as marginalized and vulnerable people). In this instance, the stakeholder mapping was led by Oxfam staff in Sudan, who had the contextual knowledge and existing local relationships and networks to ensure an inclusive and representative list of invitees. Another aspect was paying specific attention to power relations and attempting to create a safe/neutral space for people to speak up and share their perspectives and idea. Prior to beginning the data collection, other preparation steps were conducted, which sought to ensure the primary data could be gathered as efficiently and effectively as possible. These included the training of Oxfam, Kabkabya Smallholders Charitable Society and Jabal Marra Charity Organization staff, the identification of a primary team of facilitators, and the selection of suitable venues, which took into consideration issues of safe access and appropriate facilities that would put participants at ease.

This assessment was structured over two 3-day workshops, one carried out in Nyala—capital of South Darfur—one in El Fasher, capital of North Darfur. The VRA brought together traditional and nontraditional stakeholders as a representative sample of the geographical location/livelihood zone where the assessments took place. The participants of the Knowledge Groups (KG) included pastoralists, agriculture farmers, traders, government officials and academic professionals, local women's farming groups and community associations, as well as marginalized voices (disabled people, women, and youth). The participatory process was led by a main facilitator and group discussions in the local dialect were overseen by several pretrained individuals from Oxfam and partners. The process consisted of five steps: building awareness of the main hazards and social issues affecting people; identifying who was most affected; gaining a better understanding of current coping strategies; exploring the variety of impacts and consequences caused by the hazards and social issues and, finally, supporting the joint development of adaptive strategies to promote resilient development (Fig. 4.4.1).

Step 1, named "Initial Vulnerability Assessment," focuses on the participants agreeing on a shortlist of priority hazards and social issues as well as identifying the most vulnerable groups. This serves as a mean of validating preexisting organizational assumptions/historical data through group work, plenary discussion, initial voting rounds and then a score card exercise assessing level of exposure and sensitivity, which produces a final, prioritized list of hazards/social issues. Once the shortlist is agreed, the KG works on a seasonal calendar to identify when these hazards/social issues occur and how they relate to other key events like planting season, harvest, religious holidays, and the school year. This builds a picture of the varied ways hazards and social issues interact with and impact upon all aspects of their lives, thus displaying the levels of vulnerability within the community.

Between Step 1 and Step 2, an activity was added named, "Identifying Coping Mechanisms," in order to gain more knowledge about the issues initially mentioned. The KG was asked to look at existing external assistance and internal coping strategies to help identify gaps and positive capacities that

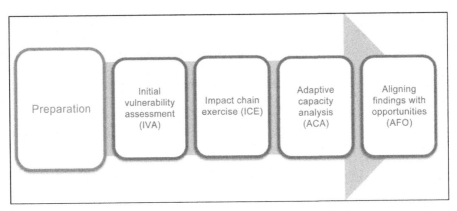

FIG. 4.4.1

The vulnerability and risk assessment process.

could be strengthened but also negative behaviors that should be discouraged. This information fed into step 3 as part of improving/scaling up adaptive capacities and discouraging negative coping strategies. It also showed where aid dependencies or lack of sustainable solutions had affected the communities' self-motivation and ownership of their own development trajectory.

Step 2, named "Impact Chain exercise," sees the KG work through flowchart-style diagrams to identify the varied and cumulative consequences for each of the shortlisted priority hazards and social issues. This activity is focused on understanding both the immediate and potential longer-term impacts and how they affect various aspects of life, e.g., income, service access, living conditions, and location and opportunities for children, etc. Subsequently, the KG looks at possible solutions and ideas to mitigate the hazard/social issue itself and reduce the impact through absorptive, adaptive, transformative mechanisms.

Step 3 take these "solutions" and develops them into more tangible ideas using the five ACCRA Characteristics of successful adaptation (ACCRA, 2012) as a basis for what would be needed to put them into practice. This exercise helps to identify:

(1) *The asset base*: The various financial, physical, natural, social, political, and human capitals necessary to best prepare a system to respond.
(2) *Institutions and entitlements*: Who the critical decision makers are to ensure equitable access and entitlement to key resources and assets.
(3) *Knowledge and information*: What information is lacking that would need to be gathered for formulating better adaptation options.
(4) *Innovation*: A key characteristic of adaptive capacity relates to the system's ability to support innovation and risk taking. Innovation can be planned, high-tech orientated, and geared toward large-scale innovations; or it can be autonomous, local-level initiatives that help innovate or adapt to the risks.
(5) *Flexible forward-thinking decision making*: Informed decision-making, transparency, and prioritization each form key elements of adaptive capacity (ACCRA, 2012).

The final task, Step 4, focuses on action planning and determining how the ideas can be taken forward. The approach allows groups that are connected through location, livelihood, or governance structures (e.g., extension officers from different departments or traders) to work together to explore how they can embed these strategies in their work and individual lives. The objective is to build a sense of shared responsibility and collective action for successful adaptation.

The results gathered during the different participatory steps were discussed and agreed with the KG at various stages during the assessment exercise thus ensuring verification was built into the process. The findings were then written up and validated/shared with the communities themselves, the immediate program team, and with external stakeholders to support further action.

4.4.4 **Results and discussion**

The Knowledge Group, as a result of the initial vulnerability assessment, came up with a long list of hazards that, through a voting system, was then reviewed in order to prioritize the hazards and social issues, which had the most significant effect on their community (Table 4.4.1).

There are clear similarities across both regions of North and South Darfur, in particular with reference to concerns over natural resource management linked to exacerbating conflict; poor infrastructure and lack of government/duty-bearer service provision. For example, the discussions highlighted a shift from the past when land was perceived as a public resource—previously negotiated at a local level through the native administration—to the current action of the government fencing off land and promoting settled farming with land ownership, which has led to a reduction in the availability of grazing land and conflicts over usage between agriculture farmers and pastoralists. Climate change impacts have heightened pressure on water resources resulting in pastoralists encroaching on settled land seeking fodder. Thus, causing farmers to harvest before crops have matured to reduce the risk of livestock forcibly entering planted fields, which damages their yield. In this way, adaptive strategies may be supporting one group's coping capacity but undermining another's, and/or hazards being dealt with in silos rather than taking a holistic perspective of building resilience for all.

The social issues identified as priority risks are deeply rooted in norms and traditions that are difficult to challenge, as communities have been practicing them for a long time and they struggle to see alternative behaviors. This is also linked to the lack of awareness about the harmfulness of some practices and/or poor legal protection and services that could prevent their adoption for example female

---

Table 4.4.1 Priority hazards and social issues as identified by the Knowledge Groups.

- Mismanagement of natural resources (water, forestry, and land)
- Climate change/droughts
- Conflict over natural resources (water, land) exacerbated by the prevalence of weapons
- Poor infrastructure and lack of basic services (roads, water, public health, Education, and government extension services)
- Endemic and epidemic diseases (human and animal)
- Lack of Early Warning system (indicators for floods, droughts, conflicts)
- Environmental pollution (mining, solid waste management)
- Inequality (gender, ethnic minorities)
- Drug abuse

genital mutilation (FGM) and early marriages. An important dimension to addressing these harmful social and cultural practices is the lack of sufficient livelihood opportunities for those who practice them, e.g., midwives performing FGM as they receive payment for such services. Drug dealing and early marriage are practices that result in individuals or families accessing high volumes of cash, which can be powerful incentives for continuing.

Another interesting point raised was the relationship between hazards and social issues and how they mutually reinforce one another and trap people in cycles of vulnerability: male youth suffering mental health issues from the psychological trauma of conflict would lead to household violence, crime and antisocial behavior that harms the community, especially women, even when peace is achieved.

By looking at the existing coping strategies, risks in this context are chronic and pervasive in nature; hence, people have been undertaking indigenous coping strategies for many decades (both negative and positive), so carrying out this step was also useful to understand what are the existing NGO/Government and community initiatives to build capacities and what support they already receive to strengthen their resilience.

Based on all the information shared during the second step, the KG was asked to develop impact chains for each of the priority hazards, and once identified the primary and secondary consequences, they were asked to think about possible solutions with a whole system approach. The KG was asked to design interventions that encompasses strategies to build both absorptive capacity (dealing with the consequences or visible challenges) and adaptive/transformative capacity (changing behaviors, laws, policies, and environmental issues that drive vulnerability).

During the completion of Step 3 of this assessment, a number of discussions centered on the application of existing adaptive strategies and support services as well as how to achieve wider behavior change and to support information management. The KG having first produced a long list of adaptive strategies that could help address the whole range of hazards and social issues, during plenary discussion chose to focus on developing ideas that could be used to address more that one of the priority hazards and social issues. While these are not new solutions in themselves, the information is extremely helpful in informing agencies like Oxfam and other key duty-bearers on what adjustments or further investments are needed to make current support more effective and have greater uptake.

Here below the main ones are reported:

- *Water Resource Management*: The KG stressed the fact that the recurrent droughts pose risks for water availability. It was noted that additional water harvesting infrastructures are needed in order to supplement water during times of scarcity. The most commonly used type of water storage solution in Sudan is called *hafir* (hand-dug dams). However, the focus should be on catchment level Integrated Water Resources Management practices, which consider that activities happening upstream have an impact on what happens downstream. Several other organizations are working on water availability and water management in the region; thus, it is vital to look for complementary interventions (UNEP, ZOA). Monitoring to support better decision making was also highlighted regarding water quality, ground-water level, rainwater variability data, and management of water points. There was an eagerness to explore sustainable solutions linked to alternative energy sources such as solar water pumps and the engagement of the private sector to invest in water service delivery. Moreover, the KG also highlighted gaps in knowledge on water needs and usage; thus, feasibility studies and climate variability assessments would need to be carried out.

- *Community-based forestry management*: Forests are both a source of income and a natural resource to be protected. The discussion focused on better-planned management and clearer policies on planting tree variety. Suggestions included planting a mix of fast-growing, local species and drought-resistant type of trees. Furthermore promoting maintenance/protection/regeneration of forests with proactive monitoring of when, what, and how much can be cut without compromising the forest, should be sought in collaboration with the Forest National Corporation (FNC) of Sudan. The KG also highlighted the significance of finding alternative livelihood opportunities, especially for women and girls who are the primary gatherers of firewood for domestic usage and as an additional source of income when the harvest fails. Providing suitable sustainable alternatives would have the additional benefit of reducing the protection risks women and girls face when gathering wood in forested areas such as sexual and gender-based violence. Gender justice approaches need to be mainstreamed and consideration should be given to alternative fuel sources and fuel usage efficiencies that would reduce the rate and scale of deforestation which is driving other climatic shocks and stresses (flooding, soil erosion, management of global carbon dioxide emissions, etc.).

- *Locally driven behavior change*: As part of the solution-focused discussions, KG members highlighted a range of suggestions that all fed into addressing social issues that had their basis within preexisting inequalities or were the legacy of harmful traditional practices. The group centered their ideas on two key aspects of this:
    - *Drug abuse* was perceived by communities as a serious issue; however, more research is needed on numbers affected, types of drugs used, and health/social impacts. Recommended actions included improved treatment services, legal frameworks to punish smugglers/sellers, and support for community and family cohesion. Particular attention should be paid to youth engagement in order to reduce this negative coping mechanism and better manage the impacts/traumas associated with conflict, especially for male youth.
    - *Harmful traditional practices*: There were different points of view on this topic particularly because while we aimed to achieve a gender balance in the social context of Sudan this was not feasible. The predominantly male nature of the KG (66%, 21) meant fewer participants raised concerns about gendered inequalities and harmful traditional practices like female genital mutilation (FGM) and early marriage. This is understandable given that they disproportionally affect women and girls. Notably, female members of the KG were vocal in raising these topics and succeeded in having them included in the assessment. As a result of this debate, the KG recognized the need for research on the scale and impact of these behaviors and the importance of raising community awareness by working with leaders, women's association, and through schools to discourage these practices. The group also recommended working with the media to support interactive theater to spread key messages.

- *Income-generating activities (IGAs) and alternative livelihood options*: The KG highlighted the need for more research/information about what other viable income-generating activities/livelihood opportunities could be available for the most vulnerable sections of the population. Finding alternative income options has the potential to offer win-win scenarios for supporting people's resilience and protecting the environment. For example, deforestation and subsequent soil erosion, desertification and the risk of flooding would be reduced if women did not have to rely solely on tree cutting and charcoal selling to earn money. Viable livelihood alternatives might also reduce the application of harmful traditional practices, like early marriage, because households would have enough funds to meet their basic needs without having to resort to dowry payments.

- *Support agricultural services*: In Darfur, there are existing packages for Agricultural Extension Services (AES) and where they are functioning, they provide beneficial support to communities, farmers, and pastoralists, but it was highlighted that the modes of delivering them could be improved. There should be an evaluation to look at which approach would be most successful in delivering high quality AES, through Community Extension Workers (CEW) or directly through the Government. Building the capacity of community agents and extension workers is also a key task and those with natural-resource based livelihoods were pushing for the AES to support them with training on improved agricultural techniques (agroecology, climate smart, Integrated Pest Management, etc.) that would boost their productivity.

- *Improve communication flow*: A key finding from this assessment was the need to increase and improve the mechanisms for inter- and intracommunity dialogue as well as the communication channels between authorities and citizens. These are important not just for knowledge exchange and increasing access to critical information such as early warning on seasonal forecasts, but it can also play a vital role in encouraging greater community engagement and participation in interventions. If people are invested in and cooperating with development interventions, the work of aid agencies and local authorities is likely to be more relevant, effective, and sustainable. Furthermore, the commitment to and business case for participatory context assessments is already well established, so supporting communication flows is embedded within this wider accountability agenda. Despite a number of agencies conducting programmatic assessments in the Darfur region, the Knowledge Group highlighted a lot of information gaps, and, notably, a real lack of continuous monitoring that would facilitate effective adaptive, flexible programming that remains relevant in the fluid context of risk and vulnerability in Darfur.

- *Governance systems and the role of the Native Administration (Ajawid)*: In several instances regarding different hazards and social issues, there was no clarity on government responsibilities and accountability. There was also confusion regarding how to access information and on how existing or new laws and by-laws are interpreted and enforced. Communication gaps remain part of this problem as well. An interesting finding was the discussion on the historical role of the *Ajawid*, and how effective the KG felt it had been, mostly regarding solving conflicts between different clans but also on other social issues, like negotiating land use and natural resource management. It was clear from the KG that many considered this local governance mechanism to be useful and that it should play an important role alongside formal governance processes in order to address localized issues and ensure there were adequate channels for community concerns to be raised to those in authority. Another reason for engaging the *Ajawid* was the KG's belief that it could be influential in tackling harmful traditional practices by encouraging an end to female genital mutilation (FGM) and early marriage. The *Ajawid* comprises respected community elders and leaders who, if they were part of campaigns and interventions, to shift cultural norms could have a significant impact on community behavior.

4.4.5 Conclusions and recommendations

Reflecting on this exercise highlighted several methodological and technical conclusions for Oxfam and our partners in Sudan, but also for Oxfam's work in fragile contexts more globally and the use of VRA approaches for context analysis.

Two important lessons around methodological approaches were that it is key to: (1) establish a participatory approach that built trust and fostering dialogue among stakeholders and then, within that, (2) create a platform for the meaningful engagement of women. The setup of the VRA, its facilitation techniques for open and equal debate, the transparent and accountable voting system and thus the fair selection of which topics were explored, was highly effective for community issues to become more visible and to enable discussion on complex and sometimes distressing problems. It helped reinforce and rebuild collaborations and raised awareness of different people's experiences within the community. It also fostered a sense of collective will and built unusual partnerships that brought together indigenous and formal technical knowledge in support of adaptive and transformative change. Having diverse actors together, the process has the additional benefit of moving the focus from the local community to a global landscape approach, which considers connections, interdependencies and how to tackle root causes rather than just treat symptoms, which is part of how Oxfam understands resilient development. The choices that were made during the stakeholder mapping, done as part of the formation of a Knowledge Group, represented a particular opportunity to create a meaningful space for women's voices to be heard and gendered social issues to gain prominence. While this is an important prerequisite action and we worked hard to ensure women's attendance, it is also important to manage expectations sensitively, remember "do no harm" principles and ensure women are not exposed to secondary threats and consequences from speaking out. A good lesson from this exercise was the positive impact of discussing the approach with male community members beforehand and assigning women staff and/or male "allies" as part of the facilitation team so that female participants could actively and safely engage in the process.

Through this process, it became increasingly clear that people's risk profile is determined by both individual and contextual factors and is a combination of hazards (natural/climate related) and social issues linked to power, good governance, and conflict. In this way, it became clear that any vulnerability and risk assessment tool needs to take a multidisciplinary approach. The findings from this exercise provided evidence that risks affect people differently and in multidimensional ways, often overlapping or reproducing one another. For any development interventions to be effective, it is vital to understand how a person's risk profile is determined, escalated, and/or perpetuated by individual and contextual factors. The VRA tool also showed that in order to achieve lasting, transformational change, a holistic strategy is needed. It is therefore necessary to build capacities and transform systems in ways that provide technological innovations alongside addressing inequalities and behaviors that exacerbate or reinforce vulnerabilities. This finding correlates with Oxfam's resilience approach, which advocates for both social change and infrastructure investment in adaptive livelihoods or risk reduction processes (Jeans et al., 2016).

The context of Darfur is one characterized by chronic long-term shocks and stresses where the humanitarian aid system has been functioning for an extended period. It has therefore been difficult to move thinking and practice toward sustainable, more resilient development ambitions. It can be particularly hard for communities where generations have become aid dependent and there is a high level of expectation around asset and in-kind distribution, to shift to approaches centered on training, start-up support or initiatives that require shared input from the community, local government, and/or the private sector. An important lesson from work in comparable contexts is the need to invest in broadening awareness, building human partnerships and collective action, and thus encouraging a more self-sufficient attitude, which is one of the foundational elements of a resilience approach (Jeans, 2017). The VRA helped highlight how, and in what ways, discussions about building capacities and supporting development pathways are conducted, is as meaningful in achieving change as the financial or physical interventions that are delivered as a result. The findings were also informative about which strategies Oxfam and other stakeholders should use in Darfur and other similar contexts. These often involve balancing support for continuing humanitarian needs with investing in sustainably implemented, adaptive and transformative measures. Doing so helps build people's capacity to cope now and in the future, regardless of what shocks or stresses they might experience. The VRA helped validate this dual approach as the discussions showed there is a strong appetite for the latter, provided people can still meet their basic needs. Establishing the enabling environment—and what immediate support is still necessary—is a critical step as agencies and other actors attempt to break the cycle of short-term emergency response by delivering both life-saving and life-changing outcomes. Working in this way is becoming increasingly critical for the aid sector, if it wants to remain relevant, responsive, and effective in reacting to the changing nature of human suffering globally. The VRA is one significant mechanism, which can help this reframing process.

Acknowledgments

We would like to extend a big thank you to the Oxfam Sudan team and our partner organizations (Kabkabya Smallholders Charitable Society and Jabal Marra Charity Organization) who co-led these assessments and continue to explore how this knowledge can guide more effective programming in the future. They are on the frontline grappling with these issues daily, working in challenging conditions to deliver life-saving and life-changing assistance. They have our respect and gratitude for all the work they do.

Disclaimer

This chapter was written to share research results, to contribute to public debate, and to invite feedback on the use of vulnerability and risk analysis methodologies. It does not necessarily reflect the policy positions of Oxfam. The views and recommendations expressed are those of the author(s) and not necessarily those of the individual organization.

References

ACCRA, 2012. The ACCRA Local Adaptive Capacity Framework. Available from: https://insights.careinternati onal. org.uk/media/k2/attachments/accra-local-adaptive-policy.pdf.

Anderson, M., Woodrow, P., 1998 [1989]. Rising From the Ashes: Development Strategies in Times of Disaster. Lynne Rienner, Boulder, CO.

BBC, 2019. Sudan Country Profile. Available from: https://www.bbc.com/news/world-africa-14094995.

Birkmann, J., et al., 2013. Framing vulnerability, risk and societal responses: the MOVE framework. Nat. Hazards 67, 193–211. https://doi.org/10.1007/s11069-013-0558-5.

Blaikie, P., Cannon, T., Davis, I., Wisner, B., 1994. At Risk: Natural Hazards, People's Vulnerability and Disasters. Routledge, London.

Blunden, J., Arndt, D., Hartfield, G., 2018. State of the Climate in 2017. Bull. Amer. Meteor. Soc. 99 (8), Si-S310. https://doi.org/10.1175/2018BAMSStateoftheClimate.I.

Bohle, H., Downing, T., Watts, M., 1994. Climate change and social vulnerability. Global Environ. Change 4 (1), 37–48.

Buchanan-Smith, M., Abdulla Fadul, A.J., Rahman, T.A., Aklilu, Y., 2012. On the Hoof Livestock Trade in Darfur. UNEP, Feinstein International Center, Tufts University. Available from: https://www.un.org/en/events/environmentconflictday/pdf/UNEP_Sudan_Tufts_Darfur_Livestock_2012.pdf.

Chambers, R., 1983. Rural Development: Putting the Last First. Longman, London.

Chiwaka, E., Yates, R., 2000. Participatory Vulnerability Analysis. ActionAid International. https://www.actionaid.org.uk/sites/default/files/doc_lib/108_1_participatory_vulnerability_analysis_guide.pdf.

Climate Data, 2019. Available from: https://en.climate-data.org/africa/sudan/south-darfur-state-1532/.

Cutter, S., 1996. Vulnerability to environmental hazards. Prog. Hum. Geogr. 20, 529–539.

Department for International Development (DFID), 2006. Reducing the risk of disasters—Helping to achieve sustainable poverty reduction in a vulnerable world. DFID Policy Paper, Department for International Development, London. Available from: http://www.preventionweb.net/files/2067_VL108502.pdf.

DFID, 1999. Sustainable Livelihood Guidance Sheets. Department for International Development, DFID, London.

Enarson, E., Morrow, B., 1997. The Gendered Terrain of Disaster: Through Women's Eyes. Praeger, New York.

Fitzpatrick, M., Young, H., 2015. The Road to Resilience: A Scoping Study for the Taadoud Transition to Development Project. Feinstein International Center.

Fullwood-Thomas, J., 2017. A practical tool for listening to the people that matter, Oxfam GB. Available from: https://views-voices.oxfam.org.uk/climate-change/2017/10/practical-tool-listening-people-matter.

Herring, S.C., Christidis, N., Hoell, A., Hoerling, M.P., Stott, P.A. (Eds.), 2019. Explaining extreme events of 2017 from a climate perspective. Bull. Am. Meteor. Soc.100 (1) , S1–S117. https://doi.org/10.1175/BAMS-ExplainingExtremeEvents2017.1.

Hilleboe, A., Sterret, C., Turnbull, M., 2013. Toward Resilience. A Guide to Disaster Risk Reduction and Climate Change Adaptation. Practical Action Publishing Ltd, Rugby.

Howard, L., Stark, A., 2018. Foreign Affairs: why civil wars are lasting longer. https://www.foreignaffairs.com/articles/syria/2018-02-27/why-civil-wars-are-lasting-longer.

ICVA, 2017. The Grand Bargain: everything you need to know, ICVA Briefing paper. https://www.agendaforhumanity.org/sites/default/files/The%20Grand%20Bargain_Everything%20You%20Need%20to%20Know%20%28ICVA%29_0.pdf.

International Federation of Red Cross and Red Crescent Societies (IFRC), 2006. What is VCA? An introduction to vulnerability and capacity assessment. Available from: https://www.ifrc.org/Global/Publications/disasters/vca/whats-vca-en.pdf.

IPCC, 2018. Summary for Policymakers. In: Masson-Delmotte, V., Zhai, P., Pörtner, H.O., Roberts, D., Skea, J., Shukla, P.R., Waterfield, T. (Eds.), Global Warming of 1.5°C. An IPCC Special Report on the Impacts of Global Warming of 1.5°C Above Pre-industrial Levels and Related Global Greenhouse Gas Emission Pathways, in the Context of Strengthening the Global Response to the Threat of Climate Change, Sustainable Development, and Efforts to Eradicate Poverty. World Meteorological Organization, Geneva, Switzerland. 32 pp. https://report.ipcc.ch/sr15/pdf/sr15_spm_final.pdf.

Jeans, H., 2017. Absorb, Adapt, Transform: Resilience Capacities. Oxfam International.https://policy-practice.oxfam.org.uk/publications/absorb-adapt-transform-resilience-capacities-620178.

Jeans, H., Thomas, S., Castillo, G., 2016. The Future Is a Choice: Oxfam Framework and Guidance for Resilience Development. Oxfam International.

Kittaneh, S., Stolk, A., 2018. Doing Nexus Differently: How Can Humanitarian and Development Actors Link or Integrate Humanitarian Action, Development, and Peace? Cooperation for Assistance and Relief Everywhere, Inc (CARE).https://www.care.org/sites/default/files/documents/care_hub_detailed_paper_doing_nexus_differently_final_sep_2018.pdf.

Knowles, S., 2011. The Disaster Experts: Mastering Risk in Modern America. University of Pennsylvania, Philadelphia.

Morchain, D., Kelsey, F., 2016. Finding Ways Together to Build Resilience: The Vulnerability and Risk Assessment Methodology. Oxfam International. ISBN 978-0-85598-673-5. http://hdl.handle.net/10546/593491.

Morchain, D., Prati, G., Kelsey, F., Ravon, L., 2015. What if gender became an essential, standard element of vulnerability assessments? Gend. Dev. 23 (3), 481–496. https://doi.org/10.1080/13552074.2015.1096620.

O'Keefe, P., Wisner, B., 1975. African drought: the state of the game. In: Richards, P. (Ed.), African Environment: Problems and Perspectives. International African Institute, London, pp. 31–39.

OCHA, 2016. Policy Development and Studies Branch, New Way of Working, Agenda for Humanity. https://www.agendaforhumanity.org/initiatives/5358.

OECD, 2016. States of Fragility 2016. http://www.oecd.org/dac/states-of-fragility-2016-9789264267213-en.htm.

Osman, A.M.K., Cohen, M.J., 2014. We No Longer Share the Land: Agricultural Change, Land, and Violence in Darfur. Oxfam International. Available from: https://policy-practice.oxfam.org.uk/publications/we-no-longer-share-the-land-agricultural-change-land-and-violence-in-darfur-315857.

Papathoma-Köhle, M., Kappes, M.S., Keiler, M., Glade, T., 2011. Physical vulnerability assessment for alpine hazards: state of the art and future needs. Nat. Hazards 58, 645–680.

Pelling, M., Obrien, K., Matyas, D., 2014. Adaptation and transformation. Clim. Change 133 (1), 113–127. Springer.

Phillips, B.D., Fordham, M., 2009. Introduction: Chapter 1. In: Phillips, B.D., Thomas, D.S.K., Fothergill, A., Blinn-Pike, L. (Eds.), Social Vulnerability to Disasters. CRC Press, Boca Raton.

Slim, H., 2017. Nexus Thinking in Humanitarian Policy: How Does Everything Fit Together on the Ground? The International Committee of the Red Cross (ICRC).https://www.icrc.org/en/document/nexus-thinking-humanitarian-policy-how-does-everything-fit-together-ground.

Susman, P., O'Keefe, P., Wisner, B., 1983. Global disasters: a radical perspective. In: Hewitt, K. (Ed.), Interpretations of Calamity. Allen and Unwin, Boston, pp. 263–283.

Turnbull, M., Turvill, E., 2012. Oxfam Participatory Capacity and Vulnerability Analysis: A practitioner's Guide. Oxfam, https://policy-practice.oxfam.org.uk/publications/participatory-capacity-and-vulnerability-analysis-a-practitioners-guide-232411.

UNDRR, 2015. United Nations Office for Disaster Risk Reduction, Sendai Framework for Disaster Risk Reduction 2015–2030. Available from: https://www.unisdr.org/we/inform/publications/43291.

UNEP, 2007. Sudan Post-Conflict Environmental Assessment. Available from: https://postconflict.unep.ch/publications/UNEP_Sudan.pdf.

UNOCHA, 2018. Sudan: Darfur Humanitarian Overview (1 April 2018). Available from: https://reliefweb.int/report/sudan/sudan-darfur-humanitarian-overview-1-april-2018.

White, G.F., 1945. Human Adjustment to Floods. University of Chicago, Department of Geography Research Paper, Chicago.

Winchester, P., 1986. Cyclone Vulnerability and Housing Policy in the Krishna Delta, South India (Doctoral Thesis). School of Development Studies, University of East Anglia, Norwich, UK.

Wisner, B., 2016. Vulnerability as concept, model, metric and tool. In: Oxford Research Encyclopaedia of Natural Hazard Science. https://doi.org/10.1093/acrefore/9780199389407.013.25. Available from: https://oxfordre.com/naturalhazardscience/view/10.1093/acrefore/9780199389407.001.0001/acrefore-9780199389407-e-25.

Wisner, B., Blaikie, P., Cannon, T., Davis, I., 2004. At Risk: Natural Hazards, People's Vulnerability and Disasters, second ed. Routledge, London.

World Bank, 2001. World Development Report 2000/2001: Attacking Poverty. Oxford University Press, New York.

World Bank, 2018. Fragility, Violence and Conflict Overview. http://www.worldbank.org/en/topic/fragilityconflictviolence/overview.

Further reading

Fullwood-Thomas, J., Facilitators Guide to Vulnerability and Risk Analysis. Available from: http://vra.oxfam.org.uk/.

RESPONSE

Understanding collective action through social media-based disaster data analytics

5.1

Andrea Garcia Tapia and Jose E. Ramirez-Marquez

Stevens Institute of Technology, Hoboken, NJ, United States

CHAPTER OUTLINE

5.1.1 Introduction

Climate change is exacerbating the magnitude of weather-related events around the world. The interaction with vulnerable and exposed natural and human systems can lead to disasters (Field et al., 2012) disrupting social order and heavily impacting the daily lives of the affected people.

Understanding Disaster Risk. https://doi.org/10.1016/B978-0-12-819047-0.00015-9

The magnitude of the disruption and the community access to financial, physical, and social resources determine the recovery process (Tapia et al., 2014). For example, several studies have shown that mortality rates after a disaster are correlated to the level of trust and to the social cohesion of the community. Higher levels of trust result in lower levels of mortality rates (Aldrich, 2012, 2017).

As of recently, most research has focused on improving early disaster warnings and strengthening existing physical infrastructure, but from our perspective social capital has been overlooked. According to the Sendai Framework Priority 1 to understand the risk at national and local level we should "enhance collaboration among people to disseminate disaster risk information through the involvement of community-based organizations and nongovernmental organizations" (UNISDR, UNO, 2015). Social media platforms could be a key component to achieve this objective.

Social cohesion is the result of three components: social capital, social inclusion, and social mobility (OECD, 2011), and social media can be a leverage to increase our social ties. In recent years, social media platforms have evolved to respond to disasters. For example:

1. Facebook has developed Safety Check.
2. Open Street Maps has a tool for crisis crowd sourcing maps.
3. Ushahidi has developed crisis mapping.
4. Google has developed Crisis Response Tools such as person finder and specialized maps.

This chapter presents a general framework to examine the emergence of collective action during the aftermath of a disruption and present Sandy hurricane in New York City (2012) as a case of study to exemplify the framework. In the case study, social media platforms such as Facebook and Twitter played a main role in the organization of disaster response. Through a mixed methods approach the content of social media accounts for the grass roots movement Occupy Sandy was analyzed.

The chapter is organized as follows: Section 5.1.2 provides a brief literature review, linking the main components of social cohesion and computational social science for disasters; Section 5.1.3 describes the methodology used to analyze communication and organization after a disaster; Section 5.1.4 describes the case of study; Section 5.1.5 describes the findings; and Section 5.1.6 summarizes the conclusions for the case study.

5.1.2 Literature review

According to the National Science Foundation and the Computing Community Consortium, Computing for Disasters is often viewed as an application domain for computing, in which data must be gathered, transmitted, transformed, and presented to stakeholders so that they can make decisions. However, one type of data is not sufficient and stakeholders must manage the situation with heterogeneous data sets that comes from different sources, arrive in different volumes, at different times, and exhibit different priorities for different phases of the disaster. Data is critical for predicting infrastructure and social vulnerabilities, preventing disasters, saving lives, mitigating damage, and minimizing recovery time. Developments in data mining, machine learning, large multimedia archiving and retrieving systems, efficient search, human-computer interfaces, and visualization and statistics enable these challenges to be overcome today. A fundamental principle of Computing for Disasters is that technology must be tailored to the human stakeholders and needed capabilities. For example, a sophisticated algorithm that produces highly optimal but counterintuitive, resource allocation may be rejected by the deciding stakeholder if the algorithm does not explain or justify recommendations. Technology is not sufficient; technology must be designed and evaluated in terms of the larger socio-technical context (Computing Community Consortium, 2012).

5.1.2.1 **Resilience**

Meerow et al. (2016) define urban resilience as an "urban system's ability to maintain or rapidly return to desired functions in the face of a disturbance, to adapt to change, and to quickly transform systems that limit current or future adaptive capacity" (Meerow et al., 2016). Resilience can be divided into three dimensions: reliability, vulnerability, and recoverability (Henry and Ramirez-Marquez, 2012; Dessavre et al., 2016). Fig. 5.1.1 shows how a disruption affects the performance function of a system and how it recovers. It is important to clarify that the recovery could have three different outcomes: returning to the same level, going to a lower level (e.g., poverty traps), or increasing the initial level also known as creative destruction or Shupmeter destruction (Tapia et al., 2014; Albala-Bertrand et al., 1993). The long-term impact of a disaster depends on the level of destruction of human, physical, and natural capital (Guha-Sapir et al., 2013). Disasters with a high level of physical infrastructure

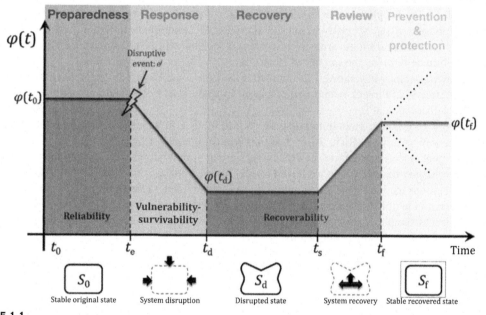

FIG. 5.1.1

System performance with a disruptive event.

Modified from Henry, D., Ramirez-Marquez, J.E., 2012. Generic metrics and quantitative approaches for system resilience as a function of time. Reliab. Eng. Syst. Saf. 99, 114–122; Dessavre, D.G., Ramirez-Marquez, J.E., Barker, K., 2016. Multidimensional approach to complex system resilience analysis. Reliab. Eng. Syst. Saf. 149, 34–43.

damage but a lower level of human capital damage (e.g., deaths and affected population) tend to recover faster than disasters with extensive damages in human capital (Albala-Bertrand et al., 1993; Skidmore and Toya, 2002). Time is a key factor in recoverability and so the rapid flow of information that social media enables can be an advantage; however, it can also be a disadvantage if it causes the spread of misinformation or rumors.

5.1.2.2 Social cohesion, collective action, and emergent response teams

There is no consensus around what comprises the elements of social cohesion. Chan et al. (2006) define it as "a state of affairs concerning both the horizontal and vertical interactions among members of society." After a review of several authors (Hulse and Stone, 2007; Jenson, 2010; Schiefer and van der Noll, 2017), the main elements of social cohesion seem to be:

1. social norms;
2. trust;
3. social relations;
4. (in)equality;
5. sense of belonging; and
6. orientation toward social good.

Other scholars have suggested that social media or web 2.0 technology enables new communication platforms that bring people together in a way that was not possible before (Penney and Dadas, 2014; Luna and Pennock, 2018), specially because it allows the information to flow horizontally instead of vertically (e.g., coming from government institutions). In other words, it assures a flat hierarchy with minimum distance between the top and the bottom.

This new technology has allowed for humanitarian crowd sourcing; one example is the Humanitarian OpenStreetMap Project, where users from all over the world can help map roads and affected areas after a disaster.

Grassroots' movements emerge to pursuit a collective goal, in the case of disaster situations they emerge as a way to survive (Rich, 2018; Stallings and Quarantelli, 1985).

An emergent response group can be defined as a group of individuals who self-organize spontaneously on a voluntary basis to act on perceived needs (Ambinder et al., 2013; Stallings and Quarantelli, 1985). Such groups tend to appear when formal groups (government) address problems inadequately or when the magnitude of the disaster is such that it overcomes the reaction capacity of the government. The most recent example of emergent groups was during the earthquake in Mexico City, where society organized rescue and reconstruction teams.

According to Stallings, there are three types of operation groups (emergent):

1. damage assessment groups;
2. operation groups; and
3. coordinating groups.

Each of them has a function in the disaster management cycle. For example, the damage assessment group provides public officials with first-hand information of the extent of the disruption, meanwhile operation groups collect and distribute food and material needed by the disaster victims. The last group functions as an impromptu citizen committees and resolves community disputes (Stallings and Quarantelli, 1985).

Over the last decade there have been multiple examples of emergent groups for disaster relief. A well-documented case is New York City after the terrorist attack on 9/11 (Voorhees, 2008). The group of volunteers was later institutionalized in what we know today as the Community Emergency Response Teams (CERT) of FEMA.

A more recent example is the 2017 earthquake of Mexico City, in which a group of data scientists collected data from social media (mainly Twitter and WhatsApp) on real time to create a map of verified information to help collect food and material for the rescue brigades.

5.1.2.3 Social media

Social media or interactive web 2.0 applications are technologies that facilitate the communication and interaction between users almost instantly, they can be used to transfer images, video, audio, or text. These applications can be used in computers, smartphones, or tablets (Luna and Pennock, 2018; Faustino et al., 2012).

Each social media application has a specific objective and hence different characteristics. The relative anonymity that reigns in interactive web 2.0 platforms increases the people's willingness to communicate with strangers (Wellman and Gulia, 1999). Fig. 5.1.2 describes the media types and examples of applications.

According to the PEW Research Center around 7 of 10 Americans use social media to connect with one another and consume news and entertainment provided by PEW Research Center (2018) (see Fig. 5.1.5). In United States, the most popular social media platform is Facebook (see Fig. 5.1.3). Most data science and computing efforts have focused on disaster response that uses information from social media, news, text messages, or telecommunication data to identify areas affected during disasters (Meier, 2015). Examples include crowd sourcing maps such as Ushahidi or Humanitarian Open Street Map Team (HOT), both of which have greatly helped emergency managers respond to emergencies. These applications have focused on mapping crises in areas of limited statehood, in which local disaster mitigation agencies have limited resources (Ahmed et al., 2019).

A common use for real-time response is text mining. Text mining can retrieve information from social media, news, or SMS regarding the development of a crisis.

For this task, machine learning algorithms such as support vector machines, logistic regression, and decision trees have been used to address the problem of identifying spam in short messages (Delany et al., 2012; Cormack et al., 2007; Gómez Hidalgo et al., 2006).

In general, applications for natural language processing have been extensively studied. Another example of natural language processing is the study by Caragea et al. (2011) that focused on correctly classifying text messages for the emergency response sector by determining a subset of features that are most informative for the target variable, either by selecting a subset of features from the entire vocabulary using feature selection with LDA, or by constructing abstract features using Feature Abstraction.

In relation with disaster mitigation recovery phase, data from Twitter has been used to assess damages or affected population (Kryvasheyeu et al., 2015), the idea behind using geo-referenced twitter intensity and damage assessment reports for predicting damages in future events. Another application to characterize human behavior during critical events is the analysis of aggregated and anonymized call detail records (CDR) captured from the mobile phone infrastructure. It has been used for predicting migration patterns in Kenya (Wesolowski and Eagle, 2010), population movements during disease outbreaks after Haiti earthquake (Bengtsson et al., 2011), and assessment of flood affected population in Tabasco, Mexico (Pastor-Escuredo et al., 2014). The CDR information has been mixed with other data

Type of social media			
Media type	**Applications type**	**Description**	**Example**
Image	Bookmark	Online pin board to share images associated with projects, goods, and services	Pinterest
	Social	Exchange of multimedia content, mainly images, and short videos and audio. Posted for short period of time	Instagram, Snapchat
Video	Video and podcasting	Platform for video sharing and comments	YouTube
Text — Short length	Microblogs	Short version of blogs. Users share contain in a limited space	Twitter, Tumblr
Text — Medium length	Social/professional networking	Environment where users communicate with others. User profile is required to establish a list of connections with whom to share information	LinkedIn, Facebook, Google +
Text — Reviews	Social rating reviews	Platform where users are able to rate and describe experiences that others can comment and share	Yelp, TripAdvisor, Google Maps
	Social bookmarking	Platform that allows users to add, edit, share bookmarks	Delicious, Google
Text — Long length	Blogs	Interactive virtual journal that allows users to express opinions, share thoughts, promote ideas, etc	WordPress, Blogger, HubPages
	Discussion forums	Online bulletin board where users are able to post questions/ concerns and expect to receive responses to the message left	bbPress, LiveJournal
	Social discovery engines & news sources	Bulletin board where users post news, links, pictures, etc. and vote for ranking their priority	Reddit, Pulse, StumbleUpon
	Wikis	Website that allows users to create, edit, and delete content. There is no defined content owner in comparison to blogs	Wikipedia, Wikispaces
Mix type	Video/ text chatting	Multimedia platform that allows users to establish dialog via text or video	Skype, WhatsApp, WeChat, Telegram, Signal

FIG. 5.1.2

Social media types.

Adapted from Faustino, J.D., Liu, B., Jin, Y., 2012. Social media during disasters: a review of the knowledge base and gaps, Final Report to Human Factors/Behavioral Sciences Division, US Department of Homeland Security, College Park, USA; Luna, S., Pennock, M.J., 2018. Social media applications and emergency management: a literature review and research agenda. Int. J. Disaster Risk Reduct. 28, 565–577.

sources such as meteorological information, socioeconomic indexes, civil protection data, and satellite images among others. The research in these areas suggests that CDR is a good source to infer mobility patterns. A similar application was made during 2011 Japan earthquake and tsunami in the project (NHK, 2011), they used the CDR information, GPS information from car navigation systems to determine the affected population and the mobility patterns. They compared this information with the census data that the Japan's government had and discovered data inconsistencies of 25%.[a] The second phase of Disaster Big Data project was to measure economic damage, in which a proxy of business connections

---

[a]The difference between the information in the census and the real-time data.

Social media usage in United States

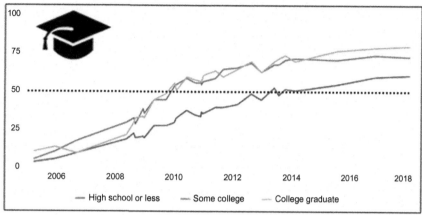

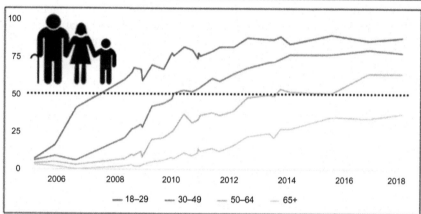

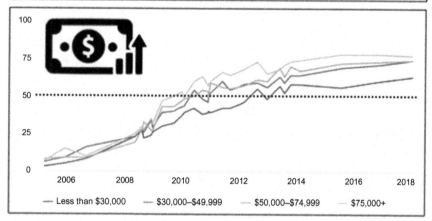

FIG. 5.1.3

Statistics per group of user.

Modified from PEW Research Center, 2018. Social Media Fact Sheet. PEW Research Center, Washington, DC. https://www. pewinternet.org/fact-sheet/social-media/.

through tax information is constructed to determine the business net and recovery plans focused on the businesses with more connections in the area.

When performing social media analysis it is important to be aware of the segments of the population represented in the social media platforms. Not every country has the same social media penetration nor every population group.

Policy recommendations based on social media analysis should not be extrapolated for general recommendations for all of the population, they should be complementary as in Varda et al. (2009).

For example, in the United States, groups with higher education (college graduates), people between 18 and 29 years of age, and people with higher income more intensively use social media platforms (see Fig. 5.1.3).

5.1.3 **Framework**

The proposed framework has four phases, from data collection to policy recommendations as described in Fig. 5.1.4. Three proxies are used as a measure of social cohesion: social relations, sense of belonging, and orientation toward the common good (see Section 5.1.2.2) encompassed in Steps 3a and 3b.

Facebook and Twitter were chosen for the following reasons:

1. *Popularity*: In the United States, 68% of the adult population uses Facebook and 24% uses Twitter. Both social media platforms are used mainly in urban areas and are widely used among different age groups and education levels (see Fig. 5.1.5).

Framework

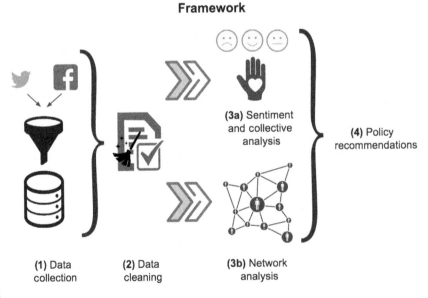

(1) Data collection
(2) Data cleaning
(3a) Sentiment and collective analysis
(3b) Network analysis
(4) Policy recommendations

FIG. 5.1.4

Framework process description.

Social media platform usage in United States	Facebook	Twitter
Total	68%	24%
Men	62%	23%
Women	74%	24%
Ages		
18–29 years	81%	40%
30–49 years	78%	27%
50–64 years	65%	19%
65+ years	41%	8%
Education		
High school or less	60%	18%
Some college	71%	25%
College graduate	77%	32%
Localization		
Urban	75%	29%
Suburban	67%	23%
Rural	58%	17%

Note: percentage difference between race categories was not significant

FIG. 5.1.5

Social media usage over time.

From PEW Research Center, survey conducted on January 3–10, 2018.

2. *Public information*: Twitter is a public platform and information can be easily accessed from their API. Facebook privacy policies have been changing. At the moment of this research, we were able to gather information from group users (private and public) and their interactions with individual users with the Netvizz app.

Fig. 5.1.5 describes the demographics of Facebook and Twitter platforms in United States (PEW Research Center, 2018). The most intensive users are concentrated in two age groups, 18–29 and 30–49 years. Younger generations are moving toward short image and video sharing platforms such as Instagram and Snapchat.

5.1.3.1 Data collection

Social media data (Facebook and Twitter) need to be collected for 6 months after the disaster since disaster short run analysis is from 1 to 6 months (Jaramillo, 2009). It is desirable to get data 1 or 2 weeks before the disaster to have a baseline before the disruption.

Data collection heavily relies on the privacy settings each social media platform has and the information available to scrap from the web. For each social media platform included in the analysis, two data sets should be collected, one for the sentiment analysis and the other for the network analysis. The data collection will be subject to the data privacy settings of the social media platform.

From Twitter, the first data set is a random sample of tweets containing the # related to the disaster (e.g., #Maria, #HurricaneHarvey). The second data set is a random sample of tweets from the official user account like "Occupy Sandy" or "FEMA." Regarding Facebook, one data set is from the official Facebook page and a second one of the comments of the resulting page. All these information can be retrieved using R, Python, and Netvizz.

5.1.3.2 Cleaning process

For cleaning the data, the following steps should be followed:

- Stop words should be removed.
- All the data should be converted to lower case.
- Punctuation should be removed.
- Links and emoticons should be removed.
- The four corpuses (two per social media platform included in the analysis) should use a stemming process.

5.1.3.2.1 Sentiment and collective analysis

Sentiment analysis is a common tool used in natural language processing (NLP), it classifies text from a message (long or short) and tells whether the underlying sentiment is positive, negative, or neutral. Recent developments in NLP field allow a broader classification of text. For the purpose of this research, we chose the more traditional approach of three categories for sentiment analysis. More detail of this process can be found in studies conducted by Pang et al. (2008) and Turney (2002). To perform the sentiment analysis to obtain the general sentiment during the disaster we use the dictionary of positive and negative words developed by Hu and Liu (2004). The sentiment analysis is used as a proxy for attachment/sense of belonging to a social entity.

To complement this sentiment analysis, a collective analysis is performed within the corpora regarding their orientation toward the common good. Collective words such as "we," "group," or "community," or disjunctive words such as "their," "my," or "others" are searched for in the corpora. Qualitative analysis of the contents regarding expressions of solidarity and emotions related to the common good are performed, looking for articulations of feelings of responsibility for the common good. The data cleaning, sentiment analysis, and collectiveness analysis can be computed with R or Python. The main question is how can disaster-relief efforts be organized through social media and mobilized through a broad network of volunteers? This framework focuses on understanding social cohesion as a function of community resilience.

5.1.3.2.2 Network analysis

The objective of the network analysis is to identify social relations and the kind of information that propagates through the network. For this we use Page Rank and Modularity algorithms.

This analysis can be done for Facebook or Twitter. The nodes in the network are posts from users (individual or group) and each interaction as links (see Fig. 5.1.6). The posts are linked to a user (could be private or public accounts). The link between nodes is the communication between the users related to a certain post, it could be a like or a comment. The size of the nodes is the number of times this post has been shared in the network.

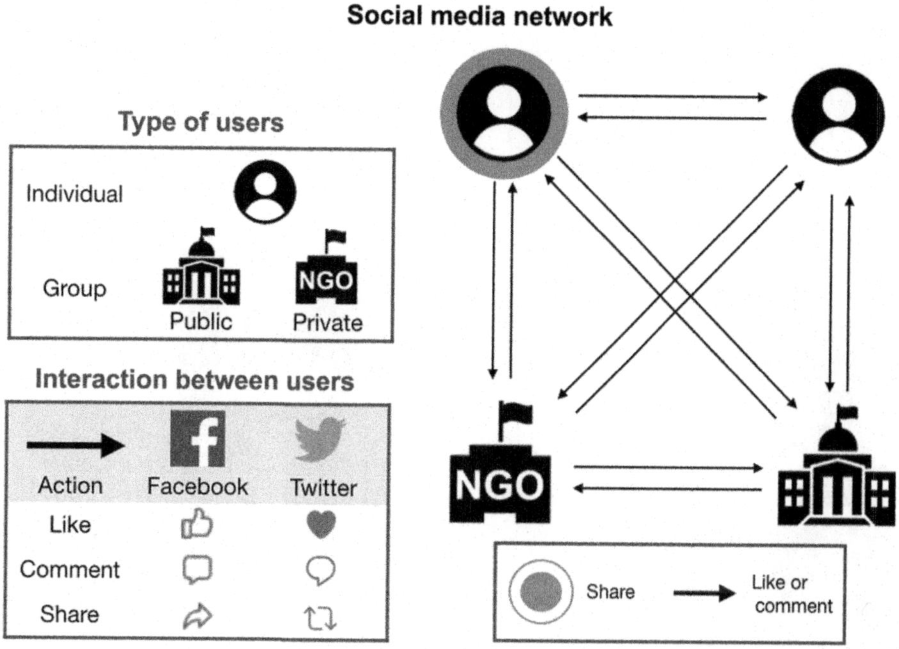

Social media network

FIG. 5.1.6

Network explanation.

Page Rank algorithm is used to find the post with more influence in the network. Page Rank algorithm ranks node "posts" depending on how often a random user following links, in this case "interaction" (likes or comments), will nonrandomly reach the original "post." Page Rank algorithm gives each post a rating of its "importance" on the network, assigning a probability that a random user will land on that post. Its computational part is based on Markov Process and more information can be found in Brin and Page (1998). The main reason to use Page Rank algorithm in the network analysis is its difficulty to manipulate their own ranking of the users.

The Modularity algorithm is used to find communities in large networks (Blondel et al., 2008), it looks for the nodes that are more densely connected together than the rest of the network. The main algorithm is a heuristic method based on modularity optimization, for more information see Newman and Girvan (2004).

Mixing Page Rank algorithm with Modularity algorithm allows to find clusters and the most important post in each community (see Fig. 5.1.7).

To layout the network, we used Force Atlas algorithm in Gephi. It allows you to manipulate the graph and shape it between a Früchterman and Rheingold's layout and Noack's LinLog (see Heymann, 2011). Force Atlas algorithm avoids overlapping on the nodes.

To create the network the program, Gephi was used and the following steps were followed:

1. Retrieve the data set of the official account with its respective comments.
2. Clean the corpus (see Section 5.1.3.2) and separate by day.

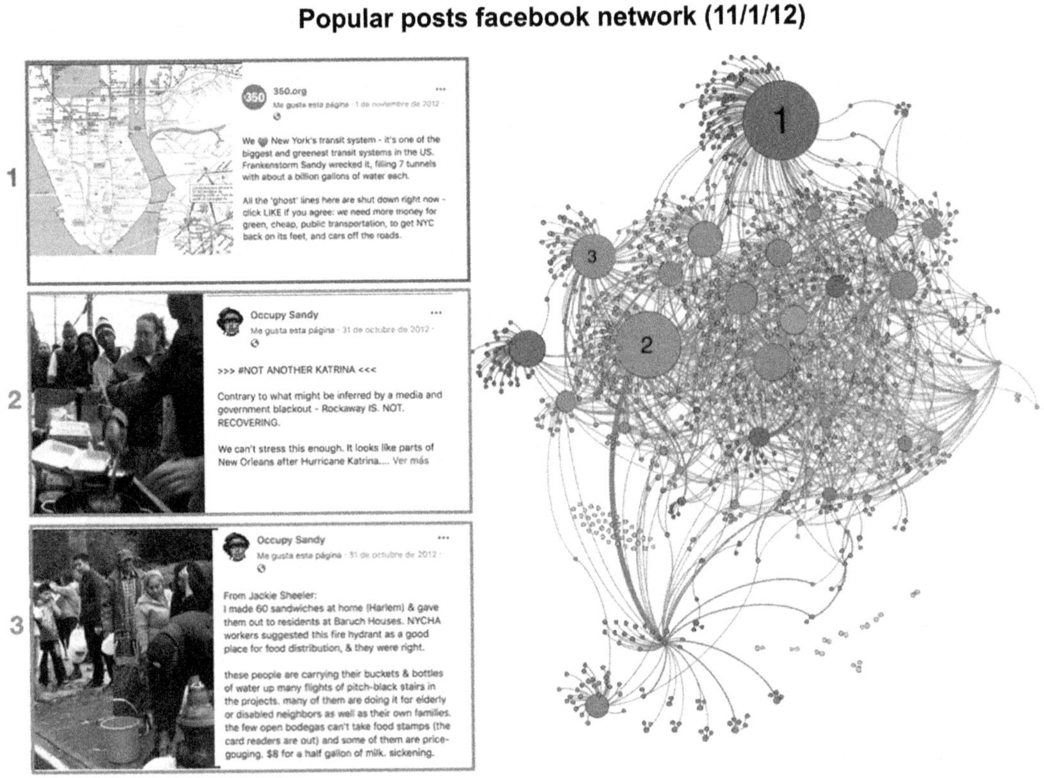

Popular posts facebook network (11/1/12)

FIG. 5.1.7

Popular posts and communities.

3. Create the network using users as nodes, edges as the links between users, and size of the node as the number of shares the post had.
4. In Gephi, apply Force Atlas algorithm to organize the network, adjust the parameter gravity to 1000.
5. Calculate the Page Rank and Modularity algorithms in Gephi.
6. Adjust the size of the nodes by Page Rank algorithm.
7. Color the nodes by Modularity algorithm.
8. Set the thickness of the edges by weight.
9. Delete users that have only one connection and users without connections.
10. Summarize and filter by times shared the most popular posts and analyze the type (link, photo, text) and content.

Fig. 5.1.7 shows the Facebook network for the November 1, 2012 and the three most popular posts. The gray communities in the lower part have thin links to the network and therefore when we zoom out the image it looks like they are unconnected communities.

5.1.3.3 **Policy recommendations**

Once the sentiment, collectiveness, and network analysis is done, summarize the findings and main type of posts shared. Policy makers will be interested in the kind of content on social media that is spread through the network. So they can disseminate important information more efficiently.

5.1.4 **Case of study: Sandy 2012**

On October 29, 2012, Hurricane Sandy made landfall, devastating extensive areas of the Atlantic basin, including substantial parts of New York and New Jersey. The death toll in the United States was 72 persons with $50 billion in property damage. Hours after Sandy had made landfall, members from the Occupy Wall Street movement used social media to spread the appeal to provide community-sourced postdisaster recovery. According to the official website and social media platforms. Occupy Sandy is a grassroots disaster relief network that emerged to provide mutual aid to communities affected by Superstorm Sandy.

They created an Occupy Sandy Facebook page, initiated the hashtag "#SandyAid" and "#OccupySandy" on Twitter and Facebook, and launched a WePay account to collect donations.

Within 4 months, Occupy Sandy had gathered 60,000 volunteers and emerged as one of the largest humanitarian actors across New York City and New Jersey. Occupy Sandy established food distribution centers and served about 10,000 meals a day in the week following the hurricane. Furthermore, the grassroots disaster-relief network coordinated "motor pools to transport construction teams and medical committees to survivors in the field" (Ambinder et al., 2013). Since the very beginning of its collective action, social media was the primary tool used by Occupy Sandy to mobilize volunteers, organize and coordinate its actions, and share information. Data were collected for the period (October 15, 2012 to March 15, 2013) for the "#OccupySandy" account. This resulted in a corpus of 12,971 entries from Twitter and includes those 2 weeks before Hurricane Sandy's landfall and up to 5 months after it. The second data set from Twitter has 2300 entries for the same period. The Facebook page "Occupy Sandy" with all the official posts resulted in 2411 entries and a second one of the comments of the page resulted in 16,814 entries. The sentiment analysis and social cohesion language analysis focused on the period between October 25, 2012 and November 30, 2012.

5.1.5 **Findings**

The analysis of the evolution of the Facebook network during the first 15 days after hurricane Sandy landfall was done organizing the network by communities using the Modularity algorithm (see Fig. 5.1.8). On average per day, there were 3227 nodes (posts), 5312 links, and 105 communities in the network. On the first day after the hurricane landed, there were 66 communities and this number increased in the following days reaching its peak on the fifth day with 282 communities, this effect faded away and ended with 20 communities on the 15th day. In Fig. 5.1.8, we colored the main 11 communities to follow them through the first 15 days after the hurricane landfall.

Network evolution Occupy sandy facebook November 1, 2012 to November 15, 2012 (cluster: by modularity, size: PageRank)

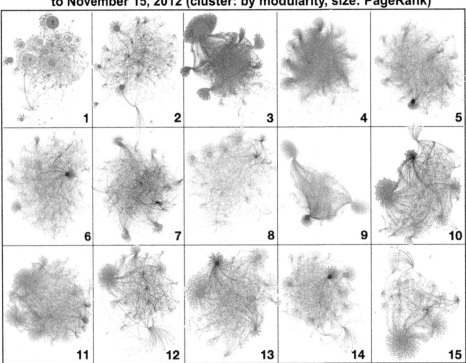

FIG. 5.1.8

Facebook network evolution (November 1, 2012 to November 15, 2012).

Regarding the type of information propagated through the network, 64% are photos followed by status (14.6%), links (17.3%), and videos (4%). The most popular posts in the first 15 days are shown in Fig. 5.1.9.

The emergence and maintenance of Occupy Sandy on social media over time was analyzed, looking for patterns regarding calls for action and organization strategies.

Fig. 5.1.7 shows an example of the most popular posts per community. The nodes are the users, the links the communication between users, and the size of the node is how many times the post was shared, as shown in Fig. 5.1.7, pictures are the most outreaching form of the post.

When analyzing the official accounts of Twitter and Facebook for Occupy Sandy it was noticed that Twitter was more active after the hurricane landed. Posts on Twitter were double the Facebook, but after a few days, activity was more constant on Facebook (see Fig. 5.1.10).

Twitter is used for immediate action calls while Facebook is used to organize long- or medium-term actions. Both the tweets and the Facebook posts show a clear pattern: most of them consisted of a description of what exactly was needed as well as the location. Some of them also included HTTP links for further, more detailed information. After 1 month, this pattern changed, the main purpose of posting on Twitter was to give information regarding where the immediate help was needed and to organize the volunteers; meanwhile Facebook was used to organize events (medium term) and propose activities.

FIG. 5.1.9

Popular posts from Facebook.

After analyzing the sentiment analysis and collectiveness analysis of the posts, the contents of the most popular posts reveal three main patterns:

- Highly engaged users demonstrated a positive sentiment on accomplishments made by Occupy Sandy.
- The majority of the posts by the official account Occupy Sandy blame formal institutions such as FEMA and Red Cross to have responded inadequately to the hurricane. In this regard, the disaster-relief network emphasized that the solution needed to be that the community stood together and took care of the disaster response by itself. For instance, in its most outreaching post on November 7, 2012, Occupy Sandy referred to an inadequate disaster-response by FEMA. In the description of the photo used in the post, the disaster-relief network urged users to "step up in FEMA's place, SHARE this photo and say #WEGOTTHIS!" (Ambinder et al., 2013). With the hashtag "#WEGOTTHIS" Occupy Sandy emphasized that the community itself could cope with the humanitarian disaster by standing together.

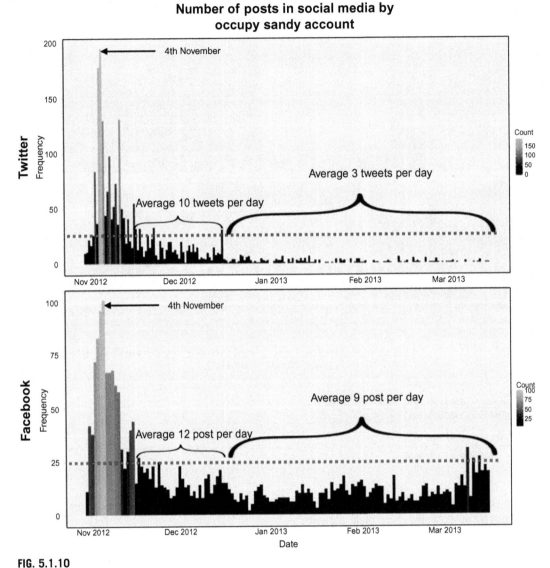

FIG. 5.1.10

Number of posts in social media.

- Occupy Sandy account engaged users to stand together as a community, as well as share their knowledge regarding specific technical issues. For instance, in the most outreaching post on November 4, 2012, Occupy Sandy shared a photo of fire-powered cell phone chargers and asked for people who knew how to make them. In most of the comments, users expressed their amazement about the power generators.

We believe the big difference from neutral posts in Twitter compared to the ones on Facebook could be explained by the privacy settings of the social media platforms.

The first of three indicators for social cohesion identified by Schiefer and van der Noll refers to social relations (Schiefer and van der Noll, 2017). To measure the extent to which Occupy Sandy users felt attached to the Occupy Sandy disaster-relief network, a sentiment analysis was conducted for both Twitter and Facebook. Their results are represented in Fig. 5.1.11. The x-axes display the dates. The y-axes indicate the number of tweets or Facebook posts. The yellow line shows the total number of tweets or Facebook posts. Positive contents are shown in blue, negative contents in red, and neutral

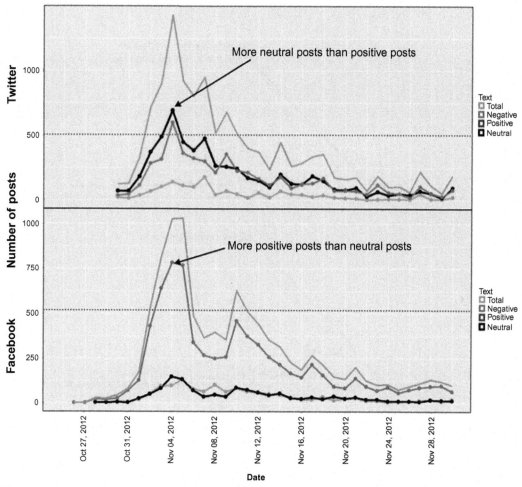

FIG. 5.1.11

Sentiment analysis from #OccupySandy.

contents are depicted by the green line. Fig. 5.1.11 shows that Facebook posts are more positive than Twitter. To complement the sentiment analysis, orientation toward the common good was measured, looking if tweets and Facebook posts included collective words compared to disjoint words such as "we," "together," "community" versus words like "you," "individual," or "single." Fig. 5.1.12 represents the results of this analysis. On the x-axes, the dates are displayed. The y-axes indicate the number of tweets or Facebook posts, respectively. The yellow lines depict the total number of tweets or Facebook posts. Neutral contents are displayed in blue, contents containing collective words in red, and the

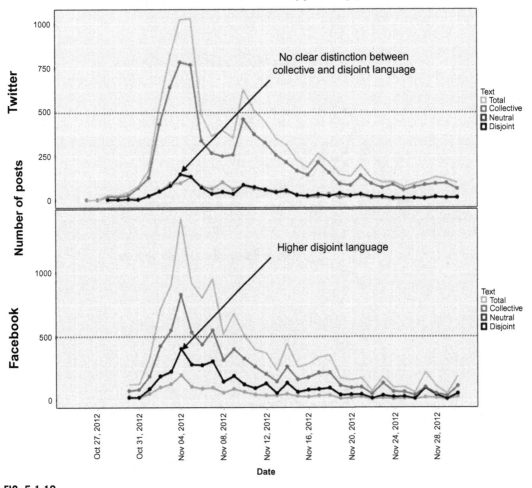

FIG. 5.1.12

Collectiveness analysis from #OccupySandy.

green line depicts contents containing disjoint words. In Twitter, there is no clear predominance of collective language, meanwhile in Facebook more disjoint language is seen (Fig. 5.1.12).

5.1.6 Conclusions

The analysis showed that social media allowed Occupy Sandy to start organizing immediately after the hurricane made landfall.

Moreover, Facebook and Twitter allowed the disaster-relief network to formulate calls for action and share information easily. Thus, Occupy Sandy kept its posts and tweets relatively brief and asked precisely for specific needs of supplies or workforce. Furthermore, the results of the analysis revealed that especially Facebook posts including photos reached a particularly high number of users who liked, shared, or commented on the post. Regarding the content of these posts, three main patterns could be observed:

1. Spreading positive news in the form of photos resulted in a particularly large amount of user reactions.
2. Occupy Sandy often blamed formal organizations for responding inadequately to the disaster.
3. In many highly outreaching posts, the disaster-relief network urged users to stand together as a community and help each other. Concerns identified during this analysis and the policy recommendations.

5.1.6.1 Concerns identified

During our study, we identified an inefficient emergency relief due to the lack of communication between public organizations and citizens resulting in frustration expressed by civil society. Social media platforms were used to coordinate and organize community-based relief.

5.1.6.2 Policy suggestions

1. Setting up official social media accounts, Twitter is more effective for short messages and Facebook is better to organize events
   - Select an intuitive # related with the event.
   - If possible create an online payment system to focus monetary help. It could be a PayPal or similar platform.

2. Organizing hubs and coordination centers
   - Localize the official coordinator centers set by the government. In case of lack of official centers use community centers, churches, or schools.
   - Set up a team in charge of the social media communication channels.

3. Register and mobilize volunteers
   - Formulate brief and precise descriptions of what supplies are needed, the amount which is needed, and the place.
   - Register every volunteer, allocate responsibilities, and set shifts.
   - Follow-up with the volunteers and make sure they are well rested.

Acknowledgments

This research has been funded in part by the National Science Foundation, through CRISP Type 2/Collaborative Research: Resilience Analytics: A Data-Driven Approach for Enhanced Interdependent Network Resilience, CMMI award number 1541165.

This research was part of the Urban Resilience Summer Project at Stevens Institute of Technology and Twente University. The authors thank Timo HartMann, BS for assistance with the data collection, this project was part of his bachelor thesis.

The authors gratefully acknowledge Professor Gregg Vesonder for sharing his pearls of wisdom with us during the course of the summer program research.

References

Ahmed, Y.A., Ahmad, M.N., Ahmad, N., Zakaria, N.H., 2019. Social media for knowledge-sharing: a systematic literature review. Telematics Inform. 37, 72–112

Albala-Bertrand, J.-M., et al., 1993. Political Economy of Large Natural Disasters: With Special Reference to Developing Countries. OUP Catalogue, Oxford University Press.

Aldrich, D.P., 2012. Social capital in post disaster recovery: towards a resilient and compassionate East Asian community. In: Economic and Welfare Impacts of Disasters in East Asia and Policy Responses, pp. 157–178.

Aldrich, D.P., 2017. In disaster recovery, social networks matter more than bottled water and batteries. https://www.citylab.com/solutions/2017/02/recovering-from-disasters-social-networks-matter-more-than-bottled-water-and-batteries/516726/.

Ambinder, E., Jennings, D.M., Blachman-Biatch, I., Edgemon, K., Hull, P., Taylor, A., 2013. The Resilient Social Network:@ OccupySandy# SuperstormSandy. Homeland Security Studies and Analysis Institute, Falls Church, VA.

Bengtsson, L., Lu, X., Thorson, A., Garfield, R., Von Schreeb, J., 2011. Improved response to disasters and outbreaks by tracking population movements with mobile phone network data: a post-earthquake geospatial study in Haiti. PLoS Med. 8 (8), e1001083.

Blondel, V.D., Guillaume, J.-L., Lambiotte, R., Lefebvre, E., 2008. Fast unfolding of communities in large networks. J. Stat. Mech. Theory Exp. 2008 (10), P10008.

Brin, S., Page, L., 1998. The anatomy of a large-scale hypertextual web search engine. Comput. Netw. ISDN Syst. 30 (1–7), 107–117.

Caragea, C., McNeese, N., Jaiswal, A., Traylor, G., Kim, H.-W., Mitra, P., Wu, D., Tapia, A.H., Giles, L., Jansen, B.J., et al., 2011. Classifying text messages for the Haiti earthquake. In: Proceedings of the 8th International Conference on Information Systems for Crisis Response and Management (ISCRAM2011), Citeseer.

Chan, J., To, H.P., Chan, E., 2006. Reconsidering social cohesion: developing a definition and analytical framework for empirical research. Soc. Indic. Res. 75 (2), 273–302.

Computing Community Consortium, 2012. Computing for Disaster Management Workshop, National Science Foundation, pp. 12–20.

Cormack, G.V., Gómez Hidalgo, J.M., Sánz, E.P., 2007. Spam filtering for short messages. In: Proceedings of the Sixteenth ACM Conference on Information and Knowledge Management. Association for Computing Machinery (ACM), New York, NY, pp. 313–320.

Delany, S.J., Buckley, M., Greene, D., 2012. SMS spam filtering: methods and data. Expert Syst. Appl. 39 (10), 9899–9908.

Dessavre, D.G., Ramirez-Marquez, J.E., Barker, K., 2016. Multidimensional approach to complex system resilience analysis. Reliab. Eng. Syst. Saf. 149, 34–43.

Faustino, J.D., Liu, B., Jin, Y., 2012. Social media during disasters: a review of the knowledge base and gaps, Final Report to Human Factors/Behavioral Sciences Division, US Department of Homeland Security, College Park, USA.

Field, C.B., Barros, V., Stocker, T.F., Dahe, Q., 2012. Managing the Risks of Extreme Events and Disasters to Advance Climate Change Adaptation: Special Report of the Intergovernmental Panel on Climate Change. Cambridge University Press.

Gómez Hidalgo, J.M., Bringas, G.C., Sánz, E.P., García, F.C., 2006. Content based SMS spam filtering. In: Proceedings of the 2006 ACM Symposium on Document Engineering, ACM, pp. 107–114.

Guha-Sapir, D., Santos, I., Borde, A., et al., 2013. The Economic Impacts of Natural Disasters. Oxford University Press.

Henry, D., Ramirez-Marquez, J.E., 2012. Generic metrics and quantitative approaches for system resilience as a function of time. Reliab. Eng. Syst. Saf. 99, 114–122.

Heymann, S., 2011. New Tutorial: Layouts in Gephi. https://gephi.wordpress.com/tag/force-atlas/.

Hu, M., Liu, B., 2004. Mining opinion features in customer reviews. In: AAAI, vol. 4, pp. 755–760.

Hulse, K., Stone, W., 2007. Social cohesion, social capital and social exclusion: a cross cultural comparison. Policy Stud. 28 (2), 109–128.

Jaramillo, C.R., 2009. Do natural disasters have long-term effects on growth? Documento CEDE No. 2009-24.

Jenson, J., 2010. Defining and Measuring Social Cohesion. vol. 1. Commonwealth Secretariat.

Kryvasheyeu, Y., Chen, H., Moro, E., Van Hentenryck, P., Cebrian, M., 2015. Performance of social network sensors during Hurricane Sandy. PLoS ONE 10 (2), e0117288.

Luna, S., Pennock, M.J., 2018. Social media applications and emergency management: a literature review and research agenda. Int. J. Disaster Risk Reduct. 28, 565–577.

Meerow, S., Newell, J.P., Stults, M., 2016. Defining urban resilience: a review. Landsc. Urban Plan. 147, 38–49.

Meier, P., 2015. Digital Humanitarians: How Big Data Is Changing the Face of Humanitarian Response. CRC Press, Boca Raton, FL.

Newman, M.E.J., Girvan, M., 2004. Finding and evaluating community structure in networks. Phys. Rev. E 69 (2), 026113.

NHK, 2011. Disaster big data. Nippon Hoso Kyokai (Japan Broadcasting Corporation). http://www.nhk.or.jp/datajournalism/about/index_en.html.

OECD, 2011. Perspectives on Global Development 2012: Social Cohesion in a Shifting World (Summary). OECD, Paris.

Pang, B., Lee, L., et al., 2008. Opinion mining and sentiment analysis. Found. Trends Inf. Retr. 2 (1–2), 1–135.

Pastor-Escuredo, D., Morales-Guzmán, A., Torres-Fernández, Y., Bauer, J.-M., Wadhwa, A., Castro-Correa, C., Romanoff, L., Lee, J.G., Rutherford, A., Frias-Martinez, V., et al., 2014. Flooding through the lens of mobile phone activity. In: Global Humanitarian Technology Conference (GHTC), 2014 IEEE, IEEE, pp. 279–286.

Penney, J., Dadas, C., 2014. (Re)tweeting in the service of protest: digital composition and circulation in the occupy wall street movement. New Media Soc. 16 (1), 74–90.

PEW Research Center, 2018. Social Media Fact Sheet, PEW Research Center, Washington, DC. https://www.pewinternet.org/fact-sheet/social-media/.

Rich, K., 2018. Girls Resist! A Guide to Activism, Leadership and Starting a Revolution. Quirk Books, Philadelphia, PA.

Schiefer, D., van der Noll, J., 2017. The essentials of social cohesion: a literature review. Soc. Indic. Res. 132 (2), 579–603.

Skidmore, M., Toya, H., 2002. Do natural disasters promote long-run growth? Econ. Inquiry 40 (4), 664–687.

Stallings, R., Quarantelli, E.L., 1985. Emergent Citizen Groups and Emergency Management. Public.

Tapia, A.G., Piña, C.M., et al., 2014. The Effect of Natural Disasters on Mexico's Regional Economic Growth: Growing Disparity or Creative Destruction? 2014.

Turney, P.D., 2002. Thumbs up or thumbs down? Semantic orientation applied to unsupervised classification of reviews. In: Proceedings of the 40th Annual Meeting on Association for Computational Linguistics, Association for Computational Linguistics, Philadelphia, PA, pp. 417–424.

UNISDR, U.N.O., 2015. Sendai framework for disaster risk reduction 2015–2030. In: Proceedings of the 3rd United Nations World Conference on DRR, Sendai, Japan, pp. 14–18.

Varda, D.M., Forgette, R., Banks, D., Contractor, N., 2009. Social network methodology in the study of disasters: issues and insights prompted by post-Katrina research. Popul. Res. Policy Rev. 28 (1), 11–29.

Voorhees, W.R., 2008. New Yorkers respond to the world trade center attack: an anatomy of an emergent volunteer organization. J. Conting. Crisis Manag. 16 (1), 3–13.

Wellman, B., Gulia, M., 1999. Net-surfers don't ride alone: virtual communities as communities. In: Networks in the Global Village: Life in Contemporary Communities. West View Press, Boulder, CO, pp. 331–366.

Wesolowski, A., Eagle, N., 2010. Parameterizing the dynamics of slums. In: 2010 AAAI Spring Symposium Series.

Exploring alternative livelihood in oil-spill impacted communities: A Nigerian perspective

5.2

Oshienemen Albert, Dilanthi Amaratunga, and Richard Haigh

University of Huddersfield, Global Disaster Resilience Centre, Huddersfield, United Kingdom

CHAPTER OUTLINE

Understanding Disaster Risk. https://doi.org/10.1016/B978-0-12-819047-0.00016-0

319

5.2.1 Introduction

The concept of livelihood, which consists of human, natural, financial, physical, and social capital, involves inherent values that accompany human existence and thus is a crucial aspect for any community. The Niger Delta communities, whose existence is strongly dependent on traditional livelihood systems, have experienced incontrovertible negative impacts of oil spills on traditional livelihood structures across decades of oil and gas production. These impacts have contributed to an increase in poverty and lowering of the standard of living in communities (Bradley, 1974; Jones and Schmitz, 2009; Picou et al., 2009); violent conflict, restiveness, and cultural damage (Aaron, 2005; Kadafa, 2012a,b; Nwachukwu and Mbachu, 2018); and environmental and mangrove damage (Lewis, 1983; Patin, 1999). In addition, a study on the effect of the Deepwater Horizon oil spill on house markets revealed that the spill caused a significant decline in home prices of between 4% and 8%, which continued until late 2015 (Cano-Urbina et al., 2019). This implies that, regardless of the region, country, or community, the impact of oil spills remains widely acknowledged. For example, Osuagwu and Olaifa (2018) and Chijioke and Ebong (2018) revealed that oil spills had significant impacts on fish hatcheries and contamination of valuable fish production, while decreasing agricultural output in Nigeria. These oil spill issues have received wide exploration (Albert et al., 2018; Bruederle and Hodler, 2019; Mogaji et al., 2018; Pan et al., 2019), but with little attention paid to the communities' alternative methods for livelihood amid the impacts (Sobrasuaipiri, 2016).

Noticeably, relatively few studies have examined the adaptation systems of oil spill communities with regards to livelihood structure, considering the depth of relationship between the natural resources of Niger Delta Communities (NDCs) and the community socioeconomic and sociocultural activities. Fewer studies have reported the impact across NDCs with qualitative focus groups in community-based investigation, which reinforces the gap this current study intends to fill in the literature. In addition, this study draws from a disaster perspective, which rarely have researchers in this field adopted for oil spill literature. Consequently, this research adapts to the Sendai Framework for Disaster Risk Reduction (SFDRR, 2015–2030): priority 1, "Understanding Disaster Risk," which is knowledge concealed at the community level. The justification behind looking at oil spills from a disaster perspective is the lack of awareness of disaster risk and the understanding that an oil spill, irrespective of how, when, or where it occurs, is a disaster, considering the number of deaths (Harrison, 2019), livelihood impact, and health and social disruption that accompany the event (Croisant and Sullivan, 2018; Gam et al., 2018). Finally, the theoretical contribution of this study encompasses the emerging themes, while exposing policy makers in disaster risk and environmental sectors to the possible and effective ways of improving community-improvised livelihood initiatives specifically for the Niger Delta region of Nigeria.

5.2.2 Oil-spill environmental hazards

Environmental hazards in the form of oil spills have increasingly contributed to livelihood disruptions, negative health impacts, cultural impairments, and psychological and ecological damage across various sectors (Gill et al., 2012; Jung et al., 2017; Pegg and Zabbey, 2013). Oil spill hazards degrade the natural environment and resources and thus decrease the environmental resource availability in the aftermath of the incidents. This contributes to grave economic losses in both developed and developing countries (Garza-Gil et al., 2006a), with difficulties in ascertaining the total of all economic losses (direct and indirect). According to Liu and Wirtz (2006), the measurement of economic loss of oil spills should encompass all categories, for example, from local stakeholders (all avenues for livelihood support that are impacted) to state and federal levels, as to how the impact affects the economy before the estimated values. Liu and Wirtz (2006) pointed out that an economic measurement that includes all the categories of direct and indirect sources of livelihood will lead to a complete estimation of the loss. In a similar vein, Grigalunas et al. (1986) mentioned that, when considering local, regional, and national effects, the market and nonmarket value cost in the distribution are vital aspects in the measurement of economic loss for those affected.

Garza-Gil et al. (2006b) revealed that there were losses of adult and juvenile shellfish and aquacultures in the infamous Galicia oil spill, generating an economic loss of 65 million Euros in the fishing sector during the year of the oil spill. Evidence from Cheong (2011) further emphasized how fishing, tourism, and agricultural communities were deeply affected by the Hebei-Spirit oil spill. Similarly, Cheong (2011) mentioned, for example, that the fishing catch declined from 7798 metric tons (m/t) in 2007 to 789 m/t in January–March 2008, which contributed to community cohesion decline and economic loss. While Agbogidi et al. (2005) revealed that there were decrease in the agricultural output of farming communities, which contributed to the decline in earning capacity due to damages caused by oil and gas production.

According to a report by CRED (2016), from a global perspective, an estimated US$154 billion economic damage was revealed from disasters in 2016. Likewise, it is estimated that an economic loss of US$306 billion has been incurred from disasters (from natural and man-made hazards) (Poledna et al., 2018). Also, a minimum of US$157 billions worth of property and infrastructure were destroyed in 2012 (Guha-Sapir et al., 2013). Quantifying the resulting economic losses to national economies is a major challenge (Poledna et al., 2018), even though statistics show substantial economic costs and losses from disasters yearly (EM-DAT, 2017). Economic losses thus differ from decade to decade and year to year, with either increases or decreases in losses and fatalities and they are highly dependent on the measurement scale and what it encompasses, as well as the approach taken for the measurement of economic losses from oil spills and/or disasters. Nonetheless, regardless of the economic losses across nations, states, and local communities, adaptation for livelihood and alternative measures are vital for subsistence in the aftermath or amid the incidents. Hence, the next section discusses adaptation, livelihood, and sustainable livelihood concepts.

5.2.3 Adaptation for livelihood

The actions taken to secure the necessities of life are a vital aspect of any community or society faced with an adverse impact of a disaster on their livelihood structure. Accordingly, the adaptation definitions commonly found in climate-change literature have slight differences, depending on the context and relative focus (Bohnert et al., 1995; Lack, 1968; Watson et al., 1996). Adaptability according to Brooks (2003) refers to *"the adjustments in a system's behaviour and characteristics that enhance its ability to cope with external stresses."* Smit et al. (2000) and Pielke (1998), from a climate-change perspective, describe adaptations as the adjustment of an ecological and socioeconomic system in response to climatic effects or the adjustment of individual, group, or institutional behaviors toward the reduction of society's vulnerability to climate conditions. Indeed, adaptation can be anticipatory or reactive, and can be planned based on differing descriptions and perceptions (Scoones, 1998). Considering the different contexts and perceptions of adaptation, this study describes adaptation as the process, action, or outcome in a system (household, community, group, sector, region, country) that allows the system to better cope with, manage, or adjust to some changing condition, stress, hazard, risk, or opportunity (Smit and Wandel, 2006). Such conditions include, but are not limited to, economic change, social change, environmental change, and cultural change. Although the concept is strongly interrelated with resilience, adaptation, vulnerability, and exposure apply not just at the local community levels but also have broad global applications in science (Smit et al., 2000; Smit and Wandel, 2006). The application of this concept scales from the community level to household and individual vulnerability to a particular stress. For example, the concepts of stress and vulnerability within the context of this paper revolve around oil spills in the human-environmental system (pollution), which have contributed to household, regional, group, and community social stress, livelihood stress, and sociocultural grief.

Adaptation is essential for vulnerable coastal communities faced with a threat to livelihood as well as health and safety across communities (Uy et al., 2011). Thus, to increase the adaptive capacity and build resilience across communities affected by the adverse impacts of change and/or oil spillages, there is a need for improvised adaptive alternative methods for livelihood support (Danoff-Burg and Conrad, 2020). Akpabio et al. (2007) arguably stated that some adaptation strategies for livelihood support across the Nigerian Niger Delta Communities are hostage taking, protest, and agitation. It is argued that such activities are practically not for livelihood support, but are instead acts of frustration to which some groups resort as a means of drawing government attention to community plights as related to the issues of marginalization, neglect, and oil and gas resource injustices (Dibua, 2005). Notably, Adekola and Mitchell (2011) and Davies (2016) emphasize that livelihoods of the Niger Delta Communities are immensely impacted negatively through the oil and gas activities, for example, gas flaring, operational noise and vibrations, and different pollution types. Such have influenced communities' quest for alternative support for livelihood as they become less resilient and more vulnerable due to the unsustainable structures available.

5.2.4 Livelihood and sustainable livelihood concept

A livelihood comprises a people's means of living, for instance, income and assets, food, and capabilities for managing the assets owned. The management of assets, whether tangible or intangible, is an integral component of livelihood. Researchers describe the complex context of livelihood using different dimensions, depending on what is intended to be achieved in a given period. According to Chambers and Conway (1991), "*A livelihood comprises the capabilities, assets (including both material and social resources) and activities required for a means of living.*" Thus, the sustainability of livelihood is dependent on the coping strategies and the resilience of the affected people. Chambers and Conway (1991) further mentioned that livelihood is socially and environmentally sustainable when it maintains or enhances the community, local, and global assets on which the livelihood entirely depends, while also benefiting other livelihoods. Further, Krantz (2001) argued that sustainable livelihood could serve as an integrating factor that allows policies to address development, sustainable resources management, and poverty eradication concurrently, given that the concept was introduced to link socioeconomic and ecological considerations in a unified and policy-relevant structure.

As discussed by Scoones (1998), sustainable livelihood is an essential aspect of the development debate; thus the achievement of sustainable development would encompass access to natural, economic, human, and social capitals combined in the pursuit of other livelihood strategies, for example, agricultural intensification and diversification of livelihood and/or migration.

However, livelihood in the Niger Delta as the case study for this chapter is strongly reliant on its natural resources and traditional systems, as presented in Table 5.2.1. Most importantly, the concept of livelihood as used in the context of this research means the indigenous socioeconomic and sociocultural activities that are based on the available resources, which serve as a means of everyday household existence. According to Erondu (2015), the Niger Delta traditional livelihood consists of any form of farming, fishing, and agricultural activities that contribute to the local economic development of a community. As evident from different scholarly studies, a majority of the Niger Delta communities' sources of livelihood revolve around the use of fauna and flora products for subsistence (Barad et al., 2020; Olawoye, 2019; Ugochukwu and Ertel, 2008). It is important to consider that a livelihood consists of any assets that can contribute to everyday subsistence of either households, groups of people, or society in general. The Niger Delta community's livelihood assets include both human and nonhuman resources on which livelihood is built and to which the people, communities, and societies have access. While most community livelihood sources largely depend on waterways, some depend significantly on available land areas (traditional industrial livelihood) for subsistence, which are a significant part of the existence of the people of the Niger Delta region. Table 5.2.1 depicts significant aspects of the Niger Delta traditional livelihood support structures.

5.2.5 Method

This study adopts a qualitative research design focusing on the NDCs impacted by oil spill environmental hazards for livelihood structures. Twenty-two community representatives were purposively selected from across profoundly impacted communities to represent the entire region at large. Seven focus group discussions were conducted concurrently to answer the question, "*What are the alternative livelihoods for oil spill and oil and gas impacted communities, and how are such alternatives invigorating the lost culture and socioeconomic conditions of the communities?*" The participants in the sample were purposively selected due to their relevant experience related to the question under investigation, having lived over 20 years within the communities in the Niger Delta, Nigeria. The participants included community representatives, those with environmental expertise, farmers, community social relations officers, and liaison officers. Participants were initially contacted by phone with a brief summary of the intended aim of the meeting before the face-to-face interviews. Unstructured and semi-structured interviews were conducted to obtain an in-depth perception of the question under investigation. As emphasized by Kvale (2008) and Rowley (2012), interviews are techniques that enable probing during a conversation, to obtain an in-depth opinion on the subject and reveal how the impacted population perceives the construct under investigation. The interview helped the researcher to explore the phenomenon in an in-depth and natural manner (Brinkmann, 2014; Qu and Dumay, 2011), while targeting the exploration of the phenomenon (Saunders et al., 2016). The result was analyzed through a content analytical approach, interpreting the data through a thematic logic by developing the interrelated interpretation within subcategories and between major categories.

Table 5.2.1 Components of Niger Delta community livelihood.

NDD traditional livelihood mechanism	Contributes to the traditional livelihood sustenance of communities (CTLSC)	Damages caused on (TLM)	Resources required to revitalize and improve (RRRI) (TLM)
Farming/agriculture	Agriculture is an important economic activity associated with Niger Delta tradition. Most traditional livelihoods across the Niger Delta are tied to agriculture. Examples: Crop cultivation, fishing, and aquaculture. Agriculture/farming links the producers and consumers of food and nonfood agricultural (farm-related) products within the region. Farming contributes to the societal activity development (e.g., food harvest festivals). Socioeconomic contributions (e.g., trade by barter) to local economic growth.	Oil and gas activities Oil spill pollution. Gas flaring	Land (free government land) and water supply (irrigation system) channels are strongly needed – Channels for increase of food securityTo stem the current slide toward poverty and food insecurity, the farming population needs to be empowered by an enhanced agricultural activity. Aquaculture practice is still at a very rudimentary level comparatively (Erondu, 2015). – Local processing industries and storage facilities are limited and still very rudimentary, and as such, the provisions of the aforementioned are requiredIncorporates dimensions such as the production, storage, processing, trade, and use of these products, the natural resource base, and the policy and regulatory environment that supports the system.
Hunting and forest resources usage	– Hunting and harvesting of forest resources from the wild contribute substantially to local economic growth. – Gathering of food through rural water and forestry resources among the rural communities constitutes viable sources of livelihoods for rural dwellers. – Niger Delta forest zones encompass an array of resources that provide wood and nontimber resources, for example, nuts, leaves/medicinal leaves, fruits, spices, barks of trees (for medicinal purposes), cane for furniture making, reed for mate making, and – Wood contributes to labor-intensive industry in different ways, such as sawmills and plymills, construction and furniture industries, drug manufacturing industries, and industrial production of useful chemicals	Oil and gas activities Oil spill pollution Gas flaring Noise and vibration	New techniques in biotechnology and phytochemical research can form the cornerstone of multimillion-naira enterprises that sustain cottage industries engaged in the collection and primary processing of such products (Onakuse and Lenihan, 2017)

Activity	Description	Factors	Recommendations
Fishing	– Hunting from the wild provides a range of products such as snails and edible animals that serves as delicacies in the food industry, and support the nutritional needs of the people across a majority of communities within the Niger Delta region. Fishing activities such as harvest of fish, oysters, periwinkles, crabs, shrimp, mullet, snapper, cichlids, tilapia, croaker, grouper, intertidal mudskippers and bonga, and mollusks including four varieties of edible common periwinkle, oyster, bloody cockle, and dog whelk are some major constituent of traditional livelihood of Niger Delta. – Harvested crustaceans, including primarily the "swimming crab" (*C. amnicola*), the purple mangrove crab (*Goniopsis peli*), the pink shrimp (*Farfantepenaeus notialis*) and the invasive tiger prawn/shrimp (*Penaeus monodon*), and different species provided an extensive economic, food, and cultural resource across many riverine Niger Delta communities. – Fishing/fish and fish products remain a significant ingredient on the global and local menu of environmental balance, an important aspect of livelihood; protein for both adults and children (Pegg and Zabbey, 2013).	Oil spill pollution Oil and gas activities	To improve the management of fisheries and aquaculture, and application of new aquaculture technologies – Provision of fishing equipment to all communities involved in fishing across the Niger Delta communities. – Provide local processing industries and storage facilities, as they are limited and still very rudimentary, hence producers of perishables and raw materials also engage in minimal and limited processing of their produce. – The power sector has failed to power this sector of the traditional economy in an up-to-date manner. The resultant effect is that much wastage and loss of potential income is experienced. This is very discouraging to fishing communities in particular and limits the number of people attracted to this traditional livelihood.
Trading	– Trading in agricultural-related and crafts products is a vital economic aspect of subsistence. – Trade by barter activities across the region during and after fishing festival activities.	Oil spill pollution Oil and gas activities Poor infrastructural facilities (e.g., poor communication and road networks)	The Niger Delta is well known for its difficult terrain that hinders free movement; therefore, the waterways become a viable channel for most community business. Waterway security means is needed for fishing population.
Crafting	The cottage craft industry is another sector of the traditional economy. These crafts include canoes, mats, baskets, and cane furniture making – All sorts of local farming implements and fishing equipment are made through the local craft industry. These include diggers, hoes, and machetes, and fishing hooks, traps, racks, and baskets.	Oil spill pollution Oil and gas activities Gas flaring	It is clear that our production system at this level lacks high-level technology. This sector has a great potential if enhanced, especially in terms of quality, designs, and aesthetics, as currently the status of the products does not meet all the standards of the international market.

5.2.6 **Study area**

The Niger Delta region of Nigeria coincides with the south-south geopolitical zones of the nation, which consisted of predominantly farmers and fishermen (agricultural activities as the main occupation) across a majority of the communities before the discovery of crude oil and production began in 1956 (Khan, 1994; Omofonmwan and Odia, 2009; Onwuka, 2005). Nigerian NDCs before the oil boom of the 1970s structurally maintained a sustainable traditional livelihood through a range of socioeconomic and sociocultural activities associated with the traditional livelihood systems (Erondu, 2015). Fig. 5.2.1 presents a Niger Delta map showing the Niger Delta constituent states therein.

The Niger Delta region consists of nine states with a population of over 30 million people, with more than 40 different ethnic groups (Ihayere et al., 2014; Omorede, 2014). The Niger Delta landscape consists of large water bodies, wetlands, and dry land, covering 70,000 km², with distinct ecological zones, coastal ridge barriers, and large spreads of mangrove forest and swamps (Ebegbulem et al., 2013; Oviasuyi and Uwadiae, 2010). In addition to a web of rivers, the landscape is dominated by rural

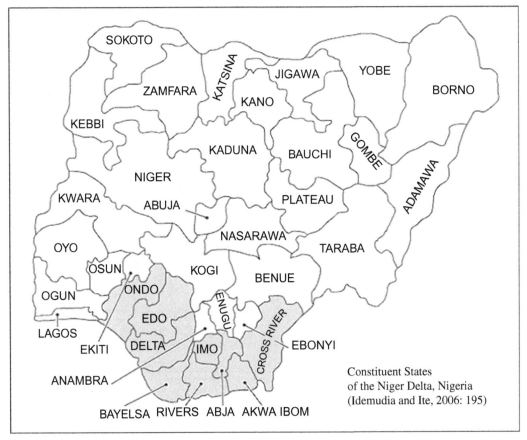

FIG. 5.2.1

Map of Nigeria showing the Niger Delta region.

communities that depend solely on their natural environment for subsistence (Ebegbulem et al., 2013). Importantly, the Niger Delta covers a large number of oil reserves located onshore, offshore, deep off-shore, and shallow offshore (Fubara et al., 2019). Petroleum products contribute substantially to the nation's treasury through resource development windfalls from oil and gas explorations and production activities within the region (Fubara et al., 2019). Oil and gas industry in Nigeria contributes to about 95% of the Nigerian foreign exchange earnings, over 60% of the country's national yearly budgetary revenue, and 15% of the gross domestic product (GDP) (Ukpong et al., 2019; NBS, 2012).

However, irrespective of the enormous contribution of the Niger Delta oil and gas production to the nation's treasury, as evidenced from various empirical studies (Odularu, 2007; Osuma et al., 2019), the region and the entire nation still have a high level of poverty, despite greater attention paid and priority attached globally to poverty reduction through the adoption of sustainable development goal frameworks (Dauda, 2019). Also, there is a major increase in obsolete learning facilities, a lack of proper health-care facilities and basic life amenities (Kadafa, 2012a,b; Oviasuyi and Uwadiae, 2010), and a massive displace-ment of labor from the original livelihood sources due to natural economic disarticulation of the oil econ-omy (Erondu, 2015; Ukpong et al., 2019). All these negative impacts and livelihood displacements have increased the quest for alternative livelihood sources across a majority of the communities within the Ni-ger Delta region (Fubara et al., 2019; Kareem et al., 2012; Ugochukwu and Ertel, 2008).

Amnesty International (2015) revealed that Rivers State and Bayelsa State are two major riverine states with a decade of high impacts of oil spills on their primary livelihood sources. The evidence from the literature fosters the justification for the chosen locations for this current study. Also, the riverine sample communities within the two states of this study represent all the communities impacted by oil spill environmental hazards across the region. According to researchers, a majority of the communities within the Niger Delta suffer similar impacts considering the similarities in primary livelihood sources, cultural beliefs, traditions, norms, and social activities embedded in the local economic structure (Abi and Nwosu, 2009; Odjuvwuederhie et al., 2006; Omofonmwan and Odia, 2009; Ugochukwu and Ertel, 2008). Thus, it is evident that most of the communities suffer physical and livelihood displacement (Opukri and Ibaba, 2008), and struggle with their livelihood and adaptation systems (Kadafa, 2012a,b). The traditional livelihood sources of the Niger Delta communities remain paramount, as over 80% of the population depends on the traditional system before the destruction by oil and gas activities (Odjuvwuederhie et al., 2006).

The study findings are analyzed using NVivo 11 software and are presented using thematic template analysis techniques and quotations from respondents. First, the data was transcribed using MS Word and then later transferred to NVivo to aid in the identification of themes and categorization, while draw. io was used to show the visual connections of the themes. The identified core results for livelihood support were: local refining of petroleum products, long-distance fishing/deep-sea fishing, distance timber production engagement, daily bricklaying casual jobs, farm and share systems, and leased farm-ing, as presented in Fig. 5.2.2.

5.2.7 **Report and discussions**

5.2.7.1 **Local refining of petroleum products**

One of the most ingenious profitable private business in the Niger Delta region is oil bunkering, as divulged by the majority of the respondents. Even though such activity is classified as an unlawful activity in Nigeria, a majority of the communities sees such act as a livelihood support mechanism

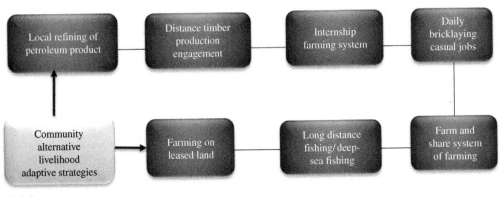

FIG. 5.2.2

Alternative livelihood adaptation strategies.

(Boniface and Samuel, 2016). Despite several studies have shown that local petroleum refining contributes to vast environmental damages and increases the threat to the Nigerian national security and national economy (Albert et al., 2018; Ingwe, 2015; Onuoha, 2008). The respondents disclosed that the oil bunkering activity is carried out on a small-scale operation at the local community level, where the petroleum product is condensed for local use and further tapped off for distribution in the local usable form.

The oil refined locally is widely used by more than 80% of households across the Niger Delta region. People use the local refined fuel as a source of electricity, and contributes to daily household upkeep. However, there is a lack of facility to conduct the activity in a more structured manner which has caused major catastrophes over the years by killing many at every slightest mistake, and contributing to the already damaged environment. A majority of the respondents emphasized that oil bunkering is an activity that would not stop in the near future except the Nigerian government and the oil and gas industry restructure their approach toward communities, by including community members in every aspect of the oil and gas business (Focus Groups 1 & 6). Respondents further apprised that oil bunkering business is conducted not just at the local community, but, involves other NGOs and sectors that engage in the operation for commercial purposes and beseeched for a sustainable livelihood support.

Sustainable livelihood communities across the region calls for a sustainable livelihood and a restructured system for local oil refineries, while asserting that oil bunkering contributes to the community household and livelihood support, irrespective of the high risk involved in the activity. This is also supported in the study conducted by Uduji and Okolo-Obasi (2019).

5.2.7.2 Long-distance fishing/deep-sea fishing

Long-distance fishing is widely adopted by communities whose immediate waterways are highly impacted by decades of oil spills without clean-up, remediation, or restoration. The interviewees whose major occupation was fishing divulged that the contamination of the waterways rendered their means of economic sustenance nonviable. Fishing has been a primary sociocultural and socioeconomic activity across most communities in the Niger Delta, with the river water serving as drinking water and used in spiritual activities. With the waterways now being green and unusable, with no living organisms due to continued pollution of different kinds, the livelihood support emanating from the waterways is greatly abridged, leaving the waterways as only a traveling channel. The majority of respondents assert that the river is valued much more than land in some areas when discussing livelihood sources, because the waterways contribute to nearly 90% of most community's and household's livelihood support through fishing. The streams and rivers are also used for cassava fermentation, swimming competitions, and canoe racing, which are sociocultural and socioeconomic activities that contribute immensely to community cohesion besides fishing activities.

This damaging effect now causes fishing communities to travel hundreds of kilometers into the Atlantic Ocean for fishing, which has increased energy use and time required for fishing and selling of fish and fish by-products. The delay due to the long distance often contributes to lowering the value of different fish species, as fish tend to drop in quality over time. This often affects the most elderly fishermen, who do not have speedboats and may use local canoes, taking hours to paddle miles for fishing activities. As asserted by a majority of respondents, the energy required for the long distances in search of fish as livelihood support compared to the catch in the usual short distance is immeasurable. The fishing population needs a hardbound engine boat, which the majority cannot afford. This has reduced the number of people who participate in long-distance fishing, thereby increasing the cost of daily living due to the limited quantities provided across the fishing communities.

Pegg and Zabbey (2013) support this assertion by establishing that the destruction of the traditional fishing method has adversely impacted the community's development, thereby calling for clean-up, remediation, and practical actions to alleviate the community's plight, while supporting livelihood strategies. It was also revealed that most community members relocate to a nearby country, such as Cameroon, for fishing activities. The relocation of community members is described by some interviewees as "forceful migration" considering the conditions that contribute to their relocation. This is supported in the research of Opukri and Ibaba (2008), which shows the contributions of oil spills and gas flarings to the diminishing fishing occupation trend of local communities, and also to the involuntary migration.

5.2.7.3 Distant timber production engagement

Timber production is one common livelihood support method adopted by most community members. Participants emphasized that most individuals prefer such activities over local oil refining and farming. As stated by different interviewees, many community members have traveled for timber production activities and never returned, which is to some extent a detriment to community development. The development of the community is very reliant on the output from the community members. Hence, if the community members continually migrate for timber production activities, some for fishing and farming activities, the sociocultural activities and development of the communities would constantly degrade, coupled with leakage from oil pipelines that habitually contributes to environmental damage (Focus Group 3). This further reduces the value of the community's activities while influencing underdevelopment, even though it contributes to household daily living.

It was strongly agreed that, irrespective of how people migrate practically for sources of livelihood, timber production activity remains a segment that contributes to household sustenance for those who engage in such activities. "*The activities around timber production are enormous*" (Focus Group 3), even though the government law stipulates "*cut down one and plant five trees*" to defeat deforestation while increasing afforestation.

5.2.7.4 **Daily bricklaying casual jobs**

Bricklaying involves craftsmen who construct brickworks, and is considered an alternative means of livelihood support in most local communities. Empirical evidence shows that a majority of youths and elderly engage in this type of work, in which they act as a laborer/assistant on construction sites (masonry or blockwork). The respondents are of the view that a job such as bricklaying cannot be disregarded, regardless of the low income it provides, when there are limited available options for livelihood support. Daily bricklaying jobs form a large part of the current alternative livelihood sources in communities within the Niger Delta region, and mostly unskilled youths engage themselves in this manner instead of in pilfering activities (Focus Groups 3 & 4). These daily bricklaying menial jobs contribute minimally to a household's survival. The people involved have stated that the job sometimes pays "*1,500, that is less than $5 per day*"; this places them in an extreme financial situation, which they perceive as better than nothing. The struggle for survival has left many with no option but to engage in anything humanly possible, as postulated by many respondents. For example, many community members have worked so strenuously at this unskilled labor that their bodies have been disfigured due to the lack of health and safety knowledge on construction sites (Focus Group 2).

Daily bricklaying jobs lack sustainability due to the nature of the activity. The majority of the respondents emphasized that, while family responsibilities linger, livelihood sources should be constant whether for the short or long term. The participants therefore call for government empowerment programs for the Niger Delta communities to improve livelihood systems while reducing conflict and the violent and restive situations which are perceived to be partially influenced by the lack of empowerment, as supported in the study of Ofem and Ajayi (2008). In line with the majority of respondents' insights, the women assert that, even though they participate in the bricklaying jobs as casual helpers, they are more economically productive in farming, emphasizing that farming is embedded in their culture and lifestyle and they therefore would engage in any available form of farming for household livelihood support.

5.2.7.5 **Farming/share system of farming**

The empirical evidence from the respondents emphasized farming as a theme that cannot be overlooked, because it forms a part of the people's tradition, culture, norms, and local economic development. The majority of the participants stated that different local farming systems have contributed to household support, irrespective of the various obstacles that often emerge with land issues.

> The farming strategy which involves sharing what is cultivated on a borrowed land 'equally' with the owner of the land is a strategy that has contributed to many household's subsistence. Otherwise, most women would have left their households for another unhealthy livelihood search (Focus group 4) even though the share system favours the owners of the land largely, people still have to engage in it for survivals (Focus group 4)

Most importantly, this farming system is widely accepted by women in the communities, especially considering that most other alternative sources, for example, bricklaying and local refining of oil, are deemed unsuitable for women within the community settings due to their cultural and traditional beliefs. Sharing crops (for example, cassava, cocoyam, yam, potatoes, and perishable crops) cultivated on a piece of borrowed land is substantially preferable to the same activities on contaminated land, which might not produce a good yield and could have numerous associated health risks (Focus Group 2).

Unavailability of land remains a major problem associated with this method of farming, as this limits the workforce, especially women, whose sole occupation is farming. The limited land as perceived by the communities affected has contributed to irregularities in which most landowners tend to harvest the cultivated crops before the original harvest period. That is why some interviewees said that internship and leased farming are more economical, beneficial, and healthier for all stakeholders involved (i.e., both the farmer and landowners).

Farm and share have caused some conflicts among individuals, even as it seems economically beneficial. Arguably, the people are of the view that most of the challenges associated with the different methods of farming can be resolved through the provision of free government land for agricultural purposes in oil and gas producing communities (Focus Groups 1 & 3).

5.2.7.6 Internship farming system

The interviewees revealed that an internship farming system is a voluntary activity in which community members relocate to a given area for some years solely for farming activities, and subsequently return to their communities with new farming techniques. This act of farming for livelihood support is carried out by a good number of farmers, not necessarily for migration purposes but with good intentions for livelihood support. According to a participant in a focus group discussion,

> ...I stayed in another village for six years immediately after the 2008, 2009 massive oil spills that took place in Bodo, I lost my canoe, farmland and left out of frustration. So, I embarked on that journey to enable me to survive with my family which yielded positively for my households (Focus group 4).

Most respondents assert that this particular farming method prevents most pregnant women from having to use contaminated waterways, which contributes to numerous deformities in newborn babies, as witnessed by women who maintained their daily activities within and around the contaminated sites. The people stated that most women who bluntly refused to stay away from contaminated sites were recorded to have given birth to deformed children, and suffered from unsightly skin diseases during the early and late periods of the disaster incidents. This particular finding is supported in the study of Atubi (2015), which revealed the people's views on how the release of hydrocarbons and other noxious materials into the atmosphere, gas combustion with the generation of intense heat, oil spills, and disposal of industrial wastes affect the fertility of the inhabitants in such a manner that it reduces fertility and increases the number of births of abnormal babies.

The interviewees stated that the internship farming strategy is an avenue to reduce the risk of coming into contact with the contaminated waterways, thus increasing the willingness of many to travel to upland communities, where there is less oil pollution, as a means to reduce the risk of death or illness. The participants asserted that most households engage in hired labor in the communities to which they relocate to support their everyday livelihood, while acquiring different cultural beliefs and lifestyles. The internship farming methods prevent many families from being exposed to air pollution, water contamination, and the strong odors that pervade affected communities daily. Respondents divulged that, regardless of how hydrocarbons affect all their livelihood sources, what brings most communities together are the common traditions of farming and fishing (Focus Groups 6 & 7).

5.2.7.7 **Leased farming**

Leased farming system is another strategy adopted by farmers whose land were forcefully damaged by oil spills or used for oil production activities. All interviewees significantly emphasized the theme "leased farming" as a livelihood strategy that has influenced the socioeconomic and sociocultural settings of the communities. According to a group called the *Periwinkle Picking Association*, leased faming remains a sustainable strategy when there is available land for the practice (Focus Groups 1 & 4). As further emphasized, *"Even though people leased these farmlands from the neighbouring communities, there is the assurance that the crops are safe as cultivated, except for criminal acts (Focus Group 1)."* Unlike the farm and share strategy by which the owner of the land sometimes harvests the crops before the harvest time. Leased farming contributes immensely to local economic growth, and most community members offer their lands for small amounts—for example, offering a piece of land for less than $5 a year—which constructively contributes to household's livelihood support. A majority of study the participants stressed that leased farming is reliable and sustainable due to the structured policy that is maintained in accordance with the individual's agreement, reinforcing the reasons why a majority of the farming population tends to pursue leased farming instead of the "farm and share" or "internship farming," irrespective of the region's terrain.

However, a fewer participants placed more emphasis on the lack of land availability and infrastructural development across the communities. Participants opine that government land should be designated and made available for any community willing to venture into agricultural activity, as the insufficient landed area have prevented agricultural practice across the region (Focus Group 1).

Consequently, the detailed insights presented in the preceding sections shows how communities defined what constitute alternative support for livelihood across communities, regardless of the obstructions associated with the activities. Empirically, most of the alternative measures contribute to socioeconomic and sociocultural development of the study's communities. However, there are different levels of contributions, given the differing nature of many alternative strategies. Therefore, the next section describes how the identified alternatives contribute to the socioeconomic and sociocultural development across the communities.

5.2.8 **Alternative livelihood contributions to socioeconomic and sociocultural development**

Drawing from the frequencies of the interview content analysis, it was revealed that most of the alternative strategies adopted for livelihood support across the Niger Delta communities contribute in a major way to household livings. The contributions cut across socioeconomic and sociocultural spectra, given that farming and fishing constitute the major intertwined sources of social and local economic activities. Empirical evidence suggests that, in addition to the fishing and farming, other alternative methods for livelihood are improvised due to various obstructions and the quest for subsistence. Fig. 5.2.3 quantifies the frequencies of participant's assertions of the identified livelihood methods' contribution to the socioeconomic and sociocultural development of communities impacted by negative oil and gas activities and oil spill environmental hazards.

The empirical evidence shows local refining of petroleum products as a main contributor to household socioeconomic livelihood irrespective of the unstructured system of the practice in the

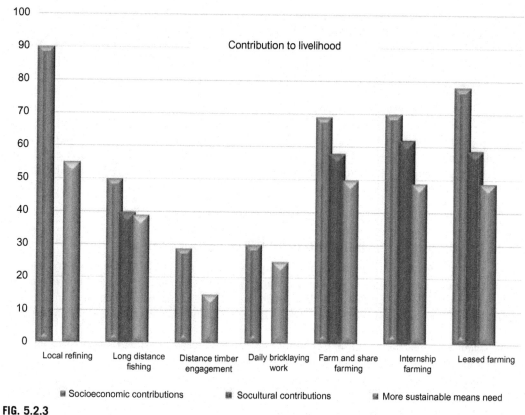

FIG. 5.2.3

Alternative livelihood contributions.

communities, followed by the farm and share system, internship farming, and leased farming. Accordingly, most of the participants frequently asserted that all different types of farming adopted by the communities serve as significant livelihood supports, regardless of the challenges that are often associated with the practice. Evidence shows about 90% contribution of local refining of petroleum product to households' socioeconomic support across communities, with 69% from farm and share, 70% from internship farming, and 78% from leased farming, with distance timber engagement and daily bricklaying jobs having the lowest percentage, about 29% and 30%, contributions to the socioeconomic sustenance of the communities. The empirical evidence further shows about 58% contribution of farm and share, 62% internship farming, and about 59% leased farming contribution to sociocultural conditions, with 0% contribution of the local refining of petroleum product, distance timber production, and daily bricklaying activities to the sociocultural conditions of the communities. Likewise, it was evident that long-distance fishing contributes about 50% to socioeconomic growth, while contributing 40% to sociocultural development.

However, while showing the important contributions of the alternatives to socioeconomic development, it was evident that the local refining of petroleum products needs a more refined, well-structured, and sustainable avenue. As such, about 55% of respondents approve a more structured practice of local refining of petroleum product to reduce environmental degradation while empowering community members. Also, evidence gathered shows that about 50% respondents to farm and share systems, 49% to internship farming, and 49% leased farming respondents approve more support for the aspect of free land provision for different agricultural practices.

Furthermore, evidence revealed that long-distance fishing further needs external support in the provision of necessary fishing tools. As such, the evidence shows 39% of the respondents attesting to the provision of simple necessary fishing tools and security across the fishing areas. Daily bricklaying also shows less emphasis on the more sustainable means of about 25% and distance timber production about 15%, although respondents maintain that such activities have increased the socioeconomic conditions of the communities within the region in the absence of the original and more convenient livelihood structure.

5.2.9 Conclusions, recommendations, and implications

The majority of the alternative methods for livelihood support show a relatively high contribution to socioeconomic and sociocultural development, while preventing the community members from direct use of contaminated land, waterways, and facilities. For instance, timber production, share farming systems, and distance fishing all enable and engage the community members resourcefully and remove individuals from the contaminated environment.

The study concludes that attention should be paid to local communities with regards to any alternative method improvised through a powerful alteration in a constructive manner, specifically the Nigerian oil and gas unique oil-producing communities and the environment. This finding is supported in the literature of Uduji and Okolo-Obasi (2019), which placed much emphasis on the engagement of the Niger Delta communities in sustainable agriculture.

However, considering that some identified alternative methods for livelihood support contribute to environmental damage, various recommendations are made.

First, government and/or policymakers should restructure the alternative methods adopted for livelihood support. For example, local oil refining should be restructured to a more organized practice in concurrence with training and retraining of the community members through awareness of disaster risk and environmental management.

A majority of households across communities succumb on a practical basis to the local refining of crude oil products, with little knowledge of the long-term risk and health impacts. This implies a lack of awareness of disaster risk at the community level, which can be mitigated through training and awareness programs of disaster risk at community levels to aid in better understanding of disaster risk, as suggested by experts during the interviews.

Such training activities would contribute to the achievement of the priority 1 of the Sendai Framework for Disaster Risk Reduction (SFDRR, 2015–2030) adopted by the United Nations. Also, considering the vulnerability of the communities across the region with a low capability of living without degrading impacts on the household's livelihood, awareness programs (predisaster risk assessment) for livelihood disaster prevention and mitigation in the local communities is recommended, to improve the readiness and expand knowledge of more environmentally friendly ways of sourcing for livelihood support.

Small modular refineries should be established at strategic communities across the Niger Delta region to influence reduction of environmental damage through the oil bunkering livelihood method.

Farming remains a major source of livelihood support and, as empirically shown previously, all the farming strategies adopted for livelihood purposes contribute to socioeconomic and cultural growth.

The government should make available free land for any oil-producing communities for agricultural purposes, considering the challenges that accompany some forms of the farming alternate methods. For example, "share farming systems" where oftentimes the landowners tend to singlehandedly harvest the cultivated crops before the agreed harvest times.

Alternative agricultural schemes should be provided to support communities who have engaged in the aforementioned farming strategies for livelihood support.

There should be an improved security agent within the river borders where communities participate in distance fishing or deep-sea fishing.

Ultimately, the communities should be equipped by making the livelihood support methods identified in this paper, and other empowerment methods during the oil and gas era, more sophisticated, to symbolize some benefits before the end of petroleum production on the land.

Other long-term empowerment strategies should be established for communities in the 21st century to prepare their people for the future.

The authors of this study believe that the adoption of these recommendations would improve the alternative methods adopted for livelihood support by the communities, and would reduce the frequencies of oil spills, environmental damage, livelihood conflicts, and resource conflicts and would enhance community livelihood empowerment.

References

Aaron, K., 2005. Perspective: big oil, rural poverty, and environmental degradation in the Niger Delta region of Nigeria. J. Agric. Saf. Health 11 (2), 127–134.

Abi, T., Nwosu, P., 2009. The effect of oil spillage on the soil of Eleme in Rivers State of the Niger Delta area of Nigeria. Res. J. Environ. Sci. 3 (3), 316–320.

Adekola, O., Mitchell, G., 2011. The Niger Delta wetlands: threats to ecosystem services, their importance to dependent communities and possible management measures. Int. J. Biodivers. Sci. Ecosyst. Serv. Manag. 7 (1), 50–68.

Agbogidi, O., Okonta, B., Dolor, D., 2005. Socio–economic and environmental impact of crude oil exploration and production on agricultural production: a case study of Edjeba and Kokori communities in Delta State of Nigeria. Glob. J. Environ. Sci. 4 (2), 171–176.

Akpabio, I.A., Okon, D.P., Inyang, E.B., 2007. Constraints affecting ICT utilization by agricultural extension officers in the Niger Delta, Nigeria. J. Agric. Educ. Ext. 13 (4), 263–272.

Albert, O.N., Amaratunga, D., Haigh, R.P., 2018. Evaluation of the impacts of oil spill disaster on communities and its influence on restiveness in Niger Delta, Nigeria. Procedia Eng. 212, 1054–1061.

Amnesty International, 2015. Niger-Delta Negligence. https://www.amnesty.org/en/latest/news/2018/03/Niger-Delta-Oil-Spills-Decoders/ Accessed July 2019.

Atubi, A.O., 2015. Effects of oil spillage on human health in producing communities of Delta State, Nigeria. Eur. J. Bus. Soc. Sci. 4 (8), 14–30.

Barad, R., Fletcher, E.K., Hillbruner, C., 2020. Leveraging existing household survey data to map livelihoods in Nigeria. World Dev. 126, 104727.

Bohnert, H.J., Nelson, D.E., Jensen, R.G., 1995. Adaptations to environmental stresses. Plant Cell 7 (7), 1099.

Boniface, Samuel, 2016. Oil bunkering activities in the Niger Delta "The Way Forward" Am. J. Eng. Res. 5 (4), 169–173.

Bradley, P.G., 1974. Marine oil spills: a problem in environmental management. Nat. Resour. J. 14, 337.

Brinkmann, S., 2014. Interview. In: Encyclopedia of Critical Psychology. Springer, pp. 1008–1010.

Brooks, N., 2003. Vulnerability, risk and adaptation: a conceptual framework. Tyndall Centre Clim. Chang. Res. 38 (38), 1–16 Working Paper.

Bruederle, A., Hodler, R., 2019. Effect of oil spills on infant mortality in Nigeria. Proc. Natl. Acad. Sci. 116 (12), 5467–5471.

Cano-Urbina, J., Clapp, C.M., Willardsen, K., 2019. The effects of the BP deepwater horizon oil spill on housing markets. J. Hous. Econ. 43, 131–156.

Chambers, R., Conway, G., 1991. Sustainable Rural Livelihood: Practical Concept for the 21st Century; 1991.

Cheong, S.-M., 2011. A social assessment of the Hebei-Spirit oil spill. GeoJournal 76 (5), 539–549.

Chijioke, B.O., Ebong, I.B., 2018. The impact of oil exploration and environmental degradation in the Niger Delta Region of Nigeria: a study of oil producing communities in Akwa Ibom State. Glob. J. Hum. Soc. Sci. Polit. Sci 18, 55–70.

CRED, 2016. Annual Disaster Statistical Review 2016: The Numbers and Trends. Retrieved from: https://reliefweb.int/report/world/annual-disaster-statistical-review-2016-numbers-and-trends.

Croisant, S., Sullivan, J., 2018. Studying the human health and ecological impacts of the deep water horizon oil spill disaster: introduction to this special issue of new solutions. New Solut. 28 (3), 410–415.

Danoff-Burg, J., Conrad, K., 2020. Lobster houses as a sustainable fishing alternative. Consilience J. Sustain. Dev. (6).

Dauda, S., 2019. The paradox of persistent poverty amid high growth: the case of Nigeria. In: Immiserating Growth: When Growth Fails the Poor. p. 250.

Davies, S., 2016. Adaptable Livelihoods: Coping with Food Insecurity in the Malian Sahel. Springer.

Dibua, 2005. Citizenship and resource control in Nigeria: the case of minority communities in the Niger Delta. Afr. Spectr. 40, 5–28.

Ebegbulem, J., Ekpe, D., Adejumo, T.O., 2013. Oil exploration and poverty in the Niger Delta Region of Nigeria: a critical analysis. Int. J. Bus. Soc. Sci. 4 (3), 279–287.

EM-DAT, 2017. OFDA/CRED International Disaster Database. Université catholique de Louvain, Brussels, Belgium. Retrieved from: https://www.emdat.be/.

Erondu, C., 2015. Enhanced traditional livelihoods: alternative to the oil economy in Niger Delta. Res. Humanit. Soc. Sci. 5 (5), 158–164.

Fubara, S.A., Iledare, O.O., Gershon, O., Ejemeyovwi, J., 2019. Natural resource extraction and economic performance of the Niger Delta Region in Nigeria. Int. J. Energy Econ. Policy 9 (4), 188.

Gam, K.B., Kwok, R.K., Engel, L.S., Curry, M.D., Stewart, P.A., Stenzel, M.R., 2018. Exposure to oil spill chemicals and lung function in deepwater horizon disaster response workers. J. Occup. Environ. Med. 60(6), e312.

Garza-Gil, M.D., Prada-Blanco, A., Vázquez-Rodríguez, M.X., 2006a. Estimating the short-term economic damages from the Prestige oil spill in the Galician fisheries and tourism. Ecological Economics 58 (4), 842–849.

Garza-Gil, M.D., Surís-Regueiro, J.C., Varela-Lafuente, M.M., 2006b. Assessment of economic damages from the prestige oil spill. Mar. Policy 30 (5), 544–551.

Gill, D., Picou, S., Ritchie, L., 2012. The Exxon Valdez and BP oil spills: a comparison of initial social and psychological impacts. Am. Behav. Sci. 56 (1), 3–23.

Grigalunas, T.A., Anderson, R.C., Brown Jr., G.M., Congar, R., Meade, N.F., Sorensen, P.E., 1986. Estimating the cost of oil spills: lessons from the Amoco Cadiz incident. Mar. Resour. Econ. 2 (3), 239–262.

Guha-Sapir, D., Santos, I., Borde, A., 2013. The Economic Impacts of Natural Disasters. Oxford University Press.

Harrison, J.A., 2019. "Down here we rely on fishing and oil": work identity and fishers' responses to the BP oil spill disaster. Sociol. Perspect. 63 (2), 333–350.

Ihayere, C., Ogeleka, D.F., Ataine, T.I., 2014. The effects of the Niger Delta oil crisis on women folks. J. Afr. Stud. Dev. 6 (1), 14–21.

Ingwe, R., 2015. Illegal oil bunkering, violence and criminal offences in Nigeria's territorial waters and the Niger Delta environs: proposing extension of informed policymaking. Inform. Econ. 19 (1), 77–86.

Jones, H.P., Schmitz, O.J., 2009. Rapid recovery of damaged ecosystems. PLoS ONE. 4(5), e5653.

Jung, D., Kim, J.-A., Park, M.-S., Yim, U.H., Choi, K., 2017. Human health and ecological assessment programs for Hebei Spirit oil spill accident of 2007: status, lessons, and future challenges. Chemosphere 173, 180–189.

Kadafa, A.A., 2012a. Environmental impacts of oil exploration and exploitation in the Niger Delta of Nigeria. Glob. J. Sci. Front. Res. Environ. Earth Sci. 12 (3), 19–28.

Kadafa, A.A., 2012b. Oil exploration and spillage in the Niger Delta of Nigeria. Civ. Environ. Res. 2 (3), 38–51.

Kareem, S.D., Kari, F., Alam, G.M., Chukwu, G.M., David, M.O., Oke, O.K., 2012. Foreign direct investment and environmental degradation of oil exploitation: the experience of Niger Delta. Int. J. Appl. Econ. Finance 6 (4), 117–126.

Khan, S.A., 1994. Nigeria: The Political Economy of Oil. Oxford University Press, Oxford, p. 10.

Krantz, L., 2001. The sustainable livelihood approach to poverty reduction. SIDA. Div. Policy Socio-Econ. Anal. 44.

Kvale, S., 2008. Doing Interviews. Sage.

Lack, D.L., 1968. Ecological Adaptations for Breeding in Birds; 1968.

Lewis, R.R., 1983. Impact of oil spills on mangrove forests. In: Biology and Ecology of Mangroves. Springer, pp. 171–183.

Liu, X., Wirtz, K.W., 2006. Total oil spill costs and compensations. Marit. Policy Manag. 33 (1), 49–60.

Mogaji, O.Y., Sotolu, A.O., Wilfred-Ekprikpo, P.C., Green, B.M., 2018. The effects of crude oil exploration on fish and fisheries of Nigerian aquatic ecosystems. In: The Political Ecology of Oil and Gas Activities in the Nigerian Aquatic Ecosystem. Elsevier, pp. 111–124.

Nigerian Bureau of Statistics (NBS), 2012. Annual Abstract of Statistics. Federal Republic of Nigeria, NBS. https://www.nigerianstat.gov.ng/ Accessed 20 December 2018.

Nwachukwu, Mbachu, 2018. The socio-cultural implications of crude oil exploration in Nigeria. In: The Political Ecology of Oil and Gas Activities in the Nigerian Aquatic Ecosystem. Academic Press, pp. 177–190.

Odjuvwuederhie, E.I., Douglason, G.O., Felicia, N.A., 2006. The effect of oil spillage on crop yield and farm income in Delta State, Nigeria. J. Cent. Eur. Agric. 7 (1), 41–48.

Odularu, G.O., 2007. Crude oil and the Nigerian economic performance. In: Oil and Gas Business. http://citeseerx.ist.psu.edu/viewdoc/download?doi=10.1.1.531.822&rep=rep1&type=pdf.

Ofem, N.I., Ajayi, A.R., 2008. Effects of youth empowerment strategies on conflict resolutions in the Niger-Delta of Nigeria: evidence from Cross River State. J. Agric. Rural Dev. 6 (1), 139–146.

Olawoye, J.E., 2019. Rural development and household livelihood: lessons and perspectives from Nigeria. Sustain. Dev. Afr. 17.

Omofonmwan, S.I., Odia, L.O., 2009. Oil exploitation and conflict in the Niger-Delta region of Nigeria. J. Hum. Ecol. 26 (1), 25–30.

Omorede, C.K., 2014. Assessment of the impact of oil and gas resource exploration on the environment of selected communities in Delta State, Nigeria. Int. J. Manag. Econ. Soc. Sci. 3 (2), 79–99.

Onuoha, F.C., 2008. Oil pipeline sabotage in Nigeria: dimensions, actors and implications for national security. Afr. Secur. Stud. 17 (3), 99–115.

Onakuse, S., Lenihan, E., 2017. Rural livelihood insecurity in Etsako East area of Edo state, Nigeria. Journal of Sustainable Development 5 (1), 2–11.

Onwuka, E., 2005. Oil extraction, environmental degradation and poverty in the Niger Delta region of Nigeria: a viewpoint. Int. J. Environ. Stud. 62 (6), 655–662.

Opukri, Ibaba, 2008. Oil induced environmental degradation and internal population displacement in the Nigeria's Niger Delta. J. Sustain. Dev. Afr. 10 (1), 173–193.

Osuagwu, E.S., Olaifa, E., 2018. Effects of oil spills on fish production in the Niger Delta. PLoS ONE. 13(10), e0205114.

Osuma, G.O., Babajide, A.A., Ikpefan, O.A., Nwuba, E.B., Jegede, P.W., 2019. Effects of global decline in oil price on the financial performance of selected deposit money banks in Nigeria. Int. J. Energy Econ. Policy 9 (3), 187.

Oviasuyi, P., Uwadiae, J., 2010. The dilemma of Niger-Delta region as oil producing states of Nigeria. J. Peace Confl. Dev. 16 (1), 10–126.

Pan, R., Wild, A.C.S.G., Drecksler, L.R., Gonçalves Jr., A.C., 2019. Environmental impact: contextualization and current reality. J. Exp. Agric. Int. 29, 1–8.

Patin, S.A., 1999. Environmental Impact of the Offshore Oil and Gas Industry. EcoMonitor Pub. East Nortport, NY. 1.

Pegg, S., Zabbey, N., 2013. Oil and water: the Bodo spills and the destruction of traditional livelihood structures in the Niger Delta. Community Dev. J. 48 (3), 391–405.

Picou, J.S., Formichella, C., Marshall, B.K., Arata, C., 2009. Community impacts of the Exxon Valdez oil spill: a synthesis and elaboration of social science research. In: Synthesis: Three Decades of Research on Socioeconomic Effects Related to Offshore Petroleum Development in Coastal Alaska. pp. 279–310.

Pielke Jr., R.A., 1998. Rethinking the role of adaptation in climate. Glob. Environ. Chang. 8 (2), 1597170.

Poledna, S., Hochrainer-Stigler, S., Miess, M.G., Klimek, P., Schmelzer, S., Sorger, J., … Linnerooth-Bayer, J., 2018. When does a disaster become a systemic event? Estimating indirect economic losses from natural disasters. arXiv preprint. arXiv:1801.09740.

Qu, S.Q., Dumay, J., 2011. The qualitative research interview. Qual. Res. Account. Manag. 8 (3), 238–264.

Rowley, J., 2012. Conducting research interviews. Manag. Res. Rev. 35 (3/4), 260–271.

Saunders, M., Lewis, P., Thornhill, A., 2016. Research Methods for Business Students. vol. 7. Pearson Education, Harlow.

Scoones, I., 1998. Sustainable rural livelihoods: a framework for analysis. In: IDS Working Paper 72. IDS, Brighton.

SFDRR, 2015–2030. Sendai Framework for Disaster Risk Reduction 2015–2030. Retrieved from: https://www.unisdr.org/files/43291_sendaiframeworkfordrren.pdf.

Smit, B., Wandel, J., 2006. Adaptation, adaptive capacity and vulnerability. Glob. Environ. Chang. 16 (3), 282–292.

Smit, B., Burton, I., Klein, R.J., Wandel, J., 2000. An anatomy of adaptation to climate change and variability. In: Societal Adaptation to Climate Variability and Change. Springer, pp. 223–251.

Sobrasuaipiri, S., 2016. Vulnerability and Adaptive Capacity in Livelihood Responses to Oil Spill in Bodo. University of Brighton, Niger Delta.

Uduji, J., Okolo-Obasi, E.N., 2019. Corporate social responsibility in Nigeria and rural youths in sustainable traditional industries livelihood in oil producing communities. J. Int. Dev. 31 (7), 658–678.

Ugochukwu, C.N., Ertel, J., 2008. Negative impacts of oil exploration on biodiversity management in the Niger De area of Nigeria. Impact Assess. Proj. Appr. 26 (2), 139–147.

Ukpong, Balcombe, Fraser, I., Areal, 2019. Preferences for mitigation of the negative impacts of the oil and gas industry in the Niger Delta region of Nigeria. Environ. Resour. Econ. 74 (2), 811–843.

Uy, N., Takeuchi, Y., Shaw, R., 2011. Local adaptation for livelihood resilience in Albay, Philippines. Environ. Hazards 10 (2), 139–153.

Watson, R.T., Zinyowera, M.C., Moss, R.H., 1996. Climate change 1995. Impacts, adaptations and mitigation of climate change: scientific-technical analyses; 1996.

Index

Note: Page numbers followed by *f* indicate figures and *t* indicate tables.